Studies in Brain and Mind

Volume 9

More information about this series at http://www.springer.com/series/6540

Bruno Mölder • Valtteri Arstila • Peter Øhrstrøm
Editors

Philosophy and Psychology of Time

Springer

Editors
Bruno Mölder
Department of Philosophy
University of Tartu
Tartu, Estonia

Peter Øhrstrøm
Department of Communication
and Psychology
Aalborg University
Aalborg, Denmark

Valtteri Arstila
Department of Behavioral Sciences and
Philosophy
University of Turku
Turku, Finland

ISBN 978-3-319-22194-6   ISBN 978-3-319-22195-3   (eBook)
Studies in Brain and Mind
DOI 10.1007/978-3-319-22195-3

Library of Congress Control Number: 2015955148

Springer Cham Heidelberg New York Dordrecht London

Printed on acid-free paper

Springer International Publishing AG Switzerland is part of Springer Science+Business Media (www.
springer.com)

Contents

Chapter 1
Introduction: Time in Philosophy and Psychology

Bruno Mölder and Valtteri Arstila

Abstract "Philosophy and Psychology of Time" comprises papers from philosophers and psychologists who work on various aspects of subjective time. In the book, the broad topic of time is examined from different aspects, divided into five parts. These main aspects are the following: the concept of time in philosophy and psychology, temporal presence, the continuity and flow of time in mind, the timing of experiences, and the relationship between time and intersubjectivity. This chapter introduces the volume and supplies a short overview of each contribution.

Time, especially the way it is experienced and the role it plays in enabling experience, has again become the focus of both philosophers and psychologists. However, it is commonplace for research in various fields to take place in isolation, without sufficient interaction between the ideas and the theoretical approaches from different disciplines. This is understandable, given the diverse objectives and methodologies of the disciplines. Notwithstanding this, the experience of time is a topic on which cross-disciplinary discussion between philosophers and psychologists would be instrumental to further progress. True, there are obstacles to such valuable interdisciplinary cooperation. But some of these obstacles are contingent, namely the use of different vocabularies and the difficulty of understanding the frameworks within which researchers of other disciplines work. Indeed, there is no straightforward translation from philosophical terms into psychological terminology and vice versa. The same applies to the theoretical frameworks which structure the debates in philosophy and psychology (which we understand broadly to encompass the study of neural processes too).

B. Mölder (✉)
Institute of Philosophy and Semiotics, University of Tartu, Ülikooli 18, 50090 Tartu, Estonia
e-mail: bruno.moelder@ut.ee

V. Arstila
Department of Behavioral Sciences and Philosophy, University of Turku, Turku, Finland
e-mail: valtteri.arstila@utu.fi

© Springer International Publishing Switzerland 2016
B. Mölder et al. (eds.), *Philosophy and Psychology of Time*, Studies
in Brain and Mind 9, DOI 10.1007/978-3-319-22195-3_1

1

This collection of papers aims to bridge this gap between philosophy and psychology. The book contains five parts, each dealing with a topic that is either central to the field in the sense that both philosophers and psychologists have shown a keen interest in it or is a topic that deserves, but has not yet received, more attention. These topics include the concept of time in philosophy and psychology, the notion of presence, continuity and flow of time in mind, the timing of experiences, as well as the relationship between time and intersubjectivity.

The book is structured in such a way that each topic is addressed both from the philosophical and the psychological angle, thus enabling an interdisciplinary exchange. The authors were encouraged to reflect upon the relevant concepts in their own and the neighbouring discipline and to write the chapters so that they can be understood by researchers from the other field. The chapters establish common ground in the respective field, but they also include a more detailed discussion which digs deeper into the topic and wherein original contributions to the discussion are made.

In what follows, we will provide a brief, general characterization of each large topic and the relevance of philosophy and psychology to each topic. This is then followed by more detailed accounts of the particular contributions.

The concept of time is conceived in philosophy and in psychology in various ways. In philosophy, the fundamental divide lies between the dynamic and the static conceptions and logics of time. Traditionally, one of the main issues has been which kind of vocabulary is more basic. On the side of psychology, there are also different concepts of time in play, but the central task is rather that of bringing out all those factors that influence our experience and/or judgements of time and temporal properties.

The topic of presence brings together issues in philosophy of mind, metaphysics of time and psychology of time. With respect to this topic, philosophers seek to describe the temporal phenomenology of consciousness and to establish which metaphysical account is in best accord with such phenomenology. Psychologists, however, are interested in the processes underpinning different levels of temporal moments that could be called "present."

The main concern with regard to subjective continuity and the flow of consciousness is the putative conflict between the seemingly continuous nature of consciousness and the discrete nature of the underlying neural processing. This is definitely a topic where conceptual analysis done by philosophers could help to clarify the matter, though an informed discussion by a neuroscientist on how discrete processing in the brain can give rise to continuous consciousness would be also needed.

The timing of experiences pertains to whether the apparent temporal order of experience matches with the objective order of events that are perceived and to the order of brain processes that underlie these experiences. Answering these questions requires careful interpretation of experimental results, and in this regard empirical input from psychologists could complement the philosophical debate.

Finally, the topic of time and intersubjectivity is one that has not been sufficiently explored. At the same time, it has often been stressed that proper timing is crucial for intersubjective action. The philosopher's role would be to clarify exactly what role time plays in coordination between persons, whereas psychologists could build specific models of the involvement of time in intersubjectivity.

Now we turn to the characterization of particular contributions to the volume.

Part I. The Concept of Time in Philosophy and Psychology The aim of the part on the concept of time is to give a general overview of the various concepts of time in philosophy and psychology, at the same time keeping an eye on the possible connections between the respective terminologies. All contributors to this part also apply their discussion to our everyday concerns about time, exemplifying how advancements in philosophy and psychology help to make sense of the elements of our daily experience.

Part I begins with a chapter by Peter Øhrstrøm, who attributes the well-known difficulty of defining time to the fact that time is so fundamental that it cannot be defined in more basic terms. Although there is no reductive account of time to be had, we can still understand time and talk about it. Øhrstrøm introduces the basic distinction between the *dynamic* (A) and the *static* (B) languages of time and proceeds with the question of which language is the fundamental one. He gives an overview of Arthur Prior's temporal logic and argues that this framework incorporates both languages. Yet, it can also be shown that the A-language is richer and more basic than the B-language. Furthermore, Øhrstrøm illustrates the usefulness of a formal approach to time by showing how it helps to make sense of common claims about time and solves age-old puzzles such as the problem of how statements about future contingent affairs could be true at present.

Samuel Baron and Kristie Miller examine the folk concept of time. They point out that any account of this concept should accommodate the point that it is "resistant to error." This means that the concept lies in the centre of the network of our practices and experiences, which are so important to us that we are not prepared to give up the concept in light of scientific developments or philosophical arguments. Given its resistance to error, Baron and Miller argue that neither the A-series nor the B-series captures the folk concept of time. In their view, the folk concept of time is a functional one, related to our perspective on the world as agents. Time as a functional concept is multiply realizable, and this explains its resistance to error: very many things can be taken to be a realization of the folk concept.

Dan Zakay approaches the concept of time from the point of view of psychology. He distinguishes between physical time, which he terms "T", and psychological time, which he understands as a subjective feeling. Zakay points out two basic dimensions of psychological time: the succession and duration of experiences. He focuses especially on duration judgements, both in that of the retrospective and prospective variety. In retrospective judgements memory plays a role, whereas prospective judgements depend on attention. Zakay explains how shifts between prospective and retrospective timing can throw light on several experiences of time and on temporal illusions familiar from our daily life.

Part II. Presence It is common to think of the "now," temporal present, as a durationless point in time that separates the past and the future. Our awareness does not appear to be confined to instants in time, however. First, most philosophers working on the topic argue that our experiences as of change, motion and other similarly temporally extended phenomena can be accounted for only if the phenomenal contents of our experiences are temporally extended. The idea that the subjectively experienced present moment is temporally extended is known as *the doctrine of the specious present*. Second, as psychologists have stressed, things that we are aware of result from processes that integrate stimuli (or the neural signals that relate to stimuli) over different periods of time. This applies to both "elementary" experiential features, e.g. colours and grouping, as well as to higher-level features required for intersubjective communication.

This subjective present, what we experience as occurring now, is the topic of the part II. It begins with a chapter by Sean Power, who separates different notions of present and contrasts them with the doctrine of the specious present. It is shown how the doctrine is compatible with some but not all metaphysical positions of time. Accordingly, depending on the aspect that one emphasizes—phenomenology or metaphysics—either phenomenology limits the plausible metaphysical views on time or metaphysical considerations force us to reconsider the doctrine and the phenomenology it is supposed to help us explain.

Marc Wittmann considers the topic from the perspective of a neuroscientist. He argues that one must separate three different notions of the subjective present. The shortest one, which he calls the *functional moment*, defines the simultaneity and temporal order of events—presumably this relates to other well-known temporal integration processes that occur within 100 ms too. Next there is the *experienced moment*, which segments and integrates temporal events into meaningful units. The experienced moment covers few seconds and enables an accurate and successful behaviour and inter-personal communication. The experienced moment also resembles best the notion of the specious present. Finally, by means of working memory related to the *mental presence*, functional moments and experienced moments become incorporated into a continuous, unitary mental state.

Part III. Continuity and Flow of Time in Mind Our consciousness is often described as a continuous stream of conscious experiences, as in Wittmann's notion of mental presence. In part III, what such a view amounts to is reflected upon. The topics of interest include the stream-likeness of consciousness as separate from the continuity of experience and the extent to which consciousness has the properties of continuity and stream-likeness. The latter topic is particularly interesting because psychological results appear to be in tension with the phenomenology—the results are sometimes taken to suggest that our experiences result from discontinuous processes and that succeeding experiences do not relate to each other in any substantial manner.

Oliver Rashbrook-Cooper approaches these issues from the philosophical tradition. Taking phenomenology as his starting point, he argues that the stream of consciousness does not boil down to having experiences succeeding each other without any apparent gap. Instead, Rashbrook-Cooper argues that the stream consists of *Phenomenal Continuity* and *Phenomenal Flow*. The first of these terms refers to the idea that the temporal boundaries of experiences are not exhibited in the phenomenology, whereas the second term refers to the idea that consciousness appears to us as a flowing phenomenon even if what we are conscious of remains the same. As a result, the gaps in experiences (which concern the contents we are conscious of) suggested by the psychological results do not threaten the stream-likeness of consciousness.

Tamas Madl, Stan Franklin, Javier Snaider and Usef Faghihi approach the topic from the perspective of cognitive neuroscience and modelling. Their focus is on the question of how succession and duration—i.e. the temporally extended events that studies of time perception are concerned with—can be perceived in the framework of the Global Workspace Theory. This issue is particularly pressing because the Global Workspace Theory is one of the major theories of consciousness and it postulates that the mechanisms behind conscious perception are discrete. They show that time perception is indeed possible in this framework by providing a computational model of the Global Workspace Theory, where time processing relies on perceptual associative memory as well as on short-term and long-term nodes grounded on sensory feature detectors.

Part IV. The Timing of Experiences In part IV, we turn to the timing of experiences. The focus is again on the theoretical issues: what determines the (apparent) time of events? The discussion thus does not concern directly the actual time when a stimulus is experienced as measured in milliseconds—something that Visual Awareness Negativity and Late Positivity aim to address—nor does it concern factors (such as intensity, location and the modality of the stimulus) that undoubtedly modulate but do not determine the time when a stimulus is experienced. Instead, at the heart of the matter lies the relationship between (i) the apparent time of experienced events and (ii) the time of the neural processes underlying the experiences of events. Do the two necessarily match or can they come apart?

Valtteri Arstila addresses this issue in the context of philosophical debate concerning postdiction effects and more precisely in relation to apparent motion. (In postdiction effects, later occurring events influence our perception of events taking place before them.) Since this debate is intimately tied up with the larger philosophical issue of the doctrine of the specious present mentioned above, he also approaches the issue of the timing of experiences from the subjective point of view: i.e. how must the apparent temporal structure of experiences be constituted if a single unified experience can represent that two events occurred at different time? In the end, Arstila suggests that the doctrine can be rejected and that the apparent time of an experienced event matches with the time of neural processes underlying the experiences.

Kielan Yarrow and Derek Arnold, in turn, approach the timing of experiences from the perspective of psychology. They focus on how to explain the results obtained from simultaneity and temporal order judgement tasks. After considering existing explanations, they emphasize that most results can be explained by assuming that the apparent time of experienced events is the same as the time of the neural processes that realize the experiences of events. This is supported by the fact that the suggested neural mechanisms underlying our performance in these tasks follow the same assumption. Thus although Yarrow and Arnold do not conclude that the time of neural processes determines the experienced simultaneity and succession of events, they argue that the competing views must be made more concrete before they can be considered to be real alternatives.

Part V. Time and Intersubjectivity Appropriate timing is central for successful on-line intersubjective action and, indeed, this has often been pointed out in accounts of social cognition. However, it has not been explored with sufficient depth and clarity. What role do time and the sense of time play in coordination between people? How could philosophy contribute to the study of this? What is the role of timing in human development? Does the development of intersubjective communication depend on temporal factors? These are some of the issues considered in part V.

Bruno Mölder explores the role of time in intersubjective processes from a philosophical perspective. For this, he relies on some tools from the philosophy of science. In particular, he generalizes Carl Craver's account of explanation in neuroscience to temporal and interactive processes between people, drawing a distinction between causally relevant conditions, constitutive components, temporal constraints and background conditions. With this set of tools, he then analyzes the mother-infant interaction and time-related phenomenological explanations that have been given for some psychopathological cases such as schizophrenia and depression. This allows for more varied and specific conclusions concerning the role of time in interaction between people.

Colwyn Trevarthen gives an all-encompassing overview of the role of time in our lives, delineating how innate motivation for intersubjective action forms the basis of our more sophisticated cognitive abilities, related to communication and culture. He shows how our social skills develop through various stages, starting from inborn capacities to imitate and share rhythms, and leading to shared culture and language. In such actions, temporal parameters related to rhythm and musicality are crucially important. Besides providing the synthesis of a wide range of developmental facts, Trevarthen discusses also the neuroscience of action and emotion, focussing especially on the importance of rhythmic movements.

We hope that this book will be of interest to both philosophers and psychologists. On the one hand, the chapters cover those key issues in the philosophy of time and mind that psychologists have also tackled. On the other hand, the volume provides an easily approachable exposition of psychological research on time that might be crucial for resolving philosophical debates, thereby extending the range of examples for the scrutiny of philosophers.

Acknowledgments The book is partly based on the talks given at the workshop that was held in the framework of the European COST Action project TD0904 TIMELY (Time In MEntaL activitY: theoretical, behavioral, bioimaging and clinical perspectives) at the University of Turku on 14th–15th August 2013. The editors acknowledge the support of the TIMELY network, which brought together time researchers from different disciplines. When preparing this volume, Bruno Mölder was also supported by the institutional research funding IUT20-5 of the Estonian Ministry of Education and Research and a grant ETF9117 from the Estonian Science Foundation.

Part I
The Concept of Time in Philosophy and Psychology

Chapter 2
The Concept of Time: A Philosophical and Logical Perspective

Peter Øhrstrøm

Abstract As pointed out by St. Augustine we cannot give a proper definition of time as such. Furthermore, conceiving time as a literal object would be highly problematic. It is, however, possible to establish a conceptual framework for meaningful discussion of the temporal aspects of reality in terms of the philosophical logic of time developed by Arthur Norman Prior. This framework is based on the view that John McTaggart's A-language is more fundamental than the B-language. Prior's view can be seen as based on some important properties of human experiences of time, and it involves the claim that it is useful to study the temporal aspects of reality in terms of so-called branching time models. This can in fact be done in several ways. It turns out that some of the most attractive and richest theories based on the ideas of branching time may be seen as formalisations of medieval and other early suggestions made by scholars such as William of Ockham and Luis de Molina. The tense-logical formalism appears to be useful wherever it is important to reason strictly regarding the temporal aspects of reality. Prior's approach gives rise to a formal language which is relevant in the context of Julius T. Frazer's hierarchical understanding of time as such and in the study of time in general. It offers a very powerful way to deal with time in a conceptually consistent, systematic and precise manner.

In any discussion concerning the notion of time it is commonplace to refer to Augustine's famous comments on the difficulties we face when answering the question, "What is time?" Augustine gave an astute formulation of the problem:

> What, then, is time? If no one asks me, I know what it is. If I wish to explain it to him who asks me, I do not know. (Augustine 1995, Book 11)

The problem seems to be that although time is an essential and well known aspect of reality, we nevertheless cannot properly define it. But why is it so difficult to provide such a definition?

P. Øhrstrøm (✉)
Department of Communication and Psychology, Aalborg University,
Rendsburggade 14, DK-9000 Aalborg, Denmark
e-mail: poe@hum.aau.dk

© Springer International Publishing Switzerland 2016
B. Mölder et al. (eds.), *Philosophy and Psychology of Time*, Studies in Brain and Mind 9, DOI 10.1007/978-3-319-22195-3_2

9

It may perhaps be argued that the reason is simply that although we refer to time again and again in our daily life, there is no precise reality which the term stands for. According to this view, the substantive "time" covers a number of very different and only rather loosely connected ideas; there is no single object or entity. If we followed this line of thought, it would not be possible to speak about the concept of time as something meaningful and precise, something that can be studied as a true element or aspect of reality. In consequence, ideas of time would have to be understood as nothing but social constructions.

There is clearly something to be said in favour of this view. The most common modern idea of time can to a large extent be seen as a product of cultural and technological development. As Lewis Mumford (2010, 15) points out, modern clocks helped "create the belief in an independent world of mathematically measurable sequences." This view of time as a literal object has been questioned by Arthur Norman Prior (1914–69), who has suggested that we simply have to drop the idea:

> But instants as literal objects … going along with the picture of time as a literal object, a sort of snake which either eats its tail or doesn't, either has ends or doesn't, either is made of separate segments or isn't, and this picture I think we must drop. (Prior 1967, 189)

On the other hand, the passage of time seems to be a crucial element of our life as human beings. And how can this be so, if time does not refer to any aspect of reality at all? In this paper, it will be argued that we can in fact speak meaningfully about the temporal aspects of reality without conceiving time as a literal object. As we shall see, it is possible to establish a conceptual framework for meaningful discussion of the temporal aspects of reality, although we cannot give a proper definition of time as such. This framework can be introduced in terms of the philosophical logic of time developed by Prior.

2.1 Prior on Time and Reality

Prior's main reason for rejecting the idea of time as a literal object is that the idea is, in his opinion, based on a mistaken view of reality. Prior (1996a, 45) claimed that: "Time is not an object, but whatever is real exists and acts in time…". According to this view, time should in fact be understood and studied on the basis of an investigation into what exists. This means that Augustine's problem should not be answered as suggested above. Although we cannot strictly-speaking define time, this does not mean that we should drop all attempts at dealing with temporal phenomena based on a careful and true analysis of reality.

Assuming that time is, after all, a meaningful concept, we have to reconsider the question: Why is it so difficult to define time? The answer may be found in reflecting on what is involved in defining time. Given that a definition of time would have to be formulated with reference to something more basic than time, the crucial problem becomes obvious. If there is nothing more fundamental in reality than time, then it follows that a proper definition of time cannot be established. However, this does not mean that time as such cannot be studied. St. Augustine's classical dilemma should not be understood solely as a denial of the possibility of explaining what time is. The real wisdom to be found in his answer is not only that time cannot be defined in terms of anything more fundamental, but rather that although this is the case, temporal awareness is in fact a crucial part of tacit human understanding. We actually have a proper understanding of time, although in most cases this knowledge can only be expressed in a rather indirect manner.

Prior gives us one very clear example of his view of temporal reality when he writes:

> I believe that what we see as a progress of events is a progress of events, a coming to pass of one thing after another, and not just a timeless tapestry with everything stuck there for good and all.... (Prior 1996b, 47–48)

In this way, Prior maintains that what we normally call "the passage of time" should be conceived as belonging to reality. What Prior rejects as a misleading idea is time as a fixed and existing structure of events, a "timeless tapestry." In Prior's opinion, we have to reject the so-called eternalism, according to which all events (past, present, and future) are equally real. The important question in this context is this: How can something exist, if it does not exist now? Prior defends the so-called presentism (see Prior 1972) according to which something exists if and only if it exists now. If something is past and does not exist now, then it does not exist (anymore). If something is future and does not exist now, then it does not exist (yet). In this way, Prior finds that the notions of time and existence are closely related.

Another crucial aspect of temporal reality is related to what Prior calls "real freedom." He sees this idea as closely bound to the belief in the reality of the passage of time:

> This belief of mine... is bound up with a belief in real freedom. One of the big differences between the past and the future is that once something has become past, it is, as it were, out of our reach—once a thing has happened, nothing we can do can make it not to have happened. But the future is to some extent, even though it is only to a very small extent, something we can make for ourselves.... if something is the work of a free agent, then it wasn't going to be the case until that agent decided that it was. (Prior 1996b, 47–48)

The main point here is that there is a very important asymmetry between the past and the future. The past is "now-unpreventable" and "out of our reach," whereas the future, at least to some extent, is open in the sense that it can be formed at least in part as a consequence of the decisions we are making now.

2.2 McTaggart and Prior on Temporal Discourse

Such reflections as on the temporal aspects of reality are essential when it comes to the study of time. Although we cannot give a simple and precise definition of time, we can in fact speak meaningfully about the temporal aspects of reality based on common human understanding. This can be done in several ways. Since the time of

John McTaggart (1866–1925), most discussions about time make use of his A- and B-language classifications. These classifications of time derive from McTaggart's famous paper, "The Unreality of Time" (1908), in which he introduced two fundamentally different ways of dealing with time, the A-series and the B-series (corresponding with what we call here, the A- and the B-language). This dynamic approach is formulated in terms of an A-language ("past," "present" and "future"). The approach is based on the view that there is a fundamental asymmetry between past and future, which is not only a feature of the human mind, but an essential aspect of reality. This view may be compared with a more static or structural understanding of time according to which the basic notions are "before," "after" and "simultaneous with" (the so-called B-language). The A-language will be used if time is seen from "within," whereas the B-language will be relevant if a temporal system is seen from the "outside" that is, from a God's eye perspective.

Prior argued that the A-language is much closer to human experience of time than the B-language. His most famous argument was first published in the paper "Thank Goodness That's Over" (1959). Later the argument was rephrased as:

> ... what we know when we know that the 1960 final examinations are over can't be just a timeless relation between dates because this isn't the thing we're pleased about when we're pleased that the examinations are over. (Prior 2003, 42)

As Prior sees it, this shows that knowledge based on temporal experience has to involve A-notions. For this reason, Prior held that the A-notions are conceptually basic, and that the B-notions are secondary.

Prior developed McTaggart's classification further, claiming that the distinction between these two languages can in fact give rise to some important philosophical positions regarding time. However, Prior also pointed out that any analysis should include the notion of temporal (or historical) necessity, that is, the notion of what is "now un-preventable." This is crucial in the account of the asymmetry between the past and the future.

In discussing which concepts should be regarded as the primary (or primitive) temporal notions regarding the understanding and description of reality, the A-notions or the B-notions, Prior argued that there are not only two possible answers to this problem. In this case Prior followed his standard procedure, according to which he first of all had to describe the positions which could in principle be held. Based on this account he would then have to explain which of the possibilities he himself would prefer, and he would have to give his reasons for this choice.

Given that the role of temporal necessity has to be taken into consideration, Prior demonstrated that we can in fact formulate four different positions, corresponding to what Prior calls "grades of tense-logical involvement":

1. The B-notions are primary regarding the understanding of the temporal aspects of reality, and the A-notions can be conceptually derived from the B-notions.
2. Ontologically, the A- and B-notions should be seen as on a par. Neither has conceptual priority over the other.

3. The A-notions are primary regarding the understanding of the temporal aspects of reality, and the B-notions can be conceptually derived from the A-notions. The basic A-language includes a notion of what is "now un-preventable."
4. The A-notions are primary regarding the understanding of the temporal aspects of reality, and the B-notions can be conceptually derived from the A-notions. The basic A-language does not include a notion of what is "now un-preventable." This modal notion can be conceptually reduced to the classical A-language of past, present and future.

These grades of tense-logical involvement may be introduced in terms of tense- and modal-logical formalisms. The formal language suggested by Prior was in fact an extension of proposition logic with three propositional operators, P, F and M. This means that if q is a proposition, then we can form new propositions using the operators:

Pq: "it has been that q"
Fq: "it will be that q"
Mq: "it is possible that q"

Given these three basic operators, it is possible to define their dual operator in this way:

Hq: "it has always been that q," defined as $H \equiv {\sim}P{\sim}$
Gq: "it will always be that q," defined as $G \equiv {\sim}F{\sim}$
Nq: "it is necessary that q," defined as $N \equiv {\sim}M{\sim}$

If one wants to go beyond the third grade of tense-logical involvement and embrace the fourth grade, then M (and N) become definable in terms of P and F (with the usual machinery of propositional logic).

It should be added that in order to deal with time in terms of tense-logic, Prior also used the so-called metric tense-operators, $P(x)$ and $F(x)$, which stand for "x time units ago it was the case that..." and "in x time units it will be the case that..."

The above formalism is A-theoretical and one of the tasks which Prior wanted to carry out was to explain how this approach relates to the B-language (static time), which is based on the assumption of a set of instants, $(TIME, <)$, ordered by a before-after relation (and a simultaneity relation), and including a formalism of $T(t,p)$ (i.e. "p is true at the time t"). In the B-language, the tense operators can be introduced in the following way:

$$T(t,Fp) \equiv_{def} \exists t_1 : t < t_1 \wedge T(t_1,p)$$
$$T(t,Pp) \equiv_{def} \exists t_1 : t_1 < t \wedge T(t_1,p)$$

Prior himself held that the A-language is conceptually primary. This means accepting at least the third grade of tense-logical involvement. After a few years of work with the development of his tense-logic, Prior decided to defend the fourth grade.

In his many papers and books, Prior demonstrated how the B-language can be constructed fully in terms of the A-language (but not vice versa). In other words, he proved that the B-language can be formally derived from the A-language.

A major point in Prior's construction of the B-language from the A-language (grades 3 and 4) is that temporal-instants may be conceived as a special class of very rich propositions. Consider any of these instant-propositions and an arbitrary other proposition, q, then the crucial property of the instant-proposition will be that it either necessarily implies q, or that it necessarily implies the negation of q, i.e. $\sim q$. Prior has demonstrated that the system of such instant-propositions will have all the formal properties required of a B-theory based on the notion of "being true at an instant." In this way, the B-language may in fact be established in terms of the A-language. (More details can be found in Prior 2003, 117 ff. and Øhrstrøm 2011). In this way, Prior (1967, 74) is able to establish the abstract idea of time. Prior argues that "time appears as a class of classes of propositions ordered by a certain relation." The classes of propositions to which Prior refers, are in fact equivalent to instant-propositions and time is conceived as the ordered set of such instant-propositions. An instant-proposition, i_1, is supposed to be before another instant proposition, i_2, if and only if i_2 necessarily implies Pi_1. In fact, it can be proved based on some rather minimal assumptions that this is the case if, and only if i_1 necessarily implies Fi_2.

Prior carried out a number of further studies of the instant-propositions understood as a special class of very rich propositions (see Prior 2003). In doing so Prior also became the founding father of so-called hybrid logic. In general, Prior's development of temporal logic and hybrid logic has given philosophers and logicians a very powerful and precise framework for further studies of the temporal aspects of reality.

2.3   Frazer's Ideas on Time and Reality

The various kinds of temporal discourse needed in order to deal with reality have also been analysed by Julius T. Frazer (1923–2010) who describes a general system of temporal discourse in which there are several levels of complexity when dealing with time and reality (Frazer 1978, 422). Frazer's levels are paraphrased below:

1. Atemporal: a world without reference to causation or temporal order.
2. Prototemporal: a world in which the only temporal references are statistical tendencies.
3. Eotemporal: a world without now, physical matter exists, time is orientable but not time-oriented.
4. Biotemporal: a world characterised by limited temporal horizons, organic present, teleological causation.
5. Nootemporal: A world in which intentionality, past/present/future, and human freedom are all meaningful notions.
6. Sociotemporal: A society referring to temporal order and organisation.

Frazer's hierarchical theory of time can be compared with both McTaggart's approach and Prior's grades. Clearly, Frazer's levels one to three may be discussed in terms of the B-language (and this language may in a certain sense not even be needed at all at the very first level). However, in order to deal with levels four to six, the A-language will be needed as well as the B-language. If levels one to six also represent the historical order of things, it may be argued that this shows that the B-language is historically fundamental as a kind of physical time, and that the appearance of the A-language should be seen as linked to the creation of biological organisms with some kind of mental or psychological awareness. However, this is not necessarily the most attractive approach to the problem of time. At the very least, we should also discuss the topic from an epistemological point of view. This means relating everything to the notion of knowledge and the process of obtaining knowledge, and it also means that Frazer's levels should not be understood as onto-logical levels. Such an approach would obviously depend on a conceptual analysis of human cognition. What does it mean that I have obtained or gained new knowledge? From a temporal point of view it clearly implies being aware of some-thing which I wasn't aware of earlier. Conceived of from this epistemological per-spective, the A-language in particular should never be completely overlooked when studying time, even if the description of the subject matter at Frazer's lower levels (one to three) does not seem to call for more than a B-language. When Frazer's hierarchical system is presented in this manner, the lower levels (one to four) have to be understood from the perspective of level five (i.e. human awareness), and the same holds for the higher and even more complex level six. In this way, the order downwards in the system (levels four, three, two, one) corresponds to an increasing abstraction according to which more and more aspects of the human perception and understanding of time (level five) are ignored.

2.4 Is Time Mind-Dependent?

Assuming Prior's third and fourth grades of tense-logical involvement it seems that time depends on human cognition. If we assume that Frazer's levels roughly corre-spond to the historical order of events in the universe, then it seems that there was a time before the first biological awareness (levels one to three). At this time, there were no human beings. But should it be assumed that this time was now-less, since the now is only possible if there are human beings or other minds to whom temporal awareness is important? If we accept Prior's analysis of the relations between the A- and B-notions in Sect. 2.2, we have to conclude that the idea of a now-less time has to be rejected. Even before the very first human mind, it would be sensible to speak of a now. The point is that we can meaningfully refer to what a human being would have experienced, if he had been there. This reference to a counterfactual observer seems in fact to be the only way in which we can meaningfully speak of these early stages of the world's history, unless we want to refer to a non-human or even divine observer. In fact the very use of the term "history" may be taken as indi-cating that we have to describe the stages in question using our temporal discourse.

In his construction of the temporal-instants needed in the B-language it is clearly Prior's strategy to present past instants as conceptually derivable from the now, which in this way turns out to be extremely rich and meaningful. In fact, not only past instants, but also the instants in the open future and even the counterfactual instants can be conceptually constructed in this manner.

It should be pointed out, however, that we cannot exclude the possibility of describing the world without any reference to our temporal awareness. The point is that we as humans cannot have any direct access to a cognition of this kind. But an atemporal cognition may certainly be possible from a divine perspective. For example, St. Thomas Aquinas seems to have held that God's knowledge is in some way outside of time. Prior discusses this possibility in his paper "The Formalities of Omniscience" (Prior 2003, 39–58), and he has convincingly argued that the kind of knowledge which is interesting and relevant from a human point of view will be conceptually related to our temporal awareness conceived in terms of the A-language. This does not prevent us from discussing and studying what was the case before the creation of mankind. Nor does it exclude us from a rational discussion of what could counterfactually have happened on the Earth, had things gone otherwise than they actually did.

It should be mentioned that Prior had a strong interest in the differences between divine and human knowledge as seen in relation to the concept of time. Indeed, much of his early work on the problems of time was motivated by his wish to find acceptable answers to the classical problem regarding the logical tension between the two Christian doctrines of divine foreknowledge and human freedom. In fact, his work on this problem strongly stimulated his development of formal tense-logic. As we shall see in the next section, he argued that the use of tense-logic may lead to a deeper understanding of the temporal aspects of reality.

2.5   Treating Time in Terms of a Tense-Logic

Prior is the founding father of modern temporal logic. He invented a formal language that made it possible to treat philosophical and conceptual problems related to time in a systematic and precise manner. Arguing for the importance of this project he wrote:

> And I think it important that people who care for rigorism and formalism should not leave the basic flux and flow of things in the hands of existentialists and Bergsonians and others who love darkness rather than light, but we should enter this realm of life and time, not to destroy it, but to master it with our techniques. (Undated note; kept in the Prior Collection, Bodleian Library, Oxford)

In terms of this tense-logical formalism, time may be discussed semantically. In order to do so, we also need a more precise idea of truth. For instance, we may discuss whether or not the proposition "Joe is drinking beer" is true. We may of course discuss who exactly Joe is. We may also discuss what it means to drink beer and what beer is. For instance, should a non-alcoholic beer be accepted as a beer? But as soon as semantical questions of that kind have been answered, we may at least in principle be able to find out by a rather simple inspection of the facts which of the two basic statements ("Joe is drinking beer" and "Joe is not drinking beer") is the true one and which of them is (right now) the false one. This seems to be based on a very basic principle, which may be seen as a version of the so-called principle of correspondence:

(C) A statement is true now if and only if the statement corresponds to facts about the present reality.

This means that the proposition "Joe is drinking beer" is true now if, and only if, it is a fact about the present reality that he is drinking beer. If that is in fact the case then it will now and at all future times be un-preventable i.e. it is and will always outside the control of Joe and everybody else that he has been drinking beer at this (present) time. Before he started to drink he may perhaps have chosen otherwise, but even when he is in fact drinking beer it seems un-preventable (i.e. outside the control of anyone) that right now he is drinking beer, although he may of course choose to stop his drinking—perhaps even very soon. Either there are facts about the present reality which makes the proposition in question true or there are no such facts. The existence of such facts may depend on what we did in the past, but it does not depend on what we are doing in the future.

Given the tense-logical formalism, Prior (1967, 74) argued that statements such as "Time will have an end," "Time is circular," "Time is continuous," etc. can be treated as meaningful without assuming "that there is some monstrous object called Time, the parts of which are arranged in such-and-such ways." The meanings which can be given to such statements arise from tense-logical formalism. This means that the statements can in fact be translated into tense-logical statements which can then be handled as any other logical problem. In Prior's words:

> Tense-logic [can be seen as] as giving the cash value of assertions about time. Postulates of the sort … can be regarded as giving the meaning of such statements as "time is continuous," "time is infinite both ways," and so forth. (Prior 1967, 74)

The idea is that a theory of time should be formulated as a tense-logical system i.e. a system with some axioms and rules of inference. The properties of time can then be explored by investigating the theorems of the system; that is, the property we want to discuss is translated into a tense-logical statement and it is then investigated whether the statement can be proved in the system.

For instance, the claim that "time is dense" corresponds to the statement, $Fq \supset FFq$ (i.e. "for any q, if q will be the case, then it will be the case that it will be the case that q"). This means that time is dense if, and only if, $Fq \supset FFq$ is a theorem in the tense-logical system we are studying. The idea behind this statement is that if something is going to happen at a later instant, t, then it will be that it will be the case, which means that there is another instant, t', which is later than the present instant and before t. In other words, this means that for any future instant, t, there will be another instant between now and t.

Another example could be the claim that "time has a last moment" which translates into the statement, $Gq \lor FGq$ (i.e. "for any q, either it will always be that q, or it will be the case that it will always be that q"). The point here is that if there is a last instant, t_{Final}, then Gq will be true at t_{Final}, no matter what q stands for, since at the last instant there is no future at all (i.e. it is an empty fulfilment). The idea here is that the first part of the disjunction, $Gq \lor FGq$, is true if we are already at the ending of time and that the other part of the disjunction is true if the ending of time is still to come.

2.6 The Idea of Branching Time

In September 1958 Prior received a letter from Saul Kripke. In this letter Kripke, only 17 years at the time, mentioned some errors in Prior's book *Time and Modality* (1957) and also suggested the use of what we now call branching time (see Ploug and Øhrstrøm 2012). Prior accepted Kripke's idea and over the following decade he developed it further in several ways. In fact he considered at least two main kinds of branching time systems. First, he considered the so-called Ockham system, which is shown in Fig. 2.1.

The idea in this kind of graphical representation is to focus on the branching moments or instants at which there is a choice between two or more alternatives. This kind of illustration may be useful but, as Prior pointed out, they are in fact no more "than handy diagrams; they need not be taken with any great metaphysical seriousness" (Prior 1967, 74).

The diagram below is named after the medieval logician William of Ockham (c. 1287–1347) who studied the logical tension between the two Christian doctrines of divine foreknowledge and human freedom. It was important for Ockham to maintain that God has complete and perfect knowledge regarding the contingent future.

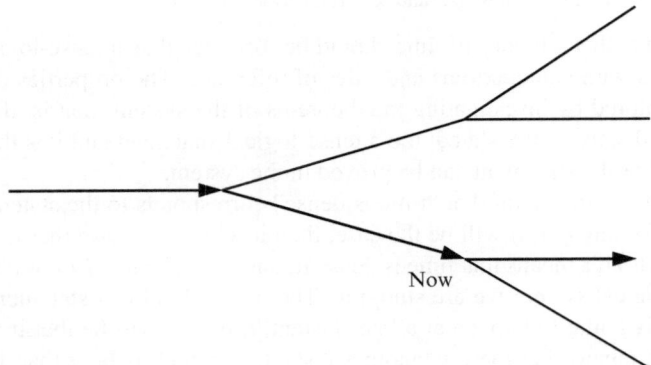

Now

Fig. 2.1 The Ockhamistic view of branching time according to which there is a chronicle now representing the true future

In the above diagram this is indicated by the arrows which represent the course of events known in advance by God. In an unpublished paper circulated in 1965 or earlier, Prior explained this idea, claiming that in the Ockhamistic system "there is a single designated line (taking one only of the possible forward routes at each fork), which might be picked out in red, representing the actual course of events" (*Postulate Sets for Tense Logic*, Bodleian Library, Oxford). This means that according to Ockham not all future possibilities have the same status. One of the possible future courses of events is privileged in the sense that it is the true one, that is, the course of events already known to God.

In his conceptual and logical analysis Prior accepted that the Ockhamistic solution is a consistent theory which should be taken into account. However, he found some consequences of the theory rather problematic. A major problem for him was that the theory means that we can by our present and future acts influence what God knew earlier. In this way, the theory seems to violate a crucial and fundamental aspect of reality, namely that the past is fixed and nothing we can do now can change the past. Indeed William of Ockham (1969) himself was aware of this problem. His response was that as humans we obviously cannot know exactly how divine omniscience works. In addition, he made a distinction between what is properly past and what is only apparently past. When I make a decision regarding tomorrow, I may influence the truth-value of a prophecy stated yesterday about what was going to happen in 2 days. However, a statement on what was yesterday going to happen 2 days later is not a statement about yesterday (but actually rather a statement about tomorrow). Such a statement is only apparently about the past. It is not about the proper past.

Another problem related to the Ockhamistic model is that if God knows what is going to happen in the contingent future and makes His knowledge known to the persons involved, then what is said becomes necessary given that God cannot be mistaken. However, Ockham himself had an answer to this problem. According to Ockham, God does not normally communicate unconditional prophecies to the persons involved. Ockham gives an example from the Old Testament referring to the prophet, Jonah, who was asked to go to Nineveh, where he should proclaim: "Forty more days, and Nineveh will be overturned" (Jonah 3.4). However, as we learn from the Bible, the citizens of Nineveh repented and the city was not overthrown at that time. But does this mean that the prophecy was in fact false when it was stated? According to Ockham, we should understand the prophecy of Jonah as presupposing the condition "unless the citizens of Nineveh repent." Obviously, this is in fact exactly how the citizens of Nineveh understood the statement of Jonah! Viewed as conditional, the prophecy may still be true. Ockham's point seems to be that although God knows the truth-values of the unconditional prophecies regarding the contingent future, He does not communicate this knowledge to human beings. In this way God's unconditional knowledge concerning the contingent future remains silent. Still, it is conceptually important that God knows the truth-value of any future contingency.

2.7   Prior's Rejection of the Ockhamistic View of Time

Prior studied the logic of the classical Ockhamistic view carefully, and he suggested a formalisation of this medieval solution. Although Prior in the end did not support this solution, his work with the model became very important, since it turns out that some other important models (including the one he himself defended) can be formally defined in terms of the Ockhamistic model.

Prior found the Ockhamistic rebuttal of determinism attractive, but in his opinion it was also rather problematic to assume that a future contingent could be true now. If a statement regarding the future is in fact true now, then there must be a strong evidence for it according to Prior. And when it comes to the contingent future, we have no such evidence. As a result Prior found it problematic to assume that one future possibility has priority over another as to which is going to be real. Although he argued that the Ockhamistic model is formally consistent, he did not find its conceptualisation of the idea of an open future acceptable. He maintained that all alternative futures should be given the same conceptual status as possibilities. Instead, he suggested and defended the so-called Peircean system. His view can be illustrated in terms of the branching time diagram shown in Fig. 2.2.

Prior argued that if a statement about the future is in fact true now, then there must be some strong evidence to support it which would also mean that the statement is now unpreventable. In terms of the problem of divine foreknowledge and human freedom he held that held that what God knew yesterday (or at any other past time) regarding the future, must (necessarily) be the case. Even in this sense we wanted to maintain the principle of the fixed past. In order to avoid determinism and maintain a belief in the reality of human freedom, he then found that he had to reject the doctrine of complete divine foreknowledge regarding the contingent future. Consider once more the person Joe who is free to decide to drink (or not to drink) beer tomorrow at noon. According to the Priorean view it cannot be true now

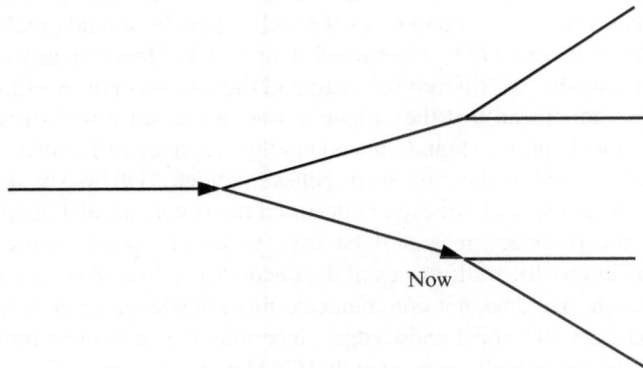

Fig. 2.2 The Priorean view of branching time according to which there is no chronicle now representing the true future. All future alternatives have the same status. Prior called this solution "The Peircean system"

neither that he is drinking beer tomorrow at noon nor that he is not drinking beer tomorrow at noon. If p is taken to stand for the proposition, "Joe is drinking beer," then the two statements regarding Joe tomorrow at noon may be formalized as $F(1)$ p and $F(1)\sim p$. According to Prior's view, both statements are false. The idea here is that since there is nothing which can make the statements true now, they must be false now. The assumption involved here are (1) that all statements are either true or false now, and (2) that a statement is only true now if there is something now which can make it true.

Prior points out, however, that it is in fact possible in a certain sense to maintain the doctrine of divine foreknowledge in a weaker sense according to which it is simply claimed that God knows everything which is true now. Since there is no truth now about Joe's relation to beer tomorrow at noon, this limited doctrine does not imply that God knows any truth concerning Joe's drinking or not drinking beer tomorrow at noon. It is interesting that Prior's view in this way has given rise to a new theological position which has been termed "open theism." This position has recently been discussed by David Jakobsen (2013).

While $F(1)p$ and $F(1)\sim p$ are both false in the Priorean system, the proposition $\sim F(1)p$ is obviously true in the system. This means that according to this view we should make a clear distinction between the following two propositions:

$F(1)\sim p$: "It will be the case tomorrow at noon that it is not the case that Joe is drinking beer."

$\sim F(1)p$: "It is not the case that it will be the case tomorrow at noon that Joe is drinking beer."

This distinction may of course be seen as difficult to handle at least within the scope of natural language. Most language users, based on common sense reasoning, would probably think that if it is not the case that Joe will be drinking beer tomorrow at noon, then it will be the case tomorrow at noon that he is not drinking beer. The distinction between $\sim F(1)p$ and $F(1)\sim p$ certainly appears to be a very high price to pay in order avoid determinism along with an acceptance of the principle of the fixed past.

2.8 Belnap's Open Future

Like Prior, Nuel Belnap et al. (2001) have criticised the classical view rejecting the idea that a proposition about the contingent future can be true now. However, they do not accept the strange Priorean distinction between $\sim F(1)p$ and $F(1)\sim p$. They have argued that it is in fact possible to maintain the validity of

$$F(1)p \lor F(1) \sim p$$

without accepting the idea of a true contingent future. This can be done by rejecting the very concept of the absolute truth-value of a proposition at a moment of time.

The idea here is that the truth-value of a proposition should depend not only on the temporal moment but also on the choice of route (or chronicle) through the branching time system. This idea had in fact been developed by Prior who termed it "Ockhamistic" (see Øhrstrøm and Hasle 1995, 203 ff.), although William of Ockham certainly would have accepted what we have called the classical view.

By making truth relative to the routes (chronicles) through the branching time system, the idea of divine foreknowledge can in principle be excluded from the theory. The various possible futures then have the same ontological status. None of them represents "the future," but they are all possible futures.

The theory suggested by Belnap et al. (2001) is elegant, and it has many followers in modern temporal logic. However, it is again a rather high price to pay in order to avoid determinism without accepting the classical view. The price in this case is that we have to drop the classical and absolute idea of truth and replace it with a more relative notion. If this worldview is accepted, the notion of truth will be rather limited. All knowledge will be conditional; i.e. truth will in principle only make sense relative to an assumed course of events. In this case there would not be any absolute truth about the contingent future which could or could not be known by anyone, and belief or a guess regarding the contingent future could neither be right nor wrong.

In terms of the example used above, this means that the statement, "Joe is going to drink beer tomorrow at noon," does not have a truth-value right now. According to the theory the same holds for the proposition, "Joe is not going to drink beer tomorrow at noon." According to Belnap et al. (2001) such propositions can only be true (or false) relative to the future course of events. It is like saying that if Joe is going to drink beer tomorrow at noon, then he is going to drink beer tomorrow at noon. This is certainly not surprising! But the loss of absolute truth is in fact a great loss. Very often we want to refer to the truth-value of contingent statements. For instance, we may be betting. Some say that tomorrow at noon Joe will be drinking beer; some may hold the opposite view. Some of the persons involved in this must be right (the winners), whereas others are wrong (the losers). We don't know now who is who, but we may of course know later.

2.9   The Idea of "The Thin Red Line" in the Theory of Time

What is so wrong about the classical view defended by William of Ockham and many other great logicians? According to Belnap et al. (2001) a model such as the one shown in Fig. 2.1 does not pay due regard to the idea of alternative possibility. After all, the notion of alternative possibilities is essential when it comes to a proper understanding of the idea of human freedom. Belnap et al. have argued that if the counterfactual possibility had been chosen, one of the possible futures at any counterfactual instant would have corresponded to what would have been the true future at that instant. This means that the model in Fig. 2.1 is far too simplistic. In order to represent the notion of the true future correctly, there should according to Belnap et al. be a "selected" future route at any choice point in the diagram.

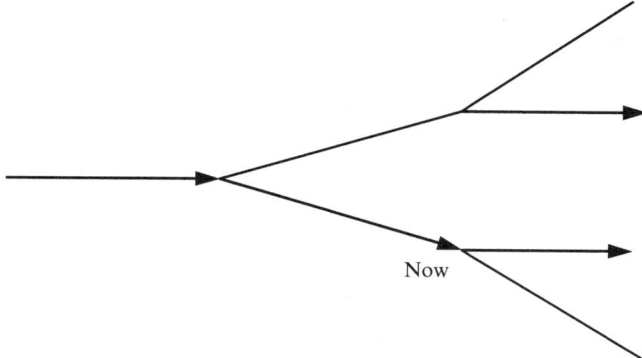

Fig. 2.3 A Molinistic model taking divine middle knowledge into account. There is a "true future" at any choice point in the diagram—even at a counterfactual instant

I believe that Belnap et al. are right and their criticism should certainly be taken into account. However their observation is definitely not new. This point was in fact well understood and defended by Luis de Molina (1535–1600), who argued that God has so-called middle knowledge. This means that God knows what any human being would freely do in any possible situation. In terms of a branching time diagram this may be illustrated in the way depicted in Fig. 2.3.

Molina pointed out that this idea of divine middle knowledge can be illustrated by an example from the New Testament:

> Woe to you, Korazin! Woe to you, Bethsaida! If the miracles that were performed in you had been performed in Tyre and Sidon, they would have repented long ago in sackcloth and ashes. (Matthew 11.23)

If we take the meaning to be that in the counterfactual situation in question the people in Tyre and Sidon would freely have repented, then this would constitute an example of divine middle knowledge. According to Molina one may in this way truly speak of the free choice of human beings even in a counterfactual situation:

> God knows that there would have been repentance in sackcloth and ashes among the Tyronians and Sidonians on the hypothesis that the wonders that were worked in Chorozain [Korazin] and Bethsaida should have been worked in Tyre and Sidon .... But because the hypothesis on which it was going to occur was not in fact actualized, this repentance never did and never will exist in reality—and yet it was a future contingent dependent on the free choice of human beings. (de Molina 1988, 116–117)

The analysis of the logical possibilities when confronting the doctrines of divine foreknowledge and human freedom shows that it is logically possible to uphold both doctrines in a consistent manner. A problem remains, namely the logical tension between the assumption regarding the true future (corresponding to the doctrine of divine foreknowledge) and the principle of the fixed past. The logical analysis shows that there are two options:

1. One can weaken the doctrine of divine foreknowledge. This can be done in Prior's way (leading to "open theism"), or in Belnap's more radical manner

according to which any truth is relative to a course of events. In both cases the price will be very high in the sense that essential parts of everyday reasoning have to be abandoned.

2. One can accept the classical idea of divine foreknowledge and drop the principle of the fixed past when it comes to God's past foreknowledge. The price is that an even higher degree of divine knowledge has to be accepted, namely that of so-called middle knowledge. In addition, we have to accept that what God knows now can be influenced by our future decisions.

It should be emphasised that although much of this discussion on time was originally formulated in a theological language, it can also be formulated in a secular language. The problem can be stated in terms of truth-values without any reference to divine knowledge. On the other hand, if we accept the so-called principle of correspondence discussed here and formulated as (C), then any discussion of truth must give rise to a discussion regarding the facts and the very nature of the present reality. It seems obvious that (1) above can be accepted without involving very much metaphysics—probably even on the basis of a purely materialistic worldview. The other possibility above, (2), seems, however, to call for an assumption regarding a deeply metaphysical nature of reality although it does not necessarily imply the classical doctrine of divine foreknowledge.

2.10 Is Time Connected?

In terms of the discussion regarding the A- and B-language as well as Prior's grades of tense-logical involvement the idea of branching may be interpreted as corresponding to the third or fourth grade if the model is assumed to include an indication of the now. But can we be certain that time is connected? Is it sufficient to represent time as just one branching time system, one tree? Why not, for instance, two unconnected trees corresponding to Fig. 2.4?

The question to consider here is whether there are possible temporal-instants which cannot be accessed from our now, instants which have never been part of a possible future. But what does it mean that an instant is possible? The answer depends on whether one defends the third or the fourth grade of tense-logical involvement. Being a defender of the fourth grade, Prior wrote:

> ... the question as to whether there are or could be unconnected time-series is a senseless one.... but these diagrams cannot represent time, as they cannot be translated into the basic non-figurative temporal language. (Prior 1967, 199)

Prior is right given the fourth grade: on this basis, there is no way to state the possibility of unconnected instants, since possibility is defined in terms of the tense-operators (P and F). However, according to the third grade, unconnected instants would in principle be possible. Clearly, the possibility that such unconnected instants should be taken into account as a part of our world-view is highly metaphysical. On the other hand, a defender of the third grade might point out that since

Fig. 2.4 The idea of an
unconnected time
according to which more
than one branching time
system is needed in order
to represent time properly

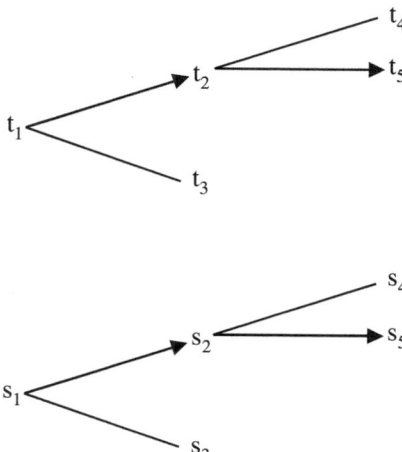

there is no strong argument against the use of an independent notion of possibility, such a notion should not be ruled out in principle. In addition, it should also be pointed out that the Ockhamistic and Molinistic versions of branching time cannot be formulated without a primitive notion of possibility. In consequence, these theories will be ruled out if we accept a world-view according to the fourth grade.

2.11   Conclusion

It has been argued that Prior's tense-logic provides a nice and useful framework for further discussions and studies of the temporal aspects of reality. This approach to time is based on the view that McTaggart's A-language is more fundamental than the B-language. It can be argued that this view can be seen as based on some important properties of human experiences of time. This does not mean that time is mind-dependent, but just that tenses are essential for a deeper understanding of the temporal aspects of reality. In fact, tense-logic seems to give rise to a formal language which is relevant in the context of Frazer's hierarchical understanding of time as such. Although some of the levels in Frazer's system do not seem to call for more than a B-language, the A-language is certainly needed at other levels.

As we have seen, it is possible to formulate a very interesting approach to the study of time based on Prior's logical ideas. Given his tense-logical systems it is in fact possible to give meaning to basic assertions about time, such as "time is dense," "time is not circular," "time is branching," "time is connected" etc. It turns out to be possible to deal with these questions in a very precise manner in terms of tense-logical formalism.

In particular, it is interesting to study the ideas of branching time. This can in fact be done in several ways. It turns out that some of the most attractive and richest theories based on the ideas of branching time may be seen as formalisations of medieval and other early suggestions made by scholars such as William of Ockham and Luis de Molina.

In the late 1970s Prior's temporal logic caught the interest of some influential computer scientists. The most important paper on the use of temporal logic in computer science from this early period was "The Temporal Logic of Programs" (1977) written by Amir Pnueli (1941–2009). Temporal logic has since become very important in computer science and in 1996 Pnueli received the Turing Award for "seminal work introducing temporal logic into computing science and for outstanding contributions to program and systems verification" (see http://amturing.acm.org). It is in fact remarkable that a theory which was originally formulated in order to deal with important challenges regarding time in theology and philosophy has found its way into the field of computer science.

Given the use of tense-logical formalism, it turns out that we can create a conceptual framework that can be applied in several (if not all) sciences and at all the levels suggested by Frazer. The formalism appears to be useful and relevant wherever it is important to reason strictly regarding the temporal aspects of reality. Although this approach does not offer a complete definition of time, it does suggest a way to deal with important aspects of temporal reality in a systematic and conceptually consistent manner. This certainly makes the tense-logical approach to time very attractive.

References

Augustine, Saint. 1955. *Confessions*. Trans. and ed. Albert C. Outler. http://www.ccel.org/ccel/Augustine/Confessions. Accessed 6 July 2014.

Belnap, Nuel, Michael Perloff, and Ming Xu. 2001. *Facing the future: Agents and choices in our indeterminist world*. Oxford: Oxford University Press.

de Molina, Luis. 1988. On divine foreknowledge Trans. Alfred J. Freddoso. Ithaca: Cornell University Press.

Frazer, Julius T. 1978. The individual and society. In *The study of time III*, ed. Julius T. Frazer, N. Lawrence, and D. Park, 419–442. New York: Springer.

Jakobsen, David. 2013. Arthur Norman Priors bidrag til metafysikken, Ph.D. dissertation. Department of Communication and Psychology, Aalborg University.

McTaggart, John M.E. 1908. The unreality of time. *Mind: A Quarterly Review of Psychology and Philosophy* 17: 456–473.

Mumford, Lewis. 2010. *Technics and civilization*. Chicago: University of Chicago Press (First published in 1934).

Øhrstrøm, Peter. 2011. Towards a common language for the discussion of time based on Prior's tense logic. In *Multidisciplinary aspects of time and time perception*, ed. Argiro Vatakis, Anna Esposito, Maria Giagkou, Fred Cummins, and Georgios Papadelis, 46–57. Berlin: Springer.

Øhrstrøm, Peter, and Per Hasle. 1995. *Temporal logic: From ancient ideas to artificial intelligence*. Dordrecht: Kluwer.

Ploug, Thomas, and Peter Øhrstrøm. 2012. Branching time, indeterminism and tense logic. *Synthese* 188(3): 367–379.

Pnueli, Amir. 1977. The temporal logic of programs. In *Proceeding SFCS '77: Proceedings of the 18th Annual Symposium on Foundations of Computer Science*, 46–57. Washington, DC: IEEE Computer Society.

Prior, Arthur N. 1957. *Time and modality.* Oxford: Oxford University Press.

Prior, Arthur N. 1959. Thank goodness that's over. *Philosophy* 34: 12–17.

Prior, Arthur N. 1967. *Past, present and future.* Oxford: Oxford University Press.

Prior, Arthur N. 1972. The notion of the present. In *The study of time*, ed. Julius T. Fraser, F.C. Haber, and C.H. Müller, 320–323. Berlin: Springer.

Prior, Arthur N. 1996a. A statement on temporal realism. In *Logic and reality: Essays on the legacy of Arthur Prior*, ed. Jack Copeland, 45–46. Oxford: Clarendon Press.

Prior, Arthur N. 1996b. Some free thinking about time. In *Logic and reality: Essays on the legacy of Arthur Prior*, ed. Jack Copeland, 47–51. Oxford: Clarendon Press.

Prior, Arthur N. 2003. In *Papers on time and tense, new edition*, ed. Per Hasle, Peter Øhrstrøm, Torben Braüner, and Jack Copeland. Oxford: Oxford University Press.

William of Ockham. 1969. *Predestination, God's foreknowledge, and future contingents* Trans. Marilyn McCord Adams and Norman Kretzmann. Indianapolis: Hackett.

...

Chapter 3
Our Concept of Time

Samuel Baron and Kristie Miller

Abstract In this chapter we argue that our concept of time is a functional concept. We argue that our concept of time is such that time is whatever it is that plays the time role, and we spell out what we take the time role to consist in. We evaluate this proposal against a number of other analyses of our concept of time, and argue that it better explains various features of our dispositions as speakers and our practices as agents.

3.1 Introduction

Recent work in the philosophy of time tends to focus on one of a number of well-entrenched metaphysical debates. Presentists and eternalists face off over questions of temporal ontology: do the past and future exist? Or does only the present exist? And if only the present exists, how do we reconcile this fact with the picture of time inherited from science, which seems to favour eternalism? Similarly, A-theorists and B-theorists lock horns over the metaphysical nature of time: is time constituted by a single, unchanging sequence of temporal instances ordered by the B-theoretic relations of earlier-than, later-than and simultaneous-with—the so-called B-series— or is time richer, constituted by a dynamical sequence of times ordered by the A-theoretic properties of pastness, presentness and futurity?

While these debates continue unabated, relatively little consideration has been given to the nature of our folk temporal concepts. What, exactly, is the structure of our everyday concept of time? And what would it take for the concept of time to go unsatisfied and thus for some brand of temporal error theory to be true? Does the folk concept favour some particular picture of temporal reality, or not? It is only by answering such questions that we can begin to develop a rigorous conceptual

S. Baron (✉)
School of Humanities, University of Western Australia
Crawley, Perth, Western Australia 6009, Australia
e-mail: samuel.baron@uwa.edu.au

K. Miller
Department of Philosophy, Quadrangle A14, University of Sydney,
Sydney, NSW 2006, Australia

© Springer International Publishing Switzerland 2016
B. Mölder et al. (eds.), *Philosophy and Psychology of Time*, Studies
in Brain and Mind 9, DOI 10.1007/978-3-319-22195-3_3

framework within which more general debates over time may be profitably developed. The goal of this chapter, then, is to analyse the folk temporal concept thereby providing the foundations for the development of such a conceptual framework. In what follows we begin by outlining—in more detail—the reasons why an analysis of the folk concept is needed and the constraints under which an analysis must be developed (Sect. 3.2). We then consider two broad strategies for analysing the folk concept already available in the literature, and find them both wanting (Sect. 3.3). In the final section, we offer our own analysis of the folk concept of time (Sect. 3.4), according to which time is a functional concept: time is as time does. We argue that this analysis better explains various features of our dispositions as speakers and our practices as agents than the accounts considered in Sect. 3.3.

Note that throughout we will be assuming a general familiarity both with current debates within the philosophy of time and with the history of the philosophy of time. Note also that we will—by and large—be taking ourselves as proxies for the folk: our own intuitions about temporal concepts will be taken as evidence of a kind that the folk concept is thus and so. We recognise that this can only be the first step in a more detailed empirical investigation into the nature of everyday temporal concepts. Still, it is not unusual for philosophers to speak for the folk as we do (being part of the folk themselves). Indeed, this is a common practice within the more general methodology of conceptual analysis in the broadly Lewisian tradition of regimenting our folk concepts. Nonetheless, it is proper to offer a promissory note to more fully determine, in the future, the extent to which the intuitions we have about our temporal concepts have ecological validity (i.e. the extent to which they are genuinely representative of folk intuitions more widely).

3.2 A Folk Concept

3.2.1 Motivation

The term "concept" means many things to many people. We cannot hope to defend a particular view about the nature of concepts in this chapter. Indeed, we think there are many legitimate views about what it is for something to be a concept, ranging from the broad notion that concepts are Fregean senses to the equally broad notion that they are mental representations. So we do not suggest that our use of the term "concept" in this chapter is the unique best one. We only suggest that there is some interesting notion that is tracked by what we mean, in this context, by "concept."

Our primary target is a *tacit folk* concept. In what follows it is assumed that something roughly in the spirit of the internalist tradition is right about the content of our concepts. It is assumed that conceptual content is exhausted by what a subject is, after relatively idealized reflection and consultation of her intuitions, disposed to say about the existence (or otherwise) of, in this case, time, across a range of possible worlds considered both actually and counterfactually.[1] An analysis of a con-

---

[1] Internalist views of this stripe are defended by, *inter alia*, Jackson (1998, 2004, 2007, 2009), Chalmers (2004), Braddon-Mitchell (2004a, b, 2005, 2009), and Pettit (2004).

cept is, on this view, a systematization of the judgments that a subject is disposed to make. A subject can, in principle, come to know the content of her concepts by coming to know this complex set of dispositions. However, since subjects frequently do not have immediate access to these dispositions, these concepts are tacit. There is no suggestion that the folk explicitly entertain these concepts or that they could easily come to know that these are the concepts they deploy.

Why *care* which concept of time the folk deploy? Surely the nature of time is best understood through an understanding of various theories in fundamental physics, while the psychology of time (i.e. the nature of temporal experience) is best understood through an understanding of theories in cognitive psychology and neuroscience. What can any understanding of the folk notion of time tell us about time that is of interest? One might well think that there is little to be gleaned from analysing a folk concept since no such analysis can shed any light on what the world is, or must be, like. After all, folk concepts are *folk* concepts for a reason; as such we should be suspicious that they are coherent, informative, or match onto any feature of reality.

This is quite right. We do not think that one can simply read features of the world off of features of our concepts. Many of our concepts are incoherent, and others, while coherent, simply do not answer to anything in the world. But that does not make an analysis of our concepts uninteresting or useless; nor does it make such an analysis unhelpful in answering questions about the world. After all, if we want to know whether, for instance, there is any free will, we first need to know what it would take for there to be free will. We can know everything there is to know about the external world: we can know all of the laws of nature, and the location of all the fundamental particles and so forth, but unless we know what the world needs to be like in order for there to be free will, we won't know whether the world's being a certain way is a way that makes it true that there is free will. We typically assume that we know what it would take for a concept to have something answering to it in the world, and proceed to examine the world to see whether or not it does. But tricky concepts, such as free will, moral responsibility, and, inter alia, time, are concepts that are sufficiently complex and difficult that we first need to do some work to figure out what our concept is like so that we can then work out whether the world is such that something answers to that concept.

To give a sense of why this matters in the current case, we think there are two reasons why an analysis of the folk concept of time is needed. First, the folk concept of time underlies a lot of debates in temporal metaphysics. As noted above, there is still heated debate about the nature of time and of temporal relations. One such debate is the debate between A-theorists and B-theorists over the existence of the A-series. The A-series, as briefly mentioned in Sect. 3.1, orders events in terms of whether they are objectively past, present or future and is such that for any event in the A-series, that event instantiates a particular irreducible A-theoretic property, (pastness, presentness or futurity), that determines its place in that series. The location of events within the A-series is thus dynamic: a set of events, E, will be present, is future, and will then become past.[2] According to the A-theory the A-series exists:

---

[2] See for instance Zimmerman (2005).

it is an objective feature of temporal reality. According to the B-theory, by contrast, there are no irreducibly tensed properties of pastness, presentness or futurity and so the A-series does not exist. Rather, all that exists is the B-series, which orders events in terms of the relations of earlier-than, later-than and simultaneous-with.[3]

A-theorists frequently contend (and some B-theorists seem to agree) that the A-theory gets something right about the way we ordinarily think about time: the A-theory is closest to summing up the ordinary person's view of temporality.[4] It also gets something right about our temporal phenomenology, or so A-theorists contend: our temporal phenomenology is as of dynamical temporal passage. Indeed, some A-theorists go so far as to hold that our ordinary concept of time is such that the A-series is, according to that concept, essential to temporality (more on this below). If that is right then either we have a powerful reason to suppose that there is, actually, an A-series given that we think that actually there is time, or we have a powerful reason to suppose that there is actually no time, if we take ourselves to have good reason to think that there is no A-series.[5]

We cannot, of course, hope to answer questions about whether the A-theory or B-theory is right simply by looking at our concepts; nor can we hope to determine, by examining our concepts, whether or not there is actually time. But it does not follow from this that there is no role for an analysis of the folk concept of time to play in debates over the nature of time. By analysing the folk concept we can determine which of the A- or B-theory better accords with our everyday thought and talk about time which some may take to constitute a kind of evidence in favour of one or other of these views. In addition, by analysing our concept it is possible to gain a better sense of the conceptual connections between our concept of time and other nearby concepts, such as persistence and causation; the metaphysics of these latter notions—for some—is thought to bear on the debate over temporal reality. Most importantly, with an analysis of the folk concept of time in hand we can go some way toward determining what the world *must be like* in order for us to conclude that there is, or is not, time. Since part of the dispute between A- and B-theorists is precisely about this issue, i.e. about the essential features of time, an analysis of the folk concept would go some way towards adjudicating an element of this dispute.

The importance of analysing the folk concept of time is not restricted to debates in metaphysics. The second reason why analysing our folk concept is useful is that an understanding of our concept of time can help to shed light on a range of contemporary scientific theories. Recent work in physics has led some physicists and philosophers to claim that time does not exist (see, for example, Barbour 1994a, b, 1999; Barbour and Isham 1999; Deutsch 1997; Rovelli 2004, 2007, 2009; Tallant

---

[3] The B-theory is typically supplemented with an account of tensed talk and thought. The A-theorist takes tensed thought and talk to pick out A-theoretic properties. The B-theorist takes tensed thought and talk to be indexical, picking out the time at which a proposition is expressed either in speech or via some doxastic state.

[4] For instance, Putnam (1967), Schmidt (2006), and Deng (2013).

[5] McTaggart (1908) and Gödel (1949).

2008, 2010). There is a number of what are now known as timeless physical theories: theories within physics according to which our world lacks a one-dimensional substructure of ordered temporal instances that provides a metric for the measure of the distance between any two time instants. Many of these theories have been offered in response to the problem of time in canonical quantum gravity. Canonical quantum gravity involves the application of standard quantization techniques to the field equations of classical general relativity. This typically involves converting general relativity into Hamiltonian form and quantizing the theory by taking pairs of configuration and momentum variables and associating with each a pair of commutative operators[6] ranging over a Hilbert space (roughly: a generalisation of a Euclidean space into higher dimensions).[7] Canonical quantization techniques, when applied to canonical quantum gravity appear to strip away the time variable entirely. The problem of time is what to say about this situation: should time be recovered post-quantization or not? Proponents of so-called timeless physical theories claim that we should not attempt to recover the time variable. We should, rather, take canonical quantum gravity at face value, as telling us that time does not exist (see Anderson 2012a, b for discussion).

What remains unclear, however, is what advocates of these timeless theories mean when they ultimately conclude that there is no time. We are happy to grant that there might be some scientific concept of time, such that *that* concept turns out not to be satisfied if a timeless physical theory is true. The question remains, however, whether our ordinary "folk" concept of time is such that, were a timeless physical theory to be true, we would conclude that there is no time. This is, we think, an important question. While it is indeed a very interesting discovery if there is nothing in our world that answers to the scientific concept of time, it would be an even more startling outcome if nothing answered to our folk concept of time. For our folk concept of time is inextricably bound up with other concepts that are central to our lives, including: concepts of agency, of rational and moral deliberation, of persistence and of causation. Accordingly, if it turns out that nothing satisfies the folk concept of time and that this is what physics tells us, then it may be that no sense can be made of the related folk notions of deliberation, causation and persistence and thus of our conception of ourselves as agents in the world. On the other hand, it may turn out that even though the folk concept of time is not satisfied, it is still possible to make sense of these other, for want of a better phrase, "timeful" notions. Either way, there is much at stake; in order to judge the extent to which we ought to fear or be sceptical of timeless physical theories, we must first know something about our folk concept of time.

---

[6] A two-place operator Rxy is commutative just when for any a and b such that Rab, Rab if and only if Rba.

[7] This account of canonical quantization is taken from Fradkin (ms., 92) . For a more detailed (and more technically demanding) overview of canonical quantum gravity, see Isham (1993). See Kuchař (1992) for a more accessible, philosophical overview.

3.2.2   Resistance to Error

Before turning to some analyses of the folk concept of time we must, first, outline an important constraint under which any such analysis must operate. Our folk concept of time—whatever its content and structure—appears to be "resistant to error" in this sense: it is unlikely that we will discover something about our world that would lead us to conclude that the concept fails actually to be satisfied. Thus to say that a concept is resistant to error is analogous to saying that a discourse in which that concept plays a role is resistant to error theory.

There are some concepts such that we set the bar very low with respect to what the world needs to be like for something to satisfy those concepts. These are concepts for which there are, epistemically speaking, many candidate satisfiers of the concept, and are such that speakers are disposed to go a long way down the list of candidates before they decide that a candidate is not a good enough deserver to satisfy that concept. Such concepts are resistant to error because there is a vast array of ways the world could be, epistemically speaking, according to which speakers are disposed to say that the concept is satisfied.

To be clear then: that a concept is resistant to error does not mean that we cannot discover that nothing actually answers to the concept. Rather, a concept is resistant to error if it is a concept that plays such a central role in our lives, and in our conception of ourselves as agents acting in the world, that there are relatively few things we could discover about our world that would make us conclude that nothing answers to the concept. Contrast, for instance, our concept of a quark with our concept of agency. There are quite likely very many things we could discover about the world that would make us conclude that there are no quarks. Now consider our concept of agency. Of course there are very "thick" agential concepts such that there might turn out to be no agents in that sense. But consider just a thin notion of agency that the folk work with: the notion according to which there are self aware beings that deliberate about what they ought, prudentially and morally, to do, and act so as to bring about the things they take themselves to have reason to want to bring about. Agents, in this sense, are self-aware deliberators and manipulators of the world around them. Agency in this thin sense, we think, is quite likely resistant to error. There are things we could discover, perhaps, that would make us conclude that there is no agency: if for instance, all of us is really just a puppet of an alien race which makes it seem to us as though we deliberate and make decisions when really we don't. But there are *relatively* few things we could discover that would make us conclude that there is no agency; for to abandon the idea that we are reasoning, deliberating, things that attempt to manipulate the world around us would be to abandon any sense of ourselves in the world at all. Indeed, it would be to abandon the experimental method entirely, since the idea that we could manipulate variables in order to track down-stream effects would be inconsistent with the idea that there are no agents in this minimal sense. To recap, the idea that a concept is resistant to error is the idea that we give the world a lot of slack when it comes to providing us something that answers to that concept. We allow that there are lots of ways the world could be, consistent with our concept being satisfied. That does not mean there are no ways the world could be, such that the concept is not satisfied. It just means that it is less likely that we will discover there are no agents, than that we will discover that there are, for instance, no quarks.

It is not our contention that one can determine, from the armchair, whether a concept is in error or not. We do think, however, that we can have reason to think that it is more, or less, likely that a particular concept is resistant to error. One way to determine whether a concept is resistant to error is to look both at the historical record of speakers' reactions to certain relevant actual discoveries and to consult our dispositions regarding various scenarios about the way the actual world might be, for all we know—that is, various epistemic possibilities—and ask whether, were we to discover that the world is that way we would conclude that nothing in our world answers to our concept. The broader the range of discoveries we could make about the world such that we continue to think that our concept is satisfied, the more resistant to error our concept is, and the less likely we are to find that the concept is in error (i.e. is unsatisfied).

We grant that there are things we could discover about the actual world that would lead us to conclude that there is no time; being in error about this particular concept is not a conceptual impossibility. Nevertheless, we think there is reason to believe that our concept of time is resistant to error. First, there is some historical evidence that supports this claim, evidence concerning the stability of speaker's dispositions with respect to their concept of time through paradigmatic shifts. For instance, consider the shift from Newtonian mechanics to relativistic mechanics that occurred in the early part of the twentieth century. The Newtonian understanding of time is one according to which there is an absolute fact of the matter about the temporal ordering of events. If two events, A and B, are related by a particular B-theoretic relation, then that is an inalienable fact about reality, and one upon which every observer should agree (assuming they have appropriate access to the evidence and, as such, are epistemically on a par). The understanding of time that we find in the special and general theories of relativity is completely different. According to relativity, there is no fact of the matter about the B-series temporal ordering of events. For any two events A and B, whether A and B are simultaneous with one another, or whether they stand at some temporal distance to one another, depends entirely on an observer's state of motion. Indeed, if, for an observer, O1, events A and B are simultaneous with one another, then there is some observer O2 for whom A is earlier than B and there is some observer O3 for whom A is later than B. Worse than this, the temporal judgements made by all three observers are on a par: there is no physical reason to suppose that O1's judgements are "more correct" than O2's or that O2's are "more correct" than O3's.

The shift from Newtonian mechanics to relativistic mechanics, then, is revolutionary: according to the latter theory events are not objectively ordered by a single B-series. What we see, rather, is a number of different B-series orderings, each of which is equally good. But things get worse: general relativity does, in fact, make use of an invariant metric of some kind and so there is a sense in which some events are—objectively—in the "past" of an observer or in an observer's "future." However, the metric used to make such determinations is one that rolls space and time together into a single four-dimensional manifold in which time is treated as a space-like dimension. The sense in which an event is "past" for every observer is that some events are at a constant spatiotemporal distance from everybody. The shift from a single ordered B-series to a multiplicity of B-series orderings and then, finally, into a spatiotemporal metric is radical indeed.

This is just one part of a much larger story about how our understanding of time has changed. The scientific view of the nature of time, as posited by physicists, has changed substantially from the ancient Greeks, through Newton, to Einstein and now to recent developments in quantum gravity. Yet despite the fact that the features needed to satisfy the folk concept of time—for instance that there is an absolute fact of the matter regarding which events are simultaneous with which other events—are not found within our best scientific theories, it is notable that the folk have never openly declared that nothing satisfies their concept of time.

This brief history of time in the sciences tells us something about the folk concept. Ordinary folk will find themselves disposed to say, across a wide range of epistemic possibilities, that if the world turns out to be that way then their concept of time is satisfied, albeit by something rather different than what they had expected. That there is *something* in our world that is a good enough deserver to count as time typically seems more certain to us than any particular view we have about the metaphysical or physical nature of time. Because the folk concept of time appears to be satisfied across a wide range of epistemic possibilities the concept is not likely to be in error.

We come now to the second reason to suppose that our concept of time is error-resistant. Temporality is one of the most entrenched, fundamental, and pivotal notions, not least in terms of our conception of ourselves as agents who *did* things in the past, and who must *decide* what to do in the future. A concept that is central to our conception of ourselves in the world is less likely to be in error than those that are less central. Plausibly, the concepts of agency, deliberation, decision, prudence, responsibility, causation and time are important concepts of this kind.

If we think of our concepts as forming an interrelated web, with some concepts more central to the web than others, then it seems likely that the concepts of deliberation, prudence, responsibility, and causation will be at the centre of that web. For it is difficult to imagine discovering that our folk concepts of deliberation, agency and causation are unsatisfied. That is not to say that we cannot imagine any particular philosophical theory of those concepts being shown to be false: we can. It is to say that we cannot imagine discovering that nothing deserves to be called agency or deliberation or causation. We think our concept of time will also be at the centre. For our concept of time is implicitly and explicitly intertwined with the central concepts of deliberation, agency, causation, prudence and causation. Just as it is difficult to see how these concepts might be unsatisfied, it is also difficult to see how we could make sense of deliberation or agency (and perhaps causation) in the absence of time, at least as that notion is understood by the folk.

It is, moreover, plausible that our concept of time is central to our phenomenology: it *seems* to us that we acted at moments previous to this one; we seem to have memories of the past, it *seems* to us that what we did in the past causally affected the way things are now; it *seems* to us that the decisions we make now will affect our future but not, in general, our past, and so on. Additionally, it seems to us that some events are earlier than, later than or simultaneous with others. Since our concept of time is inextricably bound up with our phenomenology of ourselves in the world, a phenomenology we cannot imagine failing to have, it is plausible that that concept is central to our conceptual web.

That a concept is central to our conceptual web does not show that the concept cannot be in error; it does not show that the concept will be satisfied no matter what features the actual world has. But it does suggest that we are more likely to accommodate discoveries we make about our world by both conceptual change at the periphery of our conceptual web, and by setting the bar relatively low in terms of what the world needs to be like to satisfy those core concepts. That's because the cost associated with taking a concept core to the conceptual web to be unsatisfied is far too high: if we give up on a core concept in this way, then we are forced to hold that a range of other core concepts are unsatisfied as well. The cost is particularly high in the case of time, since the concept of time is central to our sense of self simply because it is central to a range of agential phenomena, such as moral and rational deliberation and causation. So taking the concept of time to be unsatisfied would, potentially, undermine the sense we have of ourselves as agents.

In sum, if a concept, C, is resistant to error then for conditionals of the form: if the actual world is ____, then nothing satisfies C, there is a limited number of ways of filling out the ____ to render the conditional true. As discussed, we think our concept of time is like this for two reasons. First, it is very resistant to paradigm shifts in science and, second, it lies at the core of a conceptual web that is central to our self-conception as agents. Any analysis of our concept of time must both be consistent with the fact that our concept of time is resistant to error, and ideally should explain what features of the content of our concept render it resistant in this manner. In the next section we examine two different candidate analyses of our concept and evaluate them according to this desideratum.

3.3 Our Concept of Time

3.3.1 A-Theoretic and B-Theoretic Analyses

The first kind of analysis we will consider is what we shall call a "one feature" analysis. These analyses typically take some feature to be essential to time, and then use this essentiality to analyse the concept; time just is, on the one feature analysis, this or that essential feature. The classic version of a one feature analysis of the folk concept of time is the analysis that appears to underpin the A-theory. According to many A-theorists, our concept of time is a concept of something that is essentially connected to dynamical change. If the A-theorist is right about this, then temporal relations must be at least partially grounded in the A-series. For only the A-series permits that time is genuinely dynamical by allowing that events change from being future, to being present to being past.

It was McTaggart (1908) who first suggested that our concept of time is a concept of something that is essentially dynamical. He then famously went on to argue that the A-series, which he took to be essential to time, is inconsistent, and hence concluded that there is no time. Gödel (1949) later made a somewhat different—though similar in spirit—argument for the conclusion that there is no time, by appealing to features of the A-series which he, too, took to be essential to time. The details of those arguments need not concern us. The point is that no one was convinced by their conclusions. There are, no doubt, some A-theorists who, like McTaggart, would conclude that there is no time were they to discover that there is no A-series; such A-theorists differ from McTaggart only in that they believe there is, actually, an A-series. We suspect, however, that such A-theorists are in a minority. Most of us, even many A-theorists who *explicitly* say that they would conclude that there is no time were they to discover that there is no A-series, would in fact continue on exactly as they do now upon such a discovery. It is our empirical speculation that upon such a discovery almost all of us would continue to feel guilt and pride at actions we take ourselves to have engaged in previously; most of us would continue to deliberate about what to do in the future; most of us would continue to form plans and intentions; most of us would continue to engage in causal reasoning, and so forth. Indeed, we doubt that the everyday life, and the everyday assertions of most folk would change at all upon the discovery that there is no A-series. At most, some A-theorists would explicitly say things like "there is no time" all the while continuing to act as if there is time. If our empirical speculation is right, then most speakers are disposed to hold that their concept of time is satisfied even if there is no A-series.

It is worth emphasising something at this juncture. A-theorists, and sometimes even B-theorists, sometimes suggest that the A-theory is a better characterisation of our folk concept of time, though of course B-theorists go on to argue that the A-series is either inconsistent, or at the very least, does not obtain in the actual world and that the B-series is a perfectly good deserver to satisfy that concept. It may be true that the folk are inclined explicitly to talk about time as though it flows; as though future events come ever closer until they become present, and then recede into the past. From this it does not follow that their concept is one according to which time is essentially dynamical. Recall that for us, the content of a concept is exhausted by what a speaker is ideally disposed to say about whether their concept of time is satisfied across a range of scenarios—i.e. epistemic possibilities considered as actual. That is consistent with a speaker assuming that time is essentially dynamical, and even with a speaker explicitly asserting that time is such. Nevertheless, if such a speaker is disposed to say that there is time if actually it turns out that there is no A-series then, for us, it follows that their concept of time is not one according to which the A-series is essential to time: such a speaker is simply wrong about their own concept. Thus if we are right and most speakers are disposed in this manner then this is enough to show that the A-series is not, according to our folk concept, essential for time.[8]

---

[8] Notice that it is consistent with what we say, here, about our concept of time, that it is metaphysically impossible for any world to seem the way our world seems, and to lack an A-series. Perhaps the A-theorist is right, and the A-series is necessary to produce our phenomenology of time. Then

Turn now to the B-theorist's one feature analysis of our concept of time. The B-theorist thinks our concept of time is a concept of something that provides an objective ordering of events via the relations of earlier than, later than, and simultaneous with: the B-theorist thinks that the B-series is essential to time.

Certainly, we think, if asked the folk are likely to say that time provides an ordering of events into earlier, later, and simultaneous sets of events such that there are measurable distances between events at different times. Quite likely folk are (or at least were) inclined to say that time not only orders events in this manner, but that there is an absolute fact of the matter regarding that order of events, and an absolute fact of the matter regarding the distance in time between events. So if we were inclined to think that the content of our folk concept is, at least in part, given by what the folk might explicitly assert about time, then we would likely think that the folk suppose the B-series to be an important feature of time. However, as already discussed, we now know that there is no absolute fact of the matter concerning the order of events, and yet—while this is widely known these days—we continue to accept that there is time; we are not even tempted to deny that time exists simply because there is no unique B-series ordering of events. It is doubtful, then, that an absolute B-ordering is essential to the folk concept.

This puts to rest a basic B-theoretic analysis of the concept of time. But there is, perhaps, a nearby analysis that does better. Rather than treating the B-series ordering as essential to time, it is the existence of an invariant metric of some kind in which the B-series plays a role that is essential for time. On this view, it is essentially something like space-time—i.e. a metric in which B-series orderings are woven together with spatial ones—that is needed to satisfy the folk concept. The folk concept continues to be satisfied in the face of the shift from Newtonian to relativistic mechanics, then, because enough of the B-series continues to live on within the framework of space-time.

Even this modified B-theoretic analysis of the concept, however, appears fraught. To see why, suppose, as some physicists suggest, that there is no B-series ordering of times at all. That is, there is not even an invariant spatiotemporal metric within which some sense can be made of a B-series. Events are not related by any invariant ordering relations at all. Anderson (2012a) calls such views *tempus nihil est* approaches to canonical quantum gravity.

To be sure, this would be a startling discovery. Nevertheless, we suggest that upon making such a discovery the folk would continue to think and talk as they had previously; they would continue to suppose that there exist events at other times; they would continue to have a phenomenology as of some events being past, and others future; they would continue to engage in deliberation about what they ought to do, and they would continue to engage in causal reasoning regarding how to bring about what they take to be desirable outcomes. Accordingly, we think that, under those circumstances folk would still not be disposed to say that there is no time, due to the second feature of the folk concept discussed above: namely, the centrality of the time concept to a web of similar timeful concepts that underpin agency. But if

---

it turns out that time is A-theoretic. But this is not a conceptual truth: our concept of time does not demand any such thing.

we are right in this piece of speculation about what the folk would do and say under such circumstances then it follows that no invariant ordering—be it a B-series, or a spatio-temporal metric composed of different B-orderings woven together into a four-dimensional manifold—is an essential feature of the folk concept of time.

Even if we are wrong about what the folk would do were they to learn that there is no invariant ordering relation over events, there is a further reason for dubiety concerning the general strategy under consideration, whereby some particular notion such as the A-series, the B-series or a spacetime metric is taken to be core to the folk concept of time. The problem is this: taking any particular ordering feature to be essential to the folk concept renders that concept extremely inflexible. The concept so construed cannot easily make sense of the manner in which the concept is resistant to error. It cannot, for instance, easily explain why it is that the shift from Newtonian to relativistic mechanics did not result in widespread error theory about time. Nor can it explain the shift from an A-theoretic conception of time to a B-theoretic conception of time more generally.

None of this, of course, shows that either the A-theory or the B-theory is false. It might easily be that although it is not part of our *concept* of time that the A-series is essential to time, it is nevertheless the case that, in our world, time is characterised by an A-series. It is just that this would be an empirical or metaphysical discovery, not a truth about our concept. *Mutatis mutandis* for the B-theory. The point, rather, is that if we take seriously the idea that an analysis of our folk concept of time must be sensitive to the fact that that concept is resistant to error, then we have reason to reject the contention that either the A-series or the B-series or, indeed, any particular invariant metric is, according to that concept, essential to time (at least if one accepts our speculation about what the folk would be disposed to do under conditions of making certain discoveries about the actual world). In what follows we move on to consider another kind of analysis of our concept of time, a conditional analysis, which is ultimately more flexible than the "one feature" analyses just considered and so promises to do better.

3.3.2   Conditional Analyses of Our Concept of Time

So-called conditional analyses of concepts came to prominence in the debate over the status of phenomenal concepts, such as the concept of a raw feel, or quale. Phenomenal concepts are, according to a conditional analysis, to be analyzed, roughly, as follows[9]:

> CA1) If there are nonphysical states of the relevant type in the actual world, phenomenal concepts are satisfied by these states, and
>
> CA2) If there are no nonphysical states of the relevant type in the actual world then phenomenal concepts are satisfied by physical states of the relevant type.[10]

Analyses of this broad type were motivated by the insight that even physicalists about the phenomenal typically find zombie worlds (minimal physical duplicates of

---

[9] The details vary depending on the particulars of the account.

[10] Stalnaker (2002), Hawthorne (2002), and Braddon-Mitchell (2003).

our world that lack phenomenal content) conceivable and hence, according to some lines of thought, should conclude that such worlds are possible. But if zombie worlds are possible then physicalism is false. Thus, physicalists need to explain why it seems so compelling that zombie worlds are conceivable, and hence possible, if in fact they are not. The conditional analysis of phenomenal concepts offered a way to do this. According to said analysis the appearance of the conceivability of zombie worlds is to be explained by the structure of our concept of a phenomenal state. The concept effectively tells us that if there were nonphysical dualistic states of a certain kind actually, these would be the phenomenal states. Since even physicalists give some non-zero credence to there actually being such states (even though they think there are not) such physicalists will seem to find zombie worlds conceivable. For such worlds are conceivable on the assumption that phenomenal states are actually non-physical states. Such worlds are not, however, conceivable if phenomenal states are physical states. Thus, the physicalist confuses the conceivability of zombie worlds given that CA1 is satisfied, with their conceivability given that CA2 is satisfied.

It is controversial exactly what makes for a conditional analysis of a concept. According to Majeed[11] we have a conditional analysis of a concept where not only does the term expressing the concept have different extensions in different contexts, but it also has different referencing-fixing conditions in different contexts. For Majeed, such analyses are ones that attribute two or more competing sets of reference-fixing conditions to a concept: what satisfies the concept, then, depends on which set of referencing-fixing conditions is the right one, and that, in turn, is determined by features of the actual world. According to Majeed, neither set of referencing-fixing conditions is privileged. Rather, the only way to determine which set of reference-fixing conditions in fact determines reference is by determining what the actual world is like.

Majeed argues that out concept of time is conditional in this way. He argues that there are two sets of reference fixing conditions associated with "time," i.e. "being ordered in an A-series" and "being ordered in a B-series" and that the right analysis of our concept of time is as follows:

T1) If there are A-properties in the actual world, and there is A-theoretic change, "time" refers to that which is ordered in an A-series.

T2) If there are no A-properties in the actual world, but there is B-theoretic change, "time" refers to that which is ordered in a B-series.

A conditional analysis of our concept of time such as this affords certain benefits. First, it goes some way towards explaining why our concept is resistant to error. For it tells us that even if there is no A-series, our concept will be satisfied so long as there is a B-series. So the analysis can explain why it is that philosophers are generally not persuaded by arguments that move from the inconsistency of the A-series to the unreality of time (such as McTaggart's famous argument, or the argument offered by Gödel). The reason why these arguments are unpersuasive is that the folk concept of time would continue to be satisfied even if there is no A-series actually,

---

[11] Paper presented at the Frontiers in the Philosophy of Time Conference, Kyoto, Japan.

simply because when there is no A-series actually that concept is satisfied according to the reference-fixing rule governing T2 in the conditional analysis.

But while Majeed's analysis is a step in the right direction, it still will not do as a general account of the folk concept of time. According to Majeed there is no a priori privileging of one set of referencing-fixing conditions over the other: they are two candidates to fix reference, either of which might, as it turns out, do the referencing fixing, depending on the nature of the world. If what is intended by Majeed's claim that the reference-fixing conditions are on a par is that speakers are equally committed to, in the case of our concept of time, T1 and T2, then this seems right. If a conditional analysis is the right analysis of our concept of time then speakers' dispositions ought to be such that if there are A-properties then time is what is ordered by the A-series, and if there is no A-series then time is what is ordered by the B-series. This is not really, however, to say that there is no a priori privileging whatsoever of one set of referencing-fixing conditions over another.

Some kind of conceptual priority is clearly built into any conditional analysis. After all, in the case of our phenomenal concepts, it seems clear that the relevant non-physical states could co-exist with the relevant physical states. The point is that the non-physical states are a *better* deserver to satisfy our concept "phenomenal state" and hence that it is *only if* there are *no* states of that kind that the physical states are what satisfies our concept of a phenomenal state (this is what we meant above when we said that the conditional concept "prioritises" the nonphysical states). Moreover, this is something that we can know a priori. We can see this clearly in the wording of the two conditionals CA1 and CA2 outlined above. In each case the antecedent mentions the presence, or absence, of the relevant non-physical states. The presence of the relevant physical states is only mentioned in the context of the non-existence of the relevant non-physical states. So there is a clear conceptual priority given to the presence of the relevant non-physical states. It is this which partly explains why, even though physicalists give low credence to such states obtaining, they still seem to find zombie worlds conceivable.

For if the concept prioritises the relevant non-physical states, then all it takes is some low credence in the actual existence of such states to render zombies conceivable. To see this more clearly, imagine the conceptual priority in the phenomenal case were reversed, to give us a conditional analysis such as the following:

CA1) If there are physical states of the relevant type in the actual world, phenomenal concepts are satisfied by these states, and
CA2) If there are no physical states of the relevant type in the actual world then phenomenal concepts are satisfied by nonphysical states of the relevant type.

If the analysis is reversed in this way, then even if one has a non-zero credence in there being nonphysical states of the relevant kind, it is hard to see why one would thereby take zombies to be conceivable. For assuming that one believes that there are physical states of the relevant kind, then one has no reason to think that zombies are possible, even if one gives some non-zero credence to there also being the relevant non-physical states.

We find the same conceptual privileging in the proposed conditional analysis of time. In effect the two conditionals implicitly tell us that it is the presence of

A-theoretic properties that is the best deserver to satisfy our concept of time. After all, the presence of A-properties is consistent with the presence of B-properties. Yet a world with A- and B-properties is, according to the analysis, one in which time is what is ordered by the A-series. It is only if there is no A-series that time is what is ordered by the B-series. In effect, then, the conditional analysis of time grants that what is ordered by the A-series is a better deserver to count as being time, given our folk concept of time, than is what is ordered by the B-series. Now, we think some A-theorists will find this a desirable outcome; it allows that what is ordered by the A-series is a better deserver to be time than what is ordered by only the B-series, yet it has the benefit that if there is no A-series, the A-theorist need not conclude that there is no time. Such an analysis might, indeed, capture the concept that some such A-theorists deploy. But we do not think it a good candidate to capture the folk concept of time.

To see why, consider the following thought experiment. Suppose that the actual world happened to be one in which half of the world had B-properties and no A-properties, and the other half had both A- and B-properties. According to the conditional analysis just offered, only half of the world would contain time.[12] Perhaps some A-theorists would embrace this conclusion. But most B-theorists would not; most B-theorists would conclude that both halves of the world contain time. Such B-theorists would either think that time is just somewhat different in the two halves of the world, or they world think that time is what is ordered by the B-series in both halves of the world, and one half of the world has an additional, metaphysically peculiar A-series that has nothing to do with time. More importantly, our guess is that ordinary folk would not be disposed to say that only half of the world has time if this were the discovery they made about our world. Certainly we think it most unlikely that those in the half of the world with only a B-series would conclude that they should move to the other hemisphere, since only by doing so can they get some time! The point here is that we do not think that the ordinary folk concept conceptually prioritizes the A-series in the manner in which the conditional analysis suggests that it does. So we are sceptical that this is the right analysis of our concept.

Even if one can make a case for the conceptual priority of the A-series over the B-series with respect to the folk concept, there is a further difficulty with the conditional analysis. As explicated above the conditional analysis is not sufficiently exhaustive. It therefore cannot do justice to the apparent flexibility of the folk concept of time, a flexibility that underlies its resistance to error. Not everyone thinks that the candidates to be time are exhausted by what is ordered by the A- or the B-series. As discussed above, some hold that time is best thought of spatiotemporally, as a space-like dimension within a four-dimensional manifold, one that cannot be characterized by any single B-series ordering. Still others hold that time is what is ordered by the C-series, which is a *symmetrical* ordering of events and so is not, strictly speaking, a B-series ordering (which is a strict total order of events). And

---

[12] With thanks to David Braddon-Mitchell in discussion.

then there are those who offer a causal theory of time, taking the time ordering to be given by a causal relation of generation (see, e.g. Tooley 1997).

The general problem for the conditional analysis, then, is that a great many more conditionals will need to be added. After T2, presumably we will have T3: if there are no A-properties and no B-properties, then time is a space-like dimension within a four-dimensional manifold across which signals can propagate at or below the speed of light. Then we will need T4, time is a C-series ordering. Then we will have T5, time is a causal relation of generation, and so on. This not only makes for a messy analysis, but there is plenty of scope for us to disagree about the various conceptual priorities: not just whether the A-series should be prioritized over the B-series, but whether both should be prioritized over the C-series, and so on down the line.

The analysis that we are about to offer is simpler and more informative. It tells us something important about time, rather than merely cataloguing a long list of things that, as it were, might be time depending on the way the world is.

3.3.3 Time Is a Functional Concept

We think that the folk concept of time is a functional concept. That is, very roughly, according to our folk concept of time, time is whatever it is that realises a particular functional role—the time role.[13] What is the time role? In what follows we consider a number of candidates until we settle on our preferred understanding of the time role. This examination of the time role will prove useful since it will allow us to note some important things about the relationship between our ordinary concept of time and the scientific concepts of time mentioned earlier.

One possibility is that the time role is the role spelled out by the function of the t-parameter in fundamental physical theory. Let us call this the physical time role. The physical time role best captures a particularly narrow, specialised concept of time as it is deployed within physics. It would be an interesting discovery, to be sure, if nothing played this role. And that is precisely what some of the aforementioned timeless physical theories suggest is the case. When physicists say that there is, according to those theories, no time, they are in effect saying that if one of the timeless physical theories is true, then the physical time role is unrealised. Even if that is the case, however, it does not follow that the folk concept of time is unsatisfied. For, we think, a timeless physical theory could be true, and yet the world seems just as it is, experientially speaking. Indeed, that must be so if a timeless physical theory is to be at all plausible, since otherwise any such theory would be flatly inconsistent with the everyday experiences we have of the world. But the world seems like it has time. Indeed, it seems as though part of what we mean when we talk about time just are the various "timey" experiences that we have: experiences as of deliberating about the future, regretting the past, reasoning about how to

---

[13] Something along these lines has been suggested by Craig Bourne (2006, 220–222) à la Lewis (1970) in the context of discussion about our concept of time.

manipulate the world around us, its feeling as though there is temporal passage, its seeming as though we have memories of the past, and so on. Since these timey experiences would still be present even if a timeless physical theory were true, we suspect that use of the folk concept of time would continue unabated.

A second reason for doubting that the physical time role is the time role corresponding to the folk concept relates back to the flexibility of the folk concept. If the physical time role were the correct role for elucidating the folk concept, then our explanation for the apparent resistance to error that our concept has would be poor. That is not to say that there would not be an explanation for some features of the resistance to error of that concept. If the physical time role is the role specified by the t-parameter of the best fundamental physical theory of a certain kind, then the physical t-role itself will change as scientific theory changes. Indeed, as long as the best theory of fundamental physics includes a t-parameter and something realises that t-parameter, the physical time role will be realised. So even the physical time concept will be somewhat resistant to error. It will not, however, be resistant to error in all of the ways that the folk time concept is resistant to error. For it is epistemically possible (and perhaps actually true) that the best fundamental physical theory posits no t-role. In such an event, trivially, nothing realises the physical time role. Yet because it is not clear that the folk concept thereby goes unsatisfied, we have reason to think that the physical time role and the time role that captures the folk concept are different. In essence, this is because there are fewer ways that the world could be, such that the folk concept of time goes unsatisfied, than there are ways the world could be such that the physical time role goes unrealised.

Moreover, one might think it unlikely that a folk concept will be structured around a role specified by scientific theory in a case in which the folk concept pre-dates the various physical theories in question. If we think there is continuity between the folk concept of time deployed many hundreds of years ago and our folk concept then the time role cannot be the physical time role. Even if we think that the folk concept includes an aspect of deference and so currently picks out the physical time role, it will not follow that the time role just is the physical time role. For suppose one thought that the folk concept, has, all along, deferred to "experts." Thus one might hold that time is whatever it is that is realised by the role that experts tell us is the time role. The current experts in question are physicists, so the time role is the physical time role. But the experts in the distant past were not physicists. Since we are not historians we are not sure what role the purported experts in fact did or would have pointed to in the past. But we are pretty confident it would not have been the physical time role as it is now understood. So even an appeal to deference that brings together the folk and physical time roles at this point in time fails to show that the time role and the physical time role are one and the same roles.

Finally, it is possible, and quite likely, actually, that different physical sciences disagree about the t-role. For instance, it seems plausible that this is the case with respect to quantum mechanics and general relativity: the thing that plays the time role in quantum mechanics does not obviously play the time role in general relativity and vice versa. That's because—very roughly—quantum mechanics appears to require an absolute time variable, something much closer to a classical conception of time from Newtonian mechanics. General relativity, by contrast, not only makes no use of an absolute time variable, it is deeply hostile to the existence of any such thing. Indeed, some models of general relativity are not even globally hyperbolic: they do not possess even a single total temporal ordering, one that orders all of the events in reality into an ordered series. So if we are going to set the content of the folk concept to the physical role, we need to ask: which one? We see no good way to answer this question without just picking one in an ad hoc fashion. Or at least any reasons we might have for selecting one physical role as the time role over another one would be based on a prior conception of time—e.g. one is more like the folk concept than the other. But that presupposes an account of the folk concept prior to the physical time role, and so the physical time role cannot be used to elucidate the folk concept.

A second possibility is that the time role is spelled out by the function of the t-parameter in the various special sciences—for instance in biology, evolutionary science, archaeology, palaeontology and so forth. Call this the special time role. Again though, similar problems arise. It seems likely that each of the special sciences will posit a somewhat different t-role. If so then there will be no special science time role, but rather, an array of different special science time roles. Even if the special sciences were intimately linked to our folk concept of time, it is hard to see on what basis we would decide that just one of these special science roles is the time role.

Even on the simplifying assumption that there is a single special science time role—either because all the special sciences posit the same role, or because we can abstract away from the particularities of each special science to discern a role that each has in common—it is still not plausible that that special time role is the time role undergirding the folk concept. For, once again, even if we discovered that nothing realises the special time role this would not obviously lead us to the view that there is no time. As with fundamental physics, in order for a given special science to be empirically adequate in a broad sense, it must not imply that we lack the experiences of the world we in fact have, experiences that seem to us to be strongly temporal. Any such science must therefore recover our timey experiences. But the existence of such experiences would be sufficient evidence, for many, that the time role is being played by something. So because the lack of anything to play the time role in a special science would not obviously lead us to cease to talk about past and future events, to cease deliberating, planning and intending; to cease reasoning about or manipulating our environment in a temporal way and so forth, we should conclude that the time role we are looking for is distinct from the special time role or, indeed, the physical time role.

While the physical and special time roles do not appear to be good candidates for explicating the folk concept of time, the discussion so far is instructive, for it points us in the right direction. A central difficulty with both the physical role and the special role is that the discovery that nothing plays either role would not obviously be the discovery that nothing plays the folk time role. So long as our timey experiences persist we have reason to think the folk role is being played by something.

This suggests that the functional role of the folk concept of time is closely connected to certain everyday ways of experiencing the world. In particular, we experience the world as being one in which we deliberate, plan, intend, and manipulate the world around us to attain certain ends we take to be desirable. We experience the world from an agential perspective. This perspective, as we have been stressing here, is deeply interconnected with our concept of, and experience of, temporality. We deliberate about events that we might bring about at times other than the one at which we currently find ourselves. We intend to act at times other than the one at which we currently find ourselves. Manipulating the world around us occurs by bringing about certain events, which, we take it, will in turn bring about other events. Thus our experience of our world is an experience as of persisting objects, most notably ourselves, other agents, and other objects. Which is to say that it is an experience as of events being located at different times, and as of different times being differentially related to one's current self via some kind of temporal ordering. It is also an experience as of certain events being causally connected. Our experience of the world is, finally, an experience not just of an ordering of events at different times but also an experience of there being a duration or distance between these events; it may even be an experience as of future events coming ever closer, and past events receding ever further away.

Let us call the experience as of deliberating our practice of deliberation; let us call the experience as of manipulating the world the practice of causal intervention; let us call the experience as of reasoning about how to manipulate the world the practice of causal reasoning; let us call the experience of acting in the world the practice of agency; let us call the experience as of existing at different times and the experience of tracking the same object at different times the practice of persistence.

These practices are all central to our way of being in the world. To be clear, however, it is not our contention that these are the only aspects of the folk concept of time. It may be that the folk concept of time is responsive to more than just deliberation, planning, intending, manipulating and so on. The folk concept may be richer by far. The point we are trying to make is that the folk concept is at least this rich and, what's more, that these aspects of the folk concept are an integral part of the everyday notion of time. Accordingly, an adequate account of time ought to be responsive to these core aspects and, as such, ought to forge a link between the practices mentioned above and a theory of those practices.

Note that by a theory, here, we mean an account of these practices that has certain features. First, we expect that any such theory will make sense of, and vindicate, these practices. That does not mean that we expect such a theory to make reasonable every instance of causal reasoning, of agency, of deliberation or of causal intervention. Rather, our suggestion is that because these practices are so central to our being in the world—they jointly constitute a large part of our being in the world—any good theory of these practices will be one that vindicates the practices themselves. That is to say that any good theory of these practices will be one that, at the very least, does the following: (a) it renders assertible a range of claimswithin the relevant discourses associated with the practices (causal, deliberation, agential) and it clearly draws a distinction between claims that are assertible in those discourses and those that are not and (b) it renders reasonable the practices in question. These are relatively minimal constraints; they do not require that a theory of the practices render the practices justified, in some deep epistemic sense; nor does it require that the theories render any particular claims in the discourses associated with the practices true. To get a feel for the difference here, consider our moral practices and the associated discourse. One way to vindicate the practices is to offer a realist theory of morality which not only renders moral discourse truth apt and true, but makes moral practices reasonable and justified. But notice that even error theorists about moral discourse will typically want to say that moral practice is reasonable and they will want to offer some account of when claims in the moral discourse are assertible and when they are not. They may do this by appealing to moral fictionalism, or moral noncognitivism, or some other view. But however they do so they, in some sense, vindicate moral practice. By parity, we think, it is even more important to vindicate the deliberative, causal and agential practices and we assume that the best theory of these practices will do just that.

With this in mind let us suppose that the best theory of deliberation and, more generally, practical reason will have a time parameter, a t-parameter. Call this the deliberative t-role. This is the role spelled out, and indeed exhausted by, the function of the t-parameter in the best theory of deliberation. Let us suppose that the best theory of causal intervention will have a t-parameter. Call this the intervention t-role. This is the role that is exhausted by the function of the t-parameter in the best theory of causal intervention. Let us suppose that the best theory of causal reasoning will have a t-parameter. Call this the causal reasoning t-role. This is the role exhausted by the function of the t-parameter in the best theory of causal reasoning. Let us suppose that the best theory of persistence will have a t-parameter. Call this the persistence t-role. This is the role exhausted by the function of the t-parameter in the best theory of our persistence. Finally, let us suppose that the best theory of agency will have a t-parameter. Call this the agential t-role. This is the role exhausted by the function of the t-parameter of our best theory of agency.

We suspect that there will be a good deal of overlap in these five t-roles; but we are happy to concede that each of these t-roles is different. What, then, is the connection between these five t-roles and the time role? It could be that just one of these t-roles is the time role, and the rest are closely related roles. Yet there seems no principled reason to suppose that to be so; particularly since if the roles are somewhat different it is conceivable that what realises the causation t-role is not what realises the deliberative t-role, which, in turn, is not what realises the persistence t-role and so forth. Since all five of these roles are clearly central to our temporal discourse it would be ad hoc to choose just one as the time role.

Our suggestion is that the time role is *the role of having all five t-roles realised*. The time role, then, is a higher-level role: it is the role of having some other set of roles realised, where, crucially, the set of lower-level roles in question are specified by the best theories of certain "timeful" phenomena that are central to our self-conception. We are inclined to think that the five t-roles, though distinct, come as a package. It is difficult to imagine that the causation t-role is realised but the persistence t-role is not. It is difficult to imagine that the deliberative t-role is realised but the causation or persistence t-roles are not. Thus we think there is relatively little danger that some, but not all, of the five t-roles will be realised. That is important; if the time role is the higher-level role of having all five t-roles realised then our concept will not be resistant to error if there is any real likelihood that even one of the t-roles might not be realised since in such an event our concept of time would be unsatisfied.

Hence on our view "time" might not pick out a single process or phenomenon. Different processes might realise each of the five t-roles, and thus the time role would be jointly realised by these five processes. But it is precisely this kind of flexibility that renders our account appealing. First, our analysis leaves it entirely open that what realises the physical time role and the special time roles is one and the same thing, and that what realises those roles is also what realises the folk time role. Thus it leaves it open that what the folk are talking about when they talk about time is what physicists are talking about when they talk about time, even though physicists and the folk are deploying somewhat different concepts. But it also leaves open that what realises the physical time role is *not* what realises the special time role. Thus it remains an open possibility that even if a timeless physical theory is true and nothing realises the physical time role, nevertheless the special time role is still realised. If that were the case then much of the special sciences would be vindicated even if the physical time role were unrealised. Of more interest to us is that our analysis renders it an open possibility that if nothing realises the physical time role (or the special time role) nevertheless the time role may still be realised. This nicely explains how physicists could be right to say "there is no time" given their concept of time, and yet be wrong to suggest that this means the folk should conclude that there is no time given the folk concept of time. Indeed, our analysis can explain why we should be pretty confident that our concept of time will be satisfied, and indeed, why we should be confident about this even if we think there is a reasonable chance that the physical time role will not be realised. Let us explain.

We can be confident that our folk concept will be realised even if the physical time role is not realised, if we can be confident that each of the t-roles will be realised even if the physical time role is not realised. There are only two circumstances in which the five t-roles could fail to be realised. The first is that our best theories of persistence, causation, deliberation and agency all include a t-role, but that role is not realised. The second is a circumstance in which the best theories of persistence, causation, deliberation and agency simply fail to include any t-role, and hence, trivially, that role fails to be satisfied. Let us consider each in turn.

If our best theories of these phenomena included a t-parameter but this parameter fails to be realised by anything then this is to say that our best theories of these phenomena are false. If our best theory of T is false, then presumably we ought to be error theorists about T. But that hardly seems likely in the case at hand: it is unlikely indeed that we will discover that there is no true theory of persistence, causation, deliberation or agency. To be sure, there might turn out to be no true theory of some metaphysically laden notions of persistence, causation, agency or deliberation. We, however, are interested in the best theory of these (relatively) ordinary notions; notions that figure in the way in which we all understand ourselves. It seems almost inconceivable that we could discover that there is no sense to be made of the idea that we plan, deliberate, intend, reason about how to bring about desirable ends, experience events as ordered, and so forth. It seems no more conceivable that we should discover this even if a timeless physical theory is true: if the physical time role is not realised. After all, any such theory must be consistent with the appearances of our world, and those appearances strikingly include the appearance of deliberation, causation, persistence and temporal phenomenology. Our best theory of said phenomenon is a best theory of those appearances: of the way things seem to us. Thus there is excellent reason to suppose that if our best theories of deliberation, persistence, causation and temporal phenomenology include a t-role, then that t-role will be realised even if the physical time role is not realised.

The other possibility is that the best theory of these five notions fails to include a t-role. But here, again, we think this most unlikely. It is difficult to see how to make sense of the ordinary notions of agency, deliberation, persistence and so forth without the appropriate theory having something like a t-parameter: after all, as agents who deliberate we are deliberating about actions that will occur at other "times"; we are deliberating about how to manipulate events at other "times" and so on. So some kind of t-role is sure to be found in the best theory of these phenomena.

Finally, our analysis explains why our concept of time is resistant to error. For our analysis remains utterly silent on what it is that plays the time role. Anything at all will do, for us, as long as the relevant role is played. Almost certainly there are, epistemically speaking, many possible candidates that could realise the time role actually; we can conceive of *any of these* being time since we can conceive of any of them realising the relevant role. On our analysis at best a number of these jointly realise the time role and thus turn out to be time. Our analysis also allows us to explain why there are many different epistemically possible scenarios under which we will say that there is time: every scenario under which something plays the time role, no matter how weird and wonderful that thing might be, is a live possibility. So there are very many ways the world could be, consistent with our concept being satisfied. In some sense this is exactly what we would expect from a functional analysis of the time concept: if time is a functional concept then it is multiply realisable. Multiple realisability, however, is exactly the kind of thing that renders a concept resistant to error.

3.4   Conclusion

Let's take stock. We have considered three putative analyses of the concept of time: a one feature analysis, a conditional analysis and a functional analysis. We have argued that it is a functional analysis that best accounts, on the one hand, for the relative flexibility in our concept and thus its resistance to error, and, on the other hand, to the centrality of the time concept to a range of other important concepts, such as causation, deliberation, persistence and prudence. If we are right, then there is much to be done. First, we must now take this analysis of the concept of time and feed it back into contemporary physical and metaphysical theories that are billed as timeless to see if such theories really do imply that nothing plays the time role. We have suggested that there is reason to doubt that such theories imply any such thing, but we recognise that a more careful study of the various timeless theories is required to fully establish this conclusion. Second, the conceptual relationship between the concept of time and the other central concepts just mentioned must be investigated more fully, so as to round out our conceptual understanding of temporality. We have begun this process but there is more to be done in, for instance, understanding the relationship between causation and time. Finally, and in a similar vein, the individual t-roles that, we have claimed, jointly constitute the higher-order time role need to be clarified. What, exactly, is the role of time in the theories of causation, persistence and prudence? What demands on time do these theories make? To answer these questions we must now return to the metaphysical, normative and epistemic debates over these various notions and refigure them through the lens of a functional approach to time.

References

Anderson, Edward. 2012a. Problem of time in quantum gravity. *Annalen der Physik* 524(12): 757–786.

Anderson, Edward. 2012b. The problem of time in quantum gravity. In *Classical and quantum gravity: Theory, analysis and applications*, ed. Vincent R. Frignanni, 1–25. New York: Nova.

Barbour, Julian. 1994a. The timelessness of quantum gravity: I. The evidence from the classical theory. *Classical and Quantum Gravity* 11(12): 2853–2873.

Barbour, Julian. 1994b. The timelessness of quantum gravity: II. The appearance of dynamics in static configurations. *Classical and Quantum Gravity* 11(12): 2875–2897.

Barbour, Julian. 1999. *The end of time*. Oxford: Oxford University Press.

Barbour, Julian, and Chris Isham. 1999. On the emergence of time in quantum gravity. In *The arguments of time*, ed. Jeremy Butterfield, 111–168. Oxford: Oxford University Press.

Bourne, Craig. 2006. *A future for presentism*. Oxford: Oxford University Press.

Braddon-Mitchell, David. 2003. Qualia and analytical conditionals. *Journal of Philosophy* 100(3): 111–135.

Braddon-Mitchell, David. 2004a. Folk theories of the third kind. *Ratio* 17(3): 277–293.

Braddon-Mitchell, David. 2004b. Masters of our meanings. *Philosophical Studies* 118(1–2): 133–152.

Braddon-Mitchell, David. 2005. The subsumption of reference. *British Journal for the Philosophy of Science* 56(1): 157–178.

Braddon-Mitchell, David. 2009. Naturalistic analysis and the a priori. In *Conceptual analysis and philosophical naturalism*, ed. David Braddon-Mitchell and Robert Nola, 23–44. Cambridge, MA: MIT Press.

Chalmers, David. 2004. Epistemic two dimensional semantics. *Philosophical Studies* 118(1–2): 153–226.

Deng, Natalja. 2013. Fine's McTaggart, temporal passage, and the A versus B-debate. *Ratio* 26(1): 19–34.

Deutsch, David. 1997. *The fabric of reality: The science of parallel universes and its implications*. London: Penguin.

Fradkin, Eduardo. ms. Chapter four: Canonical quantization. In *General field theory*.

Gödel, Kurt. 1949. An example of a new type of cosmological solutions of Einstein's field equations of gravitation. *Review of Modern Physics* 21: 447–450.

Hawthorne, John. 2002. Advice for physicalists. *Philosophical Studies* 108: 17–52.

Isham, Chris. 1993. Canonical quantum gravity and the problem of time. *Integrable Systems, Quantum Groups, and Quantum Field Theories* 409: 157–287.

Jackson, Frank. 1998. *From metaphysics to ethics: A defence of conceptual analysis*. Oxford: Oxford University Press.

Jackson, Frank. 2004. Why we need A-intensions. *Philosophical Studies* 118(1–2): 257–277.

Jackson, Frank. 2007. Reference and description from the descriptivists' corner. *Philosophical Books* 48(1): 17–26.

Jackson, Frank. 2009. A priori biconditionals and metaphysics. In *Conceptual analysis and philosophical naturalism*, ed. David Braddon-Mitchell and Robert Nola. Cambridge, MA: MIT Press.

Kuchař, Karel V. 1992. Time and interpretations of quantum gravity. In *Proceedings of the 4th Canadian conference on general relativity and relativistic astrophysics*, ed. G. Kunstatter, D. Vincent, and J. Williams, 1–91. Singapore: World Scientific.

Lewis, David. 1970. How to define theoretical terms. *The Journal of Philosophy* 67(13): 427–446.

McTaggart, John M.E. 1908. The unreality of time. *Mind* 17(68): 457–474.

Pettit, Philip. 2004. Descriptivism, rigidified and anchored. *Philosophical Studies* 118(1–2): 323–338.

Putnam, Hilary. 1967. Time and physical geometry. *The Journal of Philosophy* 64(8): 240–247.

Rovelli, Carlo. 2004. *Quantum gravity*. Cambridge: Cambridge University Press.

Rovelli, Carlo. 2007. The disappearance of space and time. In *The ontology of spacetime*, ed. Dennis Dieks, 25–36. Amsterdam: Elsevier.

Rovelli, Carlo. 2009. Forget time. ArXiv: 0903.3832. http://fr.arxiv.org/abs/0903.3832. Accessed 04 June 2014.

Schmidt, Martin. 2006. On the impossibility of hybrid time in a relativistic setting. *Kriterion* 20: 29–36.

Stalnaker, Robert. 2002. What is it like to be a zombie. In *Conceivability and possibility*, ed. Tamar S. Gendler and John Hawthorne. Oxford: Oxford University Press.

Tallant, Jonathan. 2008. What is it to "B" a relation? *Synthese* 162: 117–132.

Tallant, Jonathan. 2010. A sketch of a presentist theory of passage. *Erkenntnis* 73(1): 133–140.

Tooley, Michael. 1997. *Time, tense and causation*. Oxford: Oxford University Press.

Zimmerman, Dean W. 2005. The A-theory of time, the B-theory of time and "Taking tense seriously". *Dialectica* 59(4): 401–457.

Chapter 4
Psychological Time

Dan Zakay

Abstract People live in a constantly changing dynamic environment. Our internal environment is also dynamic and is characterized by biological and mental processes that are in constant flux. Without these changes in both the external and internal environments, life on earth would not exist. The dimension along which all these changes occur is called "time." Without dwelling on its exact nature, "time" can be represented by a clock and is a useful notion that provides a good explanation for physical phenomena in our external environment. Like other organisms, humans must be able to relate to "time" to survive and adjust to the external environment. This presupposes that information about "time" is conveyed and perceived. Since no known human perceptual system is dedicated to "time," subjective temporal experiences are likely to compensate for this lack. Specifically, these subjective experiences may be based on internal changes in events as reflected by internal clocks or memory processes. These changes are monotonically correlated with "time" and thus can provide useful information about its passage. Psychological time is a subjective feeling which is related to the temporal experiences. Nevertheless, psychological time differs from "time," because it is non-linear and because it is dependent on the nature of events occurring within a time period. The correspondence between psychological time and "time," though imperfect, is enough to enable reality testing and normal cognitive and social functioning. This chapter discusses and analyzes psychological time, its functions and nature.

4.1 Introduction

Time is a crucial notion. No understanding of human behavior and adaptation to our dynamic environment can be complete without it. But what is time?

People can sense the passage of time (Merchant et al. 2013). We can express the feeling that time is speeding by or creeping along. Yet, the mechanisms underpin-

---

This chapter is dedicated to the memory of Iris Levin, a great time researcher

D. Zakay (✉)
School of Psychology, The Interdisciplinary Center Herzliya, Herzliya 46150, Israel
e-mail: zakay.dan@idc.ac.il

© Springer International Publishing Switzerland 2016
B. Mölder et al. (eds.), *Philosophy and Psychology of Time*, Studies in Brain and Mind 9, DOI 10.1007/978-3-319-22195-3_4

ning time remain unclear. The nature of time and the ways to measure it have intrigued thinkers since Antiquity, and still puzzle us today when contemplating for instance the laws of modern physics (Levin and Wilkening 1989). Some physicists and philosophers have claimed that time does not exist, but others, including Baron and Miller (Chap. 3 of this volume), argue that there may be a scientific concept of time. They posit that temporality is crucial to our conception of our place in the world. Øhrstrøm (Chap. 2 of this volume) writes that time is not an object, but whatever is real exists and acts in time. This leads directly to the paradox that we can feel something which might not exist or exists in a way that we do not understand. Saint Augustine (354–430 AD), best expressed this paradox in his famous saying: "What then is time? If no one asks me, I know what it is, but if I wish to explain to him who asks me, I do not know." (Augustine 1955, Book 11).

Baron and Miller (Chap. 3 of this volume) suggest that the folk concept of time is a functional concept that is closely connected to certain everyday ways of experiencing the world. Here it will be argued that psychological time is a subjective feeling that corresponds to the poorly defined notion of physical time, which will be referred to here as "T". For the human observer, "T" is a dimension along which external events like day and night take place. Psychological time, however, is a product of the mind more than a reflection of natural chronometric order (Hughes and Trautmann 1995), and reflects our need to relate to "T".

This chapter does not attempt to delve into the existence of "T". It also accepts the existence of feelings and temporal experiences that we call psychological time, which is subjective and somehow corresponds to "T". The aim of the chapter is to explore the nature and characteristics of psychological time.

4.2 Psychological Time

Psychological time refers to a sense of the passage of "T" and temporal experiences related to succession, duration, simultaneity, pace and the order of perceived external and internal events. For example we might feel that a certain interval lasts longer than another or occurred before or after another.

What are the origins of psychological time? Is time perception a valid concept?

In order to answer this question we must discuss the meaning of perception. One perspective of perception would require the existence of a dedicated sensory and perceptual system aimed at perceiving an identified type of physical energy (Coren et al. 1999). Accordingly, there must be a sense organ which can receive the physical energy and translate it to neural activity. This neural activity should reach a specific brain area which can interpret the incoming neural activity and create a respective perceptual experience. An example would be color perception. This is not the case with "T". We can't identify any external energy which conveys information about "T", nor can we identify any specific sensory or perceptual system dedicated for "T". The brain areas which are involved with the perception of "T" are also not well defined. On the other hand, a broader definition of a perceptual system will be: any system which enables the internal representation of the external environment so that adaptive behavior becomes possible (Zakay and Bentwich 1997). This broad definition does not require that all the elements included in the narrow definition will exist. Take for example face recognition. This is a crucial ability without which humans will not be able to behave adaptively. Nevertheless, face recognition is not built on a sensory and perceptual organs dedicated to perceive a specific type of a physical energy, etc. Face recognition are built on complex neural and computational processes which are conducted in several brain areas and are based on complex neural inputs (Bruce and Young 1986). There is no doubt that face recognition is part of our perceptual system. So is "T". As said by Gibson (1975) events are perceivable but time is not. The passage of "T" and durations of intervals are derived by computational and judgmental processes which are based on different neural inputs. In some respect, psychological time is similar to other perceptual dimensions like color and sound intensity (Zakay et al. 2014a). In terms of the function of psychological time, there is no doubt that it is crucial the representation of the temporal aspects of the external environment. To conclude, it might be that the term "time perception" is not accurate and maybe it should be considered as metaphorical. However, the overall perspective which sees time perception as part of the overall human perceptual system is justified.

Evolution did not grant humans with a full perceptual system dedicated for the perception of "T", and one can question the reason for such an "evolutionary flaw". A possible answer, which is presented here just as a "food for thought," is that a dedicated perceptual system is not needed, since humans and other organisms are "psychological clocks." Any activity of ours, every heartbeat, movement or spoken word, any change in our mood or in our cognitive processing activity enables us to experience the flow of "T" across events. To a considerable extent, human behavior expresses time rather than being based on explicit representation of it (Michon 1990).

Most researchers agree that there is no single neurobiological locus that serves as the core (master) clock in the human brain (Merchant et al. 2013). Attempts to base psychological time on biological cycles and pacemakers has not yielded conclusive results. Aschoff (1985) found that estimates of short intervals in the range of seconds and minutes were unrelated to any aspect of the circadian timing system. Other researchers (e.g. Wearden 1995) have tried and failed to correlate psychological time with the daily cycle of body temperature or with the cycles of certain brain waves. Whereas judgments of very brief intervals of less than a second might be explained solely by neural net activities (Merchant et al. 2013), judgment of intervals longer than 1 s cannot be accounted for by neural activity alone. It seems that psychological time is a product of a concert of both biological and cognitive processes.

4.3 Psychological Time and Objective Time ("T")

Despite some correspondence which exists between psychological time and "T", there are several prominent differences between the two notions.

One important issue has to do with the linearity of time. Time is commonly conceived as being linear, in constant motion, moving from past thorough present to the future (the time arrow). This view is in the basis of the belief that all events happen in a linear sequence and that the direction is only one way: forward (Birx 2009). The empirical fact that physical events occur in only one time direction is codified as the second law of thermodynamics, which requires that the entropy of a system, a measure of disorder, must always increase, or at least remain constant when that system is isolated from the rest of its environment (Birx 2009).

Psychological time, on the other hand, is not necessarily linear. There can be pauses in our sense of the flow of "T", like in the case of day hallucinations (Bentall 1990). Psychological time does not have a clear "time arrow" from the past to the future, since humans can dream, hallucinate and imagine time flowing from the future to the past, or being engaged in mental activities like "time travel." Factors like mood and certain mental conditions including autism and schizophrenia, affect psychological time as well (Merchant et al. 2013).

Psychological time is certainly dependent and influenced by the nature of events occurring within an interval, as illustrated by temporal illusions (to be discussed later).

The difference between psychological time and "T" has been illustrated in isolation studies (e.g. Aschoff 1985) in which humans are placed in a "Time free" environments without cues about how much physical time has elapsed. Under such conditions psychological time deviates significantly from "T".

It is agreed that "T" is not influenced by the nature of events occurring within it. For example, a clock will measure same durations for similar intervals regardless of whether or not during a target interval there was war or peace, or whether or not feelings like sadness or joy prevailed. This is not the case with psychological time. Intervals of same clock time will be perceived differently in each one of the above mentioned situations. This is due to the mechanisms and processes which underlie the formation of temporal feelings.

4.4   Psychological Time as a Feeling

As stated earlier, psychological time is exhibited as a subjective feeling. This should be differentiated from knowledge about time or from logical operations about time like calculations and reasoning. For example, knowing that a basketball game endures about 40 min, is not psychological time since it is not a temporal experience. Calculation leading to the outcome that 20 min of the game elapsed and therefore 20 min are remaining until the end of the game is also not psychological time. However, while attending the game and without using a watch, one can feel that the game is too long or too short or that its pace is too slow or fast. These feelings are examples of psychological time. Another example of knowledge and reasoning about time is that of autobiographical time. Friedman (2004) indicated that people can know how much time elapsed since a certain event in their past by inference, but this is not a temporal feeling by itself. Damasio et al. (2000) investigated the neural basis of emotions and feelings. They found that brain areas such as the somatosensory cortices and the upper brainstem nuclei are engaged in feeling emotions. These areas are involved in the mapping and/or regulation of internal organism states. This indicates the close relationship between emotion and homeostasis. The findings also lend support to the idea that the subjective process of feeling emotions is partly grounded in dynamic neural maps, which represent several aspects of the organism's continuously changing internal state.

Zajonc (1980) suggests that affective judgments may be fairly independent of, and proceed in time, the sorts of perceptual and cognitive operations commonly assumed to be the basis of these affective judgments. Affective reactions to stimuli are often the very first reactions of the organism. Furthermore, affective reactions can occur without extreme perceptual and cognitive encoding. Zajonc concluded that affect and cognition are under the control of separate and partially independent systems that can influence each other in a variety of ways. This view is compatible with our view about the origin of psychological time.

4.5 Dimensions of Psychological Time and Its Origin

Temporal experiences and feelings refer to several dimensions including duration, succession, simultaneity, tempo and order in time. Two dimensions, however, form the building blocks of our temporal experiences—succession and duration (Wittmann and Paulus 2008). Because of its centrality we shall focus here on these two.

The perception of succession refers to the sequential characteristics of events and their temporal order. The perception of duration refers to the time interval subjectively experienced between two events or the persistence of an event over time.

The taxonomy of elementary temporal experiences derives from these two basic dimensions and comprises the perceptual phenomena of simultaneity and temporal order.

As was already explained, no sense organ which is dedicated to the sensation and perception of time is known. No kind of information directly reflecting "T" was identified so far. Some researchers have entertained the possibility that time tags are encoded and become part of memory and that these tags can be used to create temporal experiences. Unfortunately, this option was not strongly supported (Hintzman and Block 1971).

If so, what is then the origin of psychological time?

It is agreed that time is not a simple entity. Instead, it is probable that a diverse group of neural mechanisms mediates temporal judgments (Eagleman 2008). In general, stimuli are initially processed by low-level sensory mechanisms that have evolved for analyzing dimensions such as spatial location and motion. As such, stimulus-driven (bottom-up) processing of low level attributes may subserve the discrimination of temporal characteristics of brief events. Temporal information concerning events that occur with longer durations may also involve concept-driven (top-down) memory and cognitive processes (Lavie and Webb 1975). Thus it seems that the origins of psychological time and the processes which underlie it are dependent on the range of the objective durations assessed. Humans are highly sensitive to temporal changes and can experience events that last less than a few milliseconds. However, in this case, the interval seems instantaneous; i.e. it has no duration. If an event or an episode persists for longer than a few milliseconds, people experience, remember and may therefore be able to judge durations (Block 1989). Only beyond the range of 100–150 ms will people be able to discriminate time intervals as being of different durations. Judgments of time periods in the range of about half a second to a few minutes tend to be fairly veridical in that judged duration is related to the actual duration in an approximately linear way, with a slope of about 1.0 (Allan 1983). For longer intervals, the experienced durations of a time period are somewhat shorter compared to their actual duration, as well as being more variable. As stated earlier, it has been hypothesized that neural-network states may be utilized to time sub-second durations without the need for a dedicated clock (Laje and Buonomano 2013). But assessments of longer time periods are based on different cognitive processes. This view is also shared by Eagleman (2008) who argues that sub-second intervals are timed automatically, but seconds, minutes and longer intervals involve cognitive processes and appear to be underpinned by entirely different neural mechanisms. Wittmann (2009) also agrees and states that a body of evidence has shown that different time perception mechanisms are associated with different timescales.

4.6 Prospective and Retrospective Timing

William James (1890) argued that ongoing estimates of duration get longer as we become more attentive to the passage of time itself, whereas duration in retrospect lengthens as a function of the "multitudinousness" of the memories which time affords. The first type of duration judgment is called "prospective" and the second "retrospective." The experience of time is termed prospective when it is related to the duration of an ongoing interval and the observer is aware of the need to judge that duration. The experience of time is retrospective when an observer is not aware of the need to judge the duration. The need only becomes apparent upon the termination of the interval. Empirical findings as well as a comprehensive meta-analysis (Block and Zakay 1997) support the differences between prospective and retrospective experiences of time. These differences indicate that prospective and retrospective timing are governed by different cognitive processes, as suggested by Fraisse (1963) who argued that immediate time judgments are based on the changes we experience and later on the changes we remember.

Retrospective duration judgment can be successfully accounted for by the Contextual Change Model (Block and Reed 1978). According to this model, when retrospective timing is required, people retrieve the contextual changes that were encoded during the target interval from memory. These contextual changes include both external changes such as the level of lighting of a room, and internal changes such as mood or the complexity of information processing. Retrospective duration judgment is a function of the amount of retrieved contextual changes. The more contextual changes are retrieved, the longer the duration is judged to be. This reflects a heuristic where the occurrence and encoding of contextual changes increase with objective time. As a result, when information processing during an interval is complex, the interval is judged to be longer in retrospect than a similar interval in which information processing was simple.

Prospective duration judgment is a function of the amount of attentional resources allocated to timing. The higher the allocated resources, the longer prospective duration judgment is. At any given moment in "time," attentional resources are split between all the concurrent tasks that need be carried out simultaneously, including timing. More attentional resources are allocated to complex tasks than to simple ones (Kahneman 1973) and for this reason, fewer attentional resources are available for timing in the former than in the latter case. Thus prospective duration judgments of the same time periods are longer when concurrent non-temporal tasks are simple than when they are complex. This finding is the opposite of the pattern observed for retrospective duration judgments (Block and Zakay 1997; Zakay and Block 1997).

4.7   Attending to Time

Early attentional models of prospective timing (e.g. Thomas and Weaver 1975; Zakay 1989) suggested that attention is focused on events and stimuli that represent changes in the internal and external environments like walking or heartbeat (see earlier the idea of "psychological clocks"). These models, however, were too vague and imprecise. A successful animal model (Gibbon et al. 1988) provided a good explanation for animals' temporal behavior. This model was based on the notion of an internal clock. Zakay and Block (1995) introduced the Attentional Gate Model (AGM) by adding an attentional gate to Church and Gibbon's model. The attentional gate notion itself was first introduced by Reeves and Sperling (1986). The gate is controlled by the amount of attentional resources and determines the number of pulses emitted by a pacemaker that can pass through the gate. The pacemaker emits the pulses continuously at a constant pace. The pulses that pass through the attentional gate are accumulated and counted in a cognitive counter. The more attentional resources are allocated for timing, the wider the attentional gate is opened, allowing for more pulses to pass through the gate and be accumulated in the counter as compared to a state in which a low amount of attentional resources are allocated to timing. In the latter state, the attentional gate is not opened wide and the number of pulses going through it that are accumulated and counted in the cognitive counter is smaller than in the first state. Prospective duration judgment is a function of the number of counted pulses, and hence will be longer when more attentional resources are allocated for timing than when fewer attentional resources are allocated.

The Dynamic Switch Model (Lejeune 1998) is another attentional model of prospective timing but does not use an attentional gate. Instead of a gate, a dynamic switch, controlled by attention, is either opened or closed at a frequency determined by the amount of attentional resources allocated to timing. With more attentional resources, the higher the frequency and the larger the number of pulses that can be accumulated and counted. The AGM and the Dynamic Switch Model are very similar and both make similar predictions (Zakay 2000).

4.8   Meaning, Temporal Relevance and Daily Temporal Experiences

In the present section we describe how the contextual-change model and attentional models of timing can be employed for explaining daily prospective and retrospective temporal experiences. But first, we should refer to the Cognitive Orientation model.

Cognitive Orientation theory is a general theory for explaining behavior which has been supported in many empirical studies (Kreitler and Kreitler 1972). According to the theory, the meaning of each referent is constantly extracted by dedicated analyzers. The referent may range from an object, an abstraction, a process, an activity or a whole situation. From the meaning of a referent the direction of behavior and the actions which should take place are evoked (Zakay and Barak 1984). The meaning of an object or an event is composed of values along meaning dimensions. There are twenty-one meaning dimensions like: The purpose or role of an object or an act, the structure, weight quantity, location, etc. of a referent, feelings or emotions that are evoked by the situation, event or an object (the referent), etc. One of the meaning dimensions is the referent's temporal qualities (Kreitler and Kreitler 1968).

We suggest that ongoing temporal experiences are built by respective processes which are based on the values assigned to the temporal meaning dimensions (the term "meaning dimensions" should not be confused with the dimensions of psychological time). During the course of time we sometimes relate to them prospectively and sometimes retrospectively. What determines the shifts between prospective and retrospective timing? Zakay (1992) suggested that the type of timing processes which is activated is determined by the importance (temporal relevance) of time in the analyzed situation we face. Temporal relevance is deducted from the temporal qualities of the meaning of the situation. When temporal relevance is high, prospective timing is activated and the executive system allocates considerable attentional resources to it. When time is not important, a retrospective state is induced. For example, enjoying ourselves on the beach at the beginning of a long vacation with no obligations will induce a retrospective state in which we will not feel the passage of "T", but when one has to complete a task within 10 min or fail, prospective timing will prevail. Zakay (2012) employed this model for providing explanations for many of our familiar daily temporal experiences. One example is that of waiting

(Osuna 1985). While waiting "T" plays a major role. Temporal relevance is high and a prospective timing process is induced. As a result the attentional gate is opened wide and the number of pulses which are accumulated and counted in the cognitive counter is also high. The feeling will be of "T" slowing down. Waiters typically look again and again at their watches, only to find out that in contrast to the feeling that a great deal of "T" has passed since the last time they looked, the clock's hands have hardly moved at all. Other familiar temporal experiences such as "time flies when we are having fun" can also be explained by the model above. While having fun, attention is not allocated to timing and hence we do not feel the passage of "T". The hands of a clock, however, continue to move and therefore when we look we are amazed to find out how much time has actually elapsed. An opposite situation was described by Loftus et al. (1987) who found that the duration of earthquakes is significantly overestimated. During earthquakes people's main wish is for the frightening event to be over, and hence they time the duration prospectively, similar to a waiting situation. Temporal illusions can also be explained by the model outlined above.

4.9 Temporal Illusions

Perceptual illusions were defined as a perception of a thing which misrepresents it, or gives it qualities not present in reality, or as distortions or incongruities between percept and reality (Zakay and Bentwich 1997).

Time perception is surprisingly prone to measurable distortions and illusions (Eagleman 2008). Indeed many states in which time perception does not faithfully represent what is regarded as objective "Time" are identified (Zakay 2009). Due to the nature of duration judgment processes (as described earlier), experiences of durations are prone to various illusions in which temporal experiences are distorted and biased. One example is known as the "empty time illusion." In a prospective duration judgment, intervals filled with complex data are judged to be shorter in duration than the same "empty" interval in which no data are processed. The opposite is true when the duration judgment is retrospective. Thus, same objective periods of "T" are translated into different psychological time values. This is also true in general when same clock time intervals are judged for its duration in a retrospective or a prospective judgment. Prospective duration judgments tend to be longer than respective retrospective ones (Block and Zakay 1997).

Another time illusion is summed up by the saying: "A watched pot never boils." In an experiment bearing out this illusion, a person is asked to watch a transparent container full of water that is placed on a flame. The person is asked to wait until the water boils. This is done prospectively. In a control condition the person is not asked to wait until the water boils (Block et al. 1980). After a certain period of time (say 90 s) the person is asked to judge how much time elapsed since the beginning of the experiment. The person who was waiting for the boiling of the water will typically judge the duration to be longer than the person who was not asked to wait until the water boils. The reason is that more attentional resources are allocated for timing when one has to wait for something to occur (the boiling of the water). Based on the Attentional Gate model (see earlier) this is well explained. Actually this is true for every waiting situation in which one has to wait for an event to occur. Another illusion, already described by Fechner as early as 1869, is called the "time order error," which is manifested in the influence of the temporal sequence of events' appearance on the estimation of their relative duration. In many cases an interval which occurred first will be perceived as longer or shorter than an interval of an identical objective duration that occurred second in time (in dependence on the respective conditions). There is still no agreement about the exact explanation of this illusion.

Gruber and Block (2013) argue that actually the temporal experience of the flow of time is an illusion by itself, because "T" is actually not continuous. However, our feeling is that "T" is flowing smoothly.

Temporal illusions demonstrate the extent to which psychological time is influenced by the nature of events occurring within "T" and the extent to which psychological time is context dependent.

Many more temporal illusions are known to influence various dimensions of psychological time (see Zakay 2009), but this is not the place to discuss them all.

4.10   The Functions of Psychological Time

In general, psychological time reflects the amount of changes experienced by a person. The changes can be perceptual, emotional, cognitive or physiological. The link between psychological time and change is exemplified by counting behavior while judging durations. For examples, when judging durations of "empty" intervals people create changes by counting or by other bodily movements. Counting is also the primary strategy generated by children without instructions, when asked to judge durations (Levin 1989). The monotonic correspondence between psychological time and "T" enables people to adapt to their physical and social environments. In addition to this vital function, psychological time fulfills some other vital functions like in the planning and performing of psychomotor activities and movements (Flanagan and Wing 1997). Psychological time also plays an important role in monitoring the durations involved in activities and comparing them to norms, thus enabling control over its regularity. An example will be meta-cognitive control during verbal communication, such as during a conversation. The temporal structure of a conversation is important and people analyze response latencies between a question and a response when interpreting the social meaning of the exchange (Boltz 2005). If the response latency is too long compared to temporal expectations, the reliability of the response is placed in doubt (Zakay et al. 2014b). Thus, duration judgment during conversations is highly significant and temporal relevance is high. This causes an activation of prospective timing during conversations, in which discrepancies between temporal expectations and actual durations of responses' latencies serve as cues for the interpretation of the quality of the communication.

Michon (1972) introduced the idea of considering time as information. He meant that temporal experiences provide information about the succession of events. Zakay (2014) elaborated this idea and argued that psychological time provides the executive system with information about the overall information processing load in the system at any specific moment. As an example, Zakay (2014) suggested that the feeling of boredom is actually a kind of a warning signal when the overall level of information processing is not optimal. Boredom is an aversive situation in which an individual experiences low information processing load. Because this is an aversive situation and people like it to end as soon as possible, temporal relevance is high and attentional resources are allocated for timing. The result is a feeling of impatience and of "T" slowing down, similar to experiences while waiting. Thus, boredom is alerting the person to the need to do something to increase the information processing load.

4.11 Conclusions

We did not resolve the St. Augustine paradox, but maybe narrowed it a little bit. The existence of both "Time" and psychological time should be admitted. It should also be admitted that each one is a different entity with some correspondence existing between the two. This correspondence is essential for adaptation to our physical and social environments. The importance of psychological time and of temporal experiences is reflected by the following clinical case of short memory syndrome. This is a rare state in which short-term memory fails to function, which prevents new information from entering the memory system. This is what happened to Clive Wearing (France 2005) who in 1985 contracted a herpes simplex virus that attacked his central nervous system and caused damage to his hippocampus. As a result his memory only lasts between 7 and 30 s. It can be said that Wearing is trapped within a "bubble" of a 30 s present. He does not have a sense of the flow of time and lacks significant temporal experiences. Of course, his quality of life is very poor. This case demonstrates the strong link between psychological time and memory, and that "T" and psychological time are two different entities (Wearing's age continues to increase), and the importance of a correspondence between "T" and psychological time. Being able to relate to "T" is essential for adaptive living. Regardless of its exact meaning, "T" shapes many aspects of our daily lives (Buhusi and Meck 2005). Being able to relate to "T" is a must for performing vital activities such as the ability to communicate or to perform actions in the right order.

To sum up, psychological time is a subjective feeling which emerges as a result of the interaction between neural-level processes and complex cognitive processing, in dependence on the objective time scale and the context. This interaction produced temporal experiences which monotonically correspond with "T". "T" is not perceived in the full sense of the notion of perception, but, adapting a broader perspective, psychological time is certainly a perceptual dimension which provides human's information processing system with several types of vital information.

Because of its importance, psychological time should be intensively studied from all relevant perspectives.

References

Allan, Lorraine G. 1983. The perception of time. *Perception & Psychophysics* 26(5): 340–354.
Aschoff, J. 1985. On the perception of time during prolonged temporal isolation. *Human Neurobiology* 4: 41–52.
Augustine, Saint. 1955. *Confessions*. Trans. and ed. Albert C. Outler. http://www.ccel.org/ccel/Augustine/Confessions. Accessed 20 Aug 2014.
Bentall, R.P. 1990. The illusion of reality: A review and integration of psychological research on hallucinations. *Psychological Bulletin* 107(1): 82–95.
Birx, H.J. (ed.). 2009. *Encyclopedia of time*. Los Angeles: Sage.
Block, Richard A. 1989. Experiencing and remembering time: Affordances, context and cognition. In *Time and human cognition: A life-span perspective*, ed. Iris Levin and Dan Zakay, 333–361. Amsterdam: Elsevier.
Block, Richard A., and M.A. Reed. 1978. Remembered duration: Evidence for a contextual-change hypothesis. *Journal of Experimental Psychology: Human Learning and Memory* 4: 656–665.
Block, Richard A., and Dan Zakay. 1997. Prospective and retrospective duration judgments. A meta-analytic review. *Psychonomic Bulletin & Review* 4(2): 184–197.
Block, Richard A., E.J. George, and M.A. Reed. 1980. A watched pot sometimes boils: a study of duration experience. *Acta Psychologica* 46: 81–94.
Boltz, Marilyn G. 2005. Temporal dimensions of conversational interaction: The role of response latencies and pauses in social impression formation. *Journal of Language and Social Psychology* 24: 103–138.
Bruce, Vicki, and Andy Young. 1986. Understanding face recognition. *British Journal of Psychology* 77(3): 305–327.
Buhusi, Catalin V., and Warren H. Meck. 2005. What makes us tick? Functional and neural mechanisms of interval timing. *Nature Reviews Neuroscience* 6: 755–765.
Coren, S., L.M. Ward, and J.T. Enns. 1999. *Sensation and perception*, 5th ed. Fort Worth: Harcourt Brace.
Damasio, Antonio R., T.J. Grabowski, A. Bechara, H. Damasio, L.L. Ponto, J. Parvizi, and R.D. Hichwa. 2000. Subcortical and cortical brain activity during the feeling of self-generated emotions. *Nature Neuroscience* 3(10): 1049–1057.
Eagleman, David M. 2008. Human time perception and its illusions. *Current Opinion in Neurobiology* 18: 131–136.
Flanagan, J.R., and A.M. Wing. 1997. The role of internal models in motion planning and control: Evidence from grip force adjustments during movements of hand-held loads. *The Journal of Neuroscience* 17(4): 1519–1528.
Fraisse, P. 1963. *The psychology of time*. New York: Harper and Row.
France, Louise. 2005. The death of yesterday. *The Observer*. January 23.
Friedman, W.J. 2004. Time in autobiographical memory. *Social Cognition* 22(5): 591–605.
Gibbon, John, Russell M. Church, Stephen Fairhurst, and Alejandro Kacelnik. 1988. Scalar expectancy theory and choice between delayed rewards. *Psychological Review* 95(1): 102–114.
Gibson, James J. 1975. Events are perceivable but time is not. In *The study of time*, vol. 2, ed. J.T. Fraser and N. Lawrence, 295–301. Berlin: Springer.
Gruber, R.P., and Richard A. Block. 2013. The flow of time as a perceptual illusion. *The Journal of Mind and Behavior* 34(1): 91–100.

Hintzman, D.L., and Richard A. Block. 1971. Repetition and memory: Evidence for a multiple-trace hypothesis. *Journal of Experimental Psychology* 88(3): 297–305.

Hughes, Diane Owen, and Thomas R. Trautmann (eds.). 1995. *Time: Histories and ethnologies.* Ann Arbor: The University of Michigan Press.

James, William. 1890. *The principles of psychology*, vol. 1. New York: Holt.

Kahneman, Daniel. 1973. *Attention and effort.* New York: Prentice Hall.

Kreitler, Shulamith, and Hans Kreitler. 1968. Dimensions of meaning and their measurement. *Psychological Reports* 23: 1307–1329.

Kreitler, Hans, and Shulamith Kreitler. 1972. The model of cognitive orientation: Towards a model of human behavior. *British Journal of Psychology* 63(1): 9–30.

Laje, Rodrigo, and Dean V. Buonomano. 2013. Robust timing and motor patterns by taming chaos in recurrent neural networks. *Nature Neuroscience* 16: 925–933.

Lavie, Peretz, and Wilse B. Webb. 1975. Time estimation in a long-term time-free environment. *American Journal of Psychology* 88: 177–186.

Lejeune, Helga. 1998. Switching or gating? The attentional challenge in cognitive models of psychological time. *Behavioural Processes* 44(2): 127–145.

Levin, Iris. 1989. Principles underlying time measurement: The development of children's constraints on counting. In *Time and human cognition: A life-span perspective*, ed. Iris Levin and Dan Zakay, 145–181. Amsterdam: Elsevier.

Levin, Iris, and F. Wilkening. 1989. Measuring time via counting. In *Time and human cognition: A life-span perspective*, ed. Iris Levin and Dan Zakay, 119–143. Amsterdam: Elsevier.

Loftus, E.F., J.W. Schooler, S.M. Boone, and D. Kline. 1987. Time went by so slowly: Overestimation of event duration by males and females. *Applied Cognitive Psychology* 1: 3–13.

Merchant, Hugo, Deborah L. Harrington, and Warren H. Meck. 2013. Neural basis of the perception and estimation of time. *Annual Review of Neuroscience* 36: 313–336.

Michon, John A. 1972. Processing of temporal information and the cognitive theory of time experience. In *The study of time*, ed. Julius T. Fraser, F.C. Haber, and C.H. Müller, 242–258. Berlin: Springer.

Michon, John A. 1990. Implicit and explicit representations of time. In *Models of psychological time*, ed. Richard A. Block, 37–54. New Jersey: Lawrence Erlbaum.

Osuna, Edgar Elias. 1985. The psychological cost of waiting. *Journal of Mathematical Psychology* 29: 82–105.

Reeves, Adam, and George Sperling. 1986. Attention gating in short-term visual memory. *Psychological Review* 93(2): 180–206.

Thomas, Ewart A.C., and Wanda B. Weaver. 1975. Cognitive processing and time perception. *Perception & Psychophysics* 17: 363–367.

Wearden, John H. 1995. Feeling the heat: Body temperature and the rate of subjective time, revisited. *The Quarterly Journal of Experimental Psychology, Section B* 48(2): 129–141.

Wittmann, Marc. 2009. Psychology and time. In *Encyclopedia of time*, ed. J.H. Birx, 1057–1064. Los Angeles: Sage.

Wittmann, Marc, and Martin P. Paulus. 2008. Decision making, impulsivity and time perception. *Trends in Cognitive Science* 12(1): 7–12.

Zajonc, Robert B. 1980. Feeling and thinking: Preferences need no inferences. *American Psychologist* 35(2): 151–175.

Zakay, Dan. 1989. Subjective time and attentional resource allocation: An integrated model of time estimation. In *Time and human cognition: A life-span perspective*, ed. Iris Levin and Dan Zakay, 365–397. Amsterdam: Elsevier.

Zakay, Dan. 1992. On prospective time estimation, temporal relevance and temporal uncertainty. In *Time, action and cognition*, ed. F. Macar, V. Pouthas, and W.J. Friedman, 109–111. Dordrecht: Kluwer Academic Publishers.

Zakay, Dan. 2000. Gating or switching? Gating is a better model of prospective timing. *Behavioural Processes* 50(1): 1–7.

Zakay, Dan. 2009. Temporal illusions. In *Encyclopedia of time*, ed. J.H. Birx, 1270–1272. Los Angeles: Sage.

Zakay, Dan. 2012. Experiencing time in daily life. *British Psychologist* 25(8): 578–582.

Zakay, Dan. 2014. Psychological time as information: The case of boredom. *Frontiers in Psychology, Perception Science* 5: 917. 1–5.

Zakay, Dan, and A. Barak. 1984. Meaning and career decision making. *Journal of Vocational Behavior* 24(1): 1–14.

Zakay, Dan, and Jonathan Bentwich. 1997. The tricks and traps of perceptual illusions. In *The mythomanias: The nature of deception and self-deception*, ed. Michael S. Myslobodsky, 73–104. New Jersey: Lawrence Erlbaum.

Zakay, Dan, and Richard A. Block. 1995. An attentional-gate model of prospective time estimation. In *Time and the dynamic control of behavior*, ed. M. Richelle, V. De Keyser, G. d'Ydewalle, and A. Vandierendonck, 167–178. Liège: Université de Liège.

Zakay, Dan, and Richard A. Block. 1997. Temporal cognition. *Current Directions in Psychological Science* 6(1): 12–16.

Zakay, Dan, Arie Bibi, and Daniel Algom. 2014a. Garner interference and temporal information processing. *Acta Psychologica* 147: 143–146.

Zakay, Dan, Dida Fleisig, and Neta David. 2014b. Prospective timing during conversations. In *Time and temporality in language and human experience*, ed. Barbara Lewandowska-Tomaszczyk and Krzysztof Kosecki, 185–197. Frankfurt am Main: Peter Lang.

Part II
Presence

Part 7
Presence

Chapter 5
Relative and Absolute Temporal Presence

Sean Enda Power

Abstract Different ways of thinking about presence can have significant conse-quences for one's thinking about temporal experience. Temporal presence can be conceived of as either absolute or relative. Relative presence is analogous to spatial presence, whereas absolute presence is not. For each of these concepts of presence, there is a theory of time which holds that this is how presence really is (and that the other concept of presence is merely derivative in some way). For the A-theory, tem-poral presence is absolute; it is a special moment in time, a time defined by events in what has been called the A-series. For the B-theory, temporal presence is relative; it is itself defined relative to moments in time, a time defined by events in the B-series. Many A-theorists (presentists) go further to claim that the present is the only real moment in time; the past and future are unreal. One can have different sets of problems depending on whether one thinks in terms of absolute presence or rela-tive presence. For example, there is the concept of the "specious" present—a dura-tion many theorists claim that we perceptually experience. It is argued in this paper that the specious present has problems given absolute presence, which it does not have given relative presence. Many of the problems are avoided by having an extended present. However, A-theory, the standard theory of time which advocates absolute presence, cannot have an extended present. Further, the best solution for absolute presence which is extended, *durational presentism*, involves denying the standard theories in the philosophy of time.

5.1 Introduction

Here is something we all experience: we perceptually[1] experience *present change*. We don't just experience things as having changed (in the past) or as about to change (in the future). We also experience things changing—now, in the present. We see

---

[1] For convenience, throughout this paper I will usually refer to perceptual experience as just "expe-rience." It could, however, also mean other kinds of experience, such as memory-experience or experiences had in acts of imagining or anticipating. The issues involving experienced change here

S.E. Power (✉)
Department of Philosophy, University College Cork, Corcaigh, Ireland
e-mail: sepower@ucc.ie

© Springer International Publishing Switzerland 2016
B. Mölder et al. (eds.), *Philosophy and Psychology of Time*, Studies in Brain and Mind 9, DOI 10.1007/978-3-319-22195-3_5

something move now: wave your hand in front of your face. Similarly, hearing, touch, even pain can be of a present change. We can hear a rapid tapping now, feel something run up our arm; a toothache can throb.

These brief descriptions of experienced change may seem obvious and uncontroversial. Yet, it is sometimes claimed that there is a puzzle, even a paradox, in such descriptions. In this paper, I argue here that this apparent puzzle comes of assuming a particular concept of presence in time. Under this concept, *absolute temporal presence*, the changes apparent to us cannot be present. Or, at least, they cannot be *really* present—only *speciously* present. Yet, I will also argue that this is not the only way to think about presence in time. Given another concept of presence, relative temporal presence, these changes can be present.

5.2 outlines the problem of experienced change and then expands on it in terms of a common concept in both psychology and philosophy of time—the *specious* present.

5.3 describes the difference between absolute and relative presence, and their relationship to extension, without focusing on the temporal variants of these concepts. Instead, the focus is on presence in a common analogue of time-space.

5.4 turns to the different concepts of time and their consequences for temporal presence.

5.5 takes those consequences and applies them to descriptions of experienced change and time.

5.6 examines a possible response from recent work on presentism—work which may allow the absolute temporal presence to be extended.

5.2 Present Change and the Specious Present

If I sweep my hand in front of my face, I see it move through a particular region of space. But this space I see cannot be traversed so rapidly by my hand that it does so in a single moment in time. This change in my hand's position, taking more than one moment to occur, to use a term by Gale (1971a), is *durational*: it has (non-zero) duration. Clearly, the same applies to experiences under any other perceptual mode (the heard tapping, the felt running and the felt throbbing pain).

The duration of this experienced change is frequently referred to as *the specious present*. The phrase has been around since the early days of psychology, originally popularised by William James. Over time, the phrase has been used to refer to a number of concepts (e.g. Power 2012). Yet, by far the most common is the concept of a perceptually experienced duration.[2]

---

do not apply to those other kinds of experiences. For example, those other experiences are not of present change, e.g. memory experience (which is of the past) or anticipation (which is of the future).

[2] The other concepts concern time-lag and the actual duration of what is perceptually experienced (neither of which need be apparent to the perceiver (Power 2012).

5.2.1 The Specious Present

The specious present is the present duration of the perceptually experienced change. As James (1918, 608) puts it, there is a "fact of our immediate experience [...] "the specious present" [...] the interval of [a duration] as a whole." Or Kelly (2005, 9): "the central idea behind [the doctrine of the specious present] is that instead of giving us a snapshot of the world at a time, perception presents us with a temporally extended window of events." Grush (2007, 3) writes of a common example—seeing the seconds' hand of a watch:

> [O]n the assumption that we can perceive motion, the content of perceptual experience, what is experienced, must include a temporal interval. The specious present doctrine [...] provides for a possible explanation of the capacity to perceive the motion of the second hand, but not the hour hand.

There are many questions one might ask about this duration of this present change. For example, one might ask: how long is it? Is it fixed or elastic? Is it the same across all kinds of experience, e.g. different sensory modes or cognitive states? (For discussion, see Wittmann 2011 and Wittmann's Chap. 6 of this volume). Although interesting, those questions will not concern us here. The question in this paper does not concern a particular value for the extent of the present change. It concerns whether or not a present change can have any extent at all.

Before going on, it is important to eliminate a possible source of confusion: one might doubt that we have any reason to think that we experience duration itself. Instead, we experience the *change* that happens over duration but change and its duration are not the same.

However, that difference does not matter here for our purposes. What matters is that the change requires the duration in order to occur. If there is too short a duration, or no duration at all, then only a stage of the change occurs. Consider an analogy with space: let us say that we see the whole of something spreading out in space, e.g. such as a front door of a house. We may not see the spatial region occupied by the door on its own. So we may claim that we do not see spatial extent when we see the door. However, it is obvious that we do see spatial extent: we see the extent filled out by the door. Similarly, when it is said that we experience duration, it is meant only that we experience the duration of change (we will return to the analogy with space).

However, the feature of experienced change most troubling for some theorists is this: how could we even experience the duration of the change?

5.2.2 Problems of the Specious Present

The first problem raised by some theorists for the idea of the specious present is this: they claim that the present is a single moment in time (for reasons to be discussed further in this paper). It is not a multiple of moments in time. Using Gale's terminology (Gale 1971a), it is *punctal*: it occupies only a single moment in time.

Yet, if what is present has no duration, and experienced changes have duration, changes cannot be present. To be more precise, if the present has no duration, these changes cannot be *wholly* present. They can overlap the present moment, of course; one of the moments constituting the change's duration is the present moment. However, the changes themselves begin or end outside that moment. Only a stage or temporal part[3] of the change can be present. The rest of the change must extend beyond the present into either the past (it has happened) or the future (it will happen).

Yet, again: experienced changes seem to be present. Look at your hand sweeping through that space before you. Do you see that motion as partially past—as *having happened*? You will *after* you have seen it but not when you are *seeing* it. Or is it partially future—as being about to happen? Perhaps you will experience that before you have seen it—but, again, not on when you are seeing it. If you see the motion, then it does not look as if that change—again, which requires duration—is anything other than present; it does not look as if it is past or future.

So: given the present cannot have duration, how can experienced changes be present?

The second related problem is this: experiencing present change requires that we experience the past and future. Whether or not this is a coherent idea, some theorists consider it counter-intuitive or strange.

Kelly, for example, struggles with the following about the specious present: it seems as if we experience duration. Yet, he argues, if we really do experience duration, it seems that we must experience some of the past or future. Kelly (2005, 211) writes:

> [It] seems to me that there are at least three devastating objections to the Specious Present Theory itself. These objections coalesce around the following three questions:
>
> 1. How can I be directly aware of something that is no longer taking place?
> 2. How can I be directly aware of a duration?
> 3. How can I be directly aware of the future?

Kelly (2005, 212) then argues that the answers to all three questions should be that we can't or, if we can, that this is a very strange position to take:

> The Specious Present Theory proposes that I am in direct perceptual contact with events that occurred in the recent past. This is at best an odd suggestion. After all, the events in the

---

[3] The existence of *temporal parts* and stages are heavily debated in one area of the philosophy of time—the constitution and mereology of persisting and changing things (e.g. people, hats, avocados). It is argued by some theorists (*endurantists*) that such things do not have stages or temporal parts; other theorists (*perdurantists*) argue that such things do—and still others that things are only those parts (*exdurantists* or *stage theorists*).

We can ignore all that and refer to the concept of a stage or temporal part without controversy in this paper. The idea of a stage or temporal part of a change is generally uncontroversial for events and changes themselves. For example, both Mellor (1998) and Lowe (2003) are opponents of temporal parts for objects (Lowe is even an A-theorist). Both accept them for events. (For good introductions to persistence theories generally, see Sider 2001; for the views of Mellor and Lowe, see Mellor 1998; Lowe 2003, respectively).

recent past are no longer occurring, and one might naturally wonder how I can be directly aware of something that is no longer taking place.

Similarly, Perrett (1999, 99) asks

[H]ow are we to make logical sense of the notion of a "present" that apparently commits us to asserting that we can presently perceive future entities before they have come into existence and past entities after they have ceased to exist?[4]

In these quotes, Kelly refers to direct perception of duration, pastness and futurity. Perrett refers to present perception of past and future entities. One might wonder if these theorists are referring to the perceptual experience of change. There is reason to treat them as the same: the puzzlement of these theorists seems undermotivated if we think of direct awareness or present perception as something representational or intentional—that is, as something akin to a thought. There are no similar questions raised with such force by those who discuss our capacity to think of the past or the future, or of how we can remember or anticipate. The issue has force if we are referring to what we seem to perceive, i.e. what we perceptually experience. And given the content otherwise of their papers, it is clear what motivates them is the perceptual experience of change.

In any case, we can summarise the puzzles of the specious present as falling into two kinds:

• How can there be a present duration?
• How can we perceptually experience the past and future?

My argument is that these two problems arise if one thinks about the experience of change in terms of absolute temporal presence. They do not arise if one thinks about it in terms of relative temporal presence.

This suggests that some theorists think in terms of absolute temporal presence. This is understandable; it is the consequence of some (at least, allegedly) intuitive theories of time, and so theorists who are not engaged in debates about the concepts of time may naturally incline towards these intuitive positions. However, these are also not the dominant concepts of time in current debate. As such, it is worth considering alternative concepts and then seeing what happens to these puzzles in the alternative light.

5.2.3 Brief Comment on Psychological and Physical Time

Before continuing, a brief comment must be made about my reference to "time." One might wonder: Is this psychological time or physical time? Am I assuming they are the same? Couldn't psychological time be different to physical time?

---

[4] For more comments on the specious present of this nature, as well as arguments defending it which are similar but not identical to those this paper, see Power (2012).

When I refer to "time" here, I assume that the "time" I refer to is both psychological time and physical time. I hold that some kind of time needs to be shared by the psychological and physical because they need to be temporally related. For example, they need to be capable of standing in causal relations and correlations. One needs to be able to say that a thought T occurs after a physical event E. This requires that T stand in temporal order with E, and so there is some common time between them.

However, one might reply that there can be important differences between physical time and psychological time. There are differences in magnitude, e.g. physical time extends over billions of years; (probably) psychological time does not. There are differences in divisibility, e.g. physical time might be a continuum (and so only abstractly divisible into infinitesimal instants); psychological time might be discontinuous (and so not only abstractly divisible).

Although this paper does concern temporal extent, it does not concern difference in magnitude or divisibility. It does not matter in this discussion if the experienced change, the specious present, or the extended present has a duration of a millisecond, a minute, hour or a year. What matters is that it has any duration at all. In that case, I am treating psychological and physical time as the same.

Finally, one might conceive of psychological time as being different to physical time because it is merely apparent or judged time. In that case, although one might call it "time," I think that it is either strictly false to call it time (it's merely apparent time or a judgement of time) and/or one is anti-realist about time.

Part of the appeal of the different concepts of time discussed here is their capacity to solve a notorious argument for the unreality of time known as McTaggart's Paradox. They do so by being realist about time (although, as we will see, not necessarily realist about the same things to do with time). As such, although this discussion may concern psychological time, e.g. when experience happens, it is not merely apparent time (which, I would argue, is not really time at all). (For a similar note about the relationship between psychological and physical time, see Power 2010.)

5.3   Absolute/Relative Presence and Extension

Let us begin by discussing the difference between absolute presence and relative presence, and its relationship to extension, in a way which does not involve time. In the next section, we begin by discussing the difference between absolute and relative. Then we apply it to presence. We finish by considering its relationship to extension.

Much of what will be said here I take to be obvious or to follow from something obvious. It should be easily grasped given folk concepts of "absolute/relative," "presence" and extension. It is the application and consequences of these concepts to *time* which may not be so obvious.

As will be evident, the best non-temporal illustrations of the concepts below are primarily spatial examples—e.g. spatial properties and spatial relations. This is not a coincidence. Many philosophers of time think of time as sharing many of the properties of space. Many of the properties of space have analogues in time. Whatever the analogue of other spatial and temporal properties, the question is whether or not we can think of temporal presence as analogous to spatial presence by being relative.[5]

5.3.1 Absolute/Relative

The difference between an absolute and relative entity (typically, a property or relation)[6] is as follows:

- F is absolute if and only if F is the case independently of all things.[7] Another way of putting it: if a property is absolute, then something has that property without it having it in *relation* to something (be that "something" itself or something else).
- F is relative if and only if F is the case depending on at least some things (to which it is relative). Another way of putting it: if a property is relative, then something has that property only in relation to something (be that "something" itself or something else).

These are simple definitions but are sufficient for the needs of this paper. And simple as they may be, they raise a number of important points when it comes to temporal presence.

5.3.1.1 Relativity Is to an Index

As used by many philosophers of time, e.g. Mellor (1998), it is common to hold that, if F is relative to some *x*, then it can also be said that F is *indexed* to *x*. If F is a relative property, then F is an *indexical* property. If F is not relative to *x*, then, also, F is not *indexed* to *x*.

---

[5] This is not just a case of temporal properties being metaphorically described using terms which literally refer to spatial properties, i.e. the idea of a "Time is Space" metaphor, as defended by Lakoff and Johnston 1980. Nor is it that time and space are, really, the same thing—space-time, as is the view in contemporary physics (e.g. Sider 2001).

[6] For convenience, I will frequently refer to properties and relations together as just properties. However, unless otherwise stated, what is discussed here can be taken to refer to both properties and relations.

[7] Note that this definition of "absolute" also implies something absolute is independent from itself. That sounds both contradictory and right: Contradictory if it means that absolute P can be the case even if P cannot. Right if it means that absolute P can be the case without needing to be so relative to P. I assume the latter meaning is trivial—an absolute electron charge need not have that charge relative to itself (indeed, I don't even know what relativity of such a property as charge would even mean here, never mind whether or not it is possible).

Generally, I refer to the distinction in terms of "relative/absolute" rather than "indexical/non-indexical." Although reference to indexicals is common amongst philosophers of time, for the purposes of this paper their differences are not useful. Further, I find the relative/absolute distinction is more familiar to those who are not philosophers of time, especially given relativistic physics (which, incidentally, also plays a significant role in the philosophy of time). There may be a difference between them but it is not obvious that it is a difference which is important here.

5.3.1.2   Relativity Can Be Universal Without Being Absolute

Note the phrase "at least some" in the definition of relativity. F might very well be relative to *everything* and yet still be relative.

We can see this with an example:

Some x is relatively *not-small*. It is not-small by being the biggest thing there is. The biggest thing there is big to anything other than itself (and so is not small to them); it is also not small to itself. Yet, this x is not absolutely *not-small*. It is not-small *relative* to everything (including itself). Absolute "not-smallness" would not be small relative to anything. It would just be not-small—and so whatever thing we pick in the world (I am not sure what it means for something to be absolutely not-small, for it would mean that x is not-small even if there was nothing to which one could compare it).

In contrast, absolute properties include:

Spatial location: Something either has a spatial location absolutely or it does not. It cannot be spatially located relative to one thing and then have no spatial location relative to other things.

Electron charge: An electron's charge is not relative to something else.

5.3.1.3   Relativity Is Neither Unreality nor Subjectivity

The relative/absolute distinction needs to be carefully distinguished from somewhat similar distinctions: unreal/real and subjective/objective. That F is either relative or absolute does not imply that F is (a) either unreal or real or (b) either subjective or objective. For reasons to do with habits of expression in contemporary metaphysics, it is important to spell this difference out.

(a) **Real/unreal** If we ask "Is A really F?" we could mean either an F that is relative or absolute. It depends on what is used to individuate or "pick out" F, as it were. Here are some non-temporal examples in which something real is relative:

   (i) *Being far away* ("far-away-ness"): A mountain can be *really* far away. This "far away-ness" is not an absolute property. The mountain is not really far away independent of everything or even relative to everything. The mountain is not far away from itself. It is really (or not really) far away relative to some things, e.g. such as the speaker's own body.

(ii) *Motion*: A train can be *really* moving or not *really* moving (and so at rest). Yet, what makes this really moving or not is not that the motion is an absolute property. The train is not moving independent of everything (or even relative to everything). The train is not moving relative to itself. It is really (or not really) moving relative to some things, e.g. such as the speaker.

Can something absolute be unreal, and vice versa? I see no particular reason why not. If there can be reference to unreal things, I take it that there can be reference to unreal *absolute* things. This plays a role in the debate concerning the metaphysics of time.[8]

(b) **Objectivity/subjectivity** I assume that: if something is subjective, then it depends in some way on *subjects*. And, anything which does not depend on subjects is objective.

This is enough to separate subjectivity from relativity. Both are kinds of dependency but something can be relative without there being any involvement of subjects. As such, just by default of not being subjective, it occurs objectively. Here are some relative objective examples:

(i) *Distance*: A mountain can be objectively distant. This distance is not a subjective property. The mountain's distance doesn't depend on a subject. If the mountain occurs in a possible world in which there never have and never will be people, it is still distant from at least some other things, e.g. anything not touching it. It is objectively not distant relative to some things, e.g. anything touching it; or itself.

(ii) *Motion*: A mountain can be *objectively* moving or not *objectively* moving (and so at rest). This motion is not an absolute property. The mountain is not really moving independent of everything or even relative to everything. The mountain is not moving relative to itself. It is really (or not really) moving relative to some things, e.g. such as the moon overhead.

---

[8] One might wonder what it means for something to be real or not. A discussion on that precise question would pull this paper too far away from its subject. Further, the debates about time only work by assuming a shared concept of reality amongst the opposing sides. E.g. the presentism/ eternalism debate turns on how things and events are real in time. McTaggart's argument is an argument about the nature of reality.

For this paper, I suggest the following approach to questions about reality as such. Unless your concept of reality depends on a concept of time (in which cases you take a side in the debates discussed in this paper), then "reality" is whatever you understand by the term. In my case, I hold the following about real things:

Real things are independent of ideas of them (*and propositions about them*). E.g. (a) If unicorns are real, then there are unicorns even if there are no ideas of unicorns (just as, if electrons are real, there are electrons even if there are no ideas of electrons). (b) If past things are real, then past things exist independently of any propositions there might be about past things. If unicorns or past things (or electrons) are not real, then there are only ideas and propositions about them (for past things, this is the presentist view).

Real things can be coherently described: It is necessary to hold this because McTaggart's arguments against the reality of time and change do so by arguing that time and change cannot be coherently described. The responses to that argument mainly try to defend the view that time and change can be coherently described.

Can something absolute be subjective and vice versa? Here we get into the complexity of the subject/object distinction. One way of understanding a "subjective" property is a property which is apparent to subjects but is not real; another way of understanding it is as property which is relative to subjects. The first understanding involves an illusion: it is unreal; it can be absolute (as discussed above). In the latter case, it might be real, but it is relative (indeed, it's just a special kind of relativity—relativity to subjects); as such, it can't be absolute.

5.3.1.4   Some Relative Properties May Be Practically and Commonly Treated as if They Are Absolute

For practical reasons, I find that I and many others frequently tend to treat some relative properties as absolute properties. If a property is the same relative to everyone, or it is useful to assume a particular index in picking out a property, then the property is treated as if it is an absolute property. Yet, it is not.

Consider motion relative to the Earth's surface. I've found that, when I ask someone if something is really moving, such as a train, the answer tends not to be that: (a) Motion is relative. (b) Something is moving relative to the train. (c) Whatever that thing is, to *it*, the train is really moving. Nor does someone deny that the train moves because the train is not moving relative to themselves (because they're on it, say). No, typically their answer depends on whether or not it is moving relative to the Earth, e.g. if it is moving relative to the landscape around it. Yet, the Earth is only another index by which we define something as relatively moving. This motion, relative to the Earth, is not *absolute* motion.[9]

5.3.2   Presence

Presence can have a number of meanings which can be relevant to our experience. One can refer to "presence" as some aspect of someone's personality, e.g. a speaker has great presence on stage. One can refer to "presence" as a property of some indescribable and supernatural something, there is an unknown presence in a haunted house. One might refer to "presence" indirectly through reference to the *non-presence* or *absence* of what is merely imagined. An example of such imagined things might an object—such as a unicorn—or an event, such as my standing beneath the Eiffel Tower (which I have never done) (e.g. Sartre 1986; McGinn 2004).[10]

---

[9] Newton did think there was a way to define absolute motion and rest—but this is one aspect of Newton's work which is no longer current in physics (Lange 2002).

[10] These uses of "presence" might involve metaphorical or indirect references to "presence" in this discussion, e.g. the supernatural "presence" refers to the sense of something being at a location in a context in which one is unable to further specify what is there (as, e.g. "entity" is often used in science fiction movies).

There is a concept of presence which is much more relevant to time than these cases. This is *spatial presence*. Many theorists treat temporal presence and spatial presence as if they are very similar. As we will see, the similarity comes mainly from the position that both spatial presence and temporal presence are relative properties.

The following about presence generally is assumed in this paper:

(a) Having presence is the same as *being present* and *having presentness*.
(b) Presence can refer to a location in time, in space and, somewhat indirectly, in relation to imagined things. I will frequently refer to presence in time as *temporal presence* and presence in space as *spatial presence*.
(c) If

- Something is temporally present, then it exists (or is happening) *now*. If it is not temporally present, it exists (or is happening) *then*.
- Something is spatially present, then it exists (or is happening) here. If it is not spatially present, then it exists (or is happening) *there*.

I assume that both temporal presence and spatial presence are common or folk concepts. Most people understand that, if something is said to be present, then it can mean that it is *here*, *now* or *both*. However, one might still wonder: is such presence relative to locations in space and time? With space the answer seems obviously that presence is so dependent. However, it is a matter of debate with time.

5.3.3 Absolute and Relative Presence

We have discussed absolute/relative properties and we have discussed presence. "Presence" looks to be a property; things and events are *present* (as, e.g. they might be coloured, or any other property). So, can we have relative presence and absolute presence?

There is an uncontroversial example of relative presence—spatial presence. Given a certain concept of time, there are important analogies between spatial presence and temporal presence; this is tenseless eternalism. There are also important relationships (it may be too much to say analogies here) between non-presence in imagination and non-presence in time; this concept of time is presentism. So let us consider the case of absolute and relative presence for space and imagination.

Applying the difference between absolute and relative to presence, we can say the following:

- To say that something is relatively present is to say that it is present depending on something or relative to something. It cannot simply be present, without further qualification.
- To say that it is absolutely present is to say that it can be present simpliciter. There is no need to define it as present *to something*.

Applying the points in the previous section about absolute/relative to presence, we get:

1. Relative presence is to an index. If x is relatively present, it is relatively present to some index i.
2. Relative presence can be universal presence (without being absolute presence).

    Consider something which is everywhere in space (e.g. as the ether was once thought to be). Being at every location, this is universally present at those locations. But this does not mean that it is absolutely present. It still needs to be present *relative to a spatial location*. To be absolutely present, there would be no need to qualify the presence this way. This unqualified presence looks implausible for space. However, it does not look implausible for time.
3. Relative presence is not identical to either unreal presence or subjective presence. Nor is absolute presence identical to real presence or objective presence.

    Again, consider spatial presence.

    That x is spatially present relative to S does not make it some kind of *unreal* spatial presence or not *really* spatially present. That is all there is to spatial presence as commonly understood. It is sufficient for x to be present at a location that it be really present (at that location).

    Nor does relative presence make x subjectively present in space. Some x can be at a location—and so spatially present at that location—in a world in which there are no subjects or subjectivity. E.g. if there are no subjects prior to life on Earth, even then x can be relatively present in space.
4. Some cases of relative presence may be practically and commonly treated as if they are cases of absolute presence.

    Examples of this are more difficult to find for presence. A close but strained example may be spatial presence according to the Earth in a geocentric possible world (where the Earth is the centre of the universe). Perhaps what we in our non-geocentric[11] world call present *relative to the Earth*—"here" if "here" on Earth; "there" if not on Earth—is actually absolute presence in a geocentric world—"here" to everything. But that is not clearly the case. Even if I held the Earth to be the real centre of the universe, I cannot see myself holding that, if I am on Mars, I am really "there" on Mars and not "here" on Mars. While on Mars, my claim that "Here is not the centre of the universe; Earth is the centre of the universe" would be a contradiction. Mars, even to those on it, is "there" because it is absolutely "there." Whatever the merits of such a view, the idea of absolute spatial presence is not plausible in contemporary thought. A better example is the subject of this paper: temporal presence.

---

[11] It should be noted that our world as described in contemporary astronomy is neither a geocentric world nor Copernicus' alternative of a *heliocentric* world. Both of those presuppose one unique centre to the universe; contemporary astronomy does not; centres are defined by gravitationally determined locations, e.g. the centre of Earth-moon system, the solar system, the Milky Way, the local cluster of galaxies.

Perhaps one might usefully conceive of these two concepts of centres in terms of absolute and relative centres (but I do not have space to consider that here).

There is another issue which must be addressed before focusing on temporal presence. This concerns the idea that something present can also be extended.

5.3.4 Extended Presence

The paper concerns experienced change, the specious present—temporal presence alleged to be extended in time. Yet, perhaps the issue is not *temporal* presence. Perhaps it is presence of *any* kind. If that is right, then there is also a problem with saying that something cannot be spatially present and spatially extended. Yet, as discussed in this section, although that combination of properties can be presented as a problem, it is a problem easily solved. Why it is easily solved serves as background to why the analogous problem might not be so easily solved for temporal presence.

Here is how we might generate the incompatibility for space. We hold the following:

(a) Spatial presence is defined as being at one place—*here*. For example, a house is spatially present on its site; it is *here* at this site.
(b) Spatial extension is defined as being at multiple places, i.e. more than one place. For example, a street occupies a multiple of sites.
(c) As such, what is spatially extended cannot be spatially present, and vice versa. What is present is at one place; what is extended is at multiple places. So, the house is not spatially extended and the street is not spatially present.

Clearly there is something wrong with this analysis. *Here* and *what* is here can spread over an area. Just because I pick out a region of space does not mean that I have to say most of it is there, and only some of it is here. The solutions are perhaps obvious.

First, there is what might be called the *partial* solution. If something is spatially extended, it is true that it cannot be *only* spatially present. A street of multiple houses cannot only be at one house. However, spatial presence does not require that kind of strict presence. It only requires that the extended thing be at that one location for some of its extent. A street can be present at a house's site by running through it.

Such an answer looks simple and it is for the street in the example. However, it will not work for anything which is only at the location where there is presence; whatever is only at that location does not extend beyond it. Take the example of the house. The house does not extend beyond the site. Unlike the street, one cannot say that the house is only at the site for some its extent. The house is at the site for all of its extent. It spreads out in space at the location of the site.

So how can we say that house is spatially extended and spatially present—that it is at multiple locations and only at that present location?

Again, for space, the solution looks simple. The problem is being generated by an assumption: What defines spatial presence is the same as what determines the spatial extent.

This explains the problem for space: All that has been said here about extent is that is an occupation of multiple locations. There is nothing more sophisticated than that (nor is there need for more sophistication, even when we turn to the subject of time). If one of those multiple locations within the spatial extent is the location which defines presence, then it is correct that the spatial extent cannot be *at* that location. That location is only one of the places over which it spreads. Again, if the street's extent is determined by its spreading over multiple houses, then it cannot be extended at one of those houses. Nor can what is at one of those houses be extended there, such as the houses themselves.

To solve this, we make the following analysis:

(a) Spatial extent is determined by a multiple of locations. Call the sum of these locations *M*. E.g. *M* is the sum of all the rooms in a house.
(b) No location in *M* is the location which defines spatial presence. E.g. there is no one room in the house which defines spatial presence.
(c) Instead, the location which defines spatial presence is *M* itself. E.g. the house defines spatial presence.

In that case, one and the same location is both spatially present and spatially extended. It is spatially extended as determined by one set of locations (the rooms) and spatially present relative to their sum (the house).

This solution may be so obvious as to make the previous analysis seem tedious and unnecessary. However, it is important to step carefully through it because of the following: we can provide this solution because we can separate what determines spatial extent from what defines spatial presence.

Importantly, spatial extent in this case is *independent* of spatial presence. Up until spatial presence was defined by *M*, it was entirely up to us whether we do define it that way. We could have defined it by one of the rooms—or indeed by anything at any spatial location. We could have chosen an entirely different location—and if we did, the house and all its rooms would not be spatially present; they would be there, not here. In all of these cases, the house would still have been spatially extended.

If it were not possible to determine spatial extent independent of spatial presence, this answer would not be available. If, for example, a particular location in space had to be one of the locations by which any extent is determined, then what is present at that location would never be extended. It would always be at one location amongst the many which make up spatial extent.

For space, it is difficult to imagine what such a necessary point could be. However, for time, it has an analogue in the present moment of A-theory and presentism. Let us turn to consider time and presence.

5.4 Temporal Presence in A-Theory and B-Theory

The concept of absolute spatial presence is not something taken seriously. For some philosophers, the concept of absolute temporal presence should be taken seriously. They argue that absolute temporal presence is necessary for our concept of time. Without it, there can be no time. Theorists who defend this position are A-theorists (or tense theorists). The opposing positions are B-theorists (or tenseless theorists): they hold that temporal presence is relative like spatial presence.

A-theory and B-theory get their names from an influential paper by the philosopher McTaggart. In it, he argues that time is unreal. The two theories come out of different ways of responding to that argument.

McTaggart (1908, 458) states that we think of time in two different ways:

> Positions in time, as time appears to us *prima facie*, are distinguished in two ways. Each position is Earlier than some, and Later than some, of the other positions. And each position is either Past, Present or Future. The distinctions of the former are permanent, while those of the latter are not. If M is ever earlier than N, it is always earlier. But an event, which is which is now present, was future and will be past.
>
> [...]
>
> For the sake of brevity I shall speak of the series of positions running from the far past through near past to the present, and then from the present to the near future and into the far future, as the A series. The series running from the earlier to the later I shall call the B series. The contents of a position in time are called events.

One might deny McTaggart's claims by asserting that one does not think of events as being in time as either the A-series of B-series. However, I assume that McTaggart's claims at this point are correct. When most people (including this author) think of things happening in time, we *do* think of them, on the one hand, as either earlier, later than or simultaneous with each other or, on the other hand, as being in the present, past and/or future.

These two concepts of events in time form the backbone of many of the current debates about the reality of time. This is in no small part because McTaggart uses them to argue that there is no such thing as real time. The different concepts of time—and, through them, concepts of temporal presence—come in part from responses to McTaggart's argument. They come from claims about what is necessary for real time and presence.

5.4.1 Two Theories of Time

McTaggart's argument has become known as McTaggart's Paradox. It need not be covered in detail here but it is important to outline some of it in order to make clear the thinking about presence.

The relevant claims by McTaggart are as follows:

1. If time is real, change in events in the A-series has to be real. There are two parts to his claim here:

(i) Without events really being in the A-series, there can be no time. The B-series might be sufficient for time by itself. However, McTaggart argues that the position of events in the B-series depends on their position in the A-series. If they are not *really* in the A-series, then they are not *really* in the B-series.

(ii) There can be no change without events being in the A-series. Change just is variation of events' positions in the A-series. There is change if, e.g. my lecturing in A changes from being present to being past. This change cannot be captured by B-series. As such, if events are only ordered in the B-series, there is neither change nor time.

2. It is impossible for there to be change in the A-series:

   (i) Positions in the A-series, what later philosophers call *A-properties*, are incompatible. It is impossible for anything to be past *and* present *and* future.

   (ii) If an event changes in the A-series, then the event has all of the A-properties. Every event is past (to a later time than when it occurs), present (to the time when it occurs) and future (to an earlier time than when it occurs).

3. Any changing event in the A-series has incompatible A-properties. It is impossible for anything real to have incompatible A-properties. So, either the events in the A-series are not real or their change is not real. In either case, change in events in the A-series cannot be real. As such, given "1" above, time cannot be real.

The variety of responses to this argument laid the foundation for various positions which defend the reality of time against this argument. The most prominent current positions are the A-theory and the B-theory (sometimes also known as *tense theory* and *tenseless theory*, respectively).

5.4.1.1   A-Theory

According to the A-theory, the A-series is how events are really laid out in time. If time is real, events really lie in the past, present and/or future. Their locations in the B-series, i.e. being earlier than/later than/simultaneous with one another, depends on their location in the A-series, i.e. in their being in the past, present and/or future. Further, in order for events to really change, they need to change in their A-properties. As such, they agree with "1" above: the A-series is necessary for change and time and also for the B-series.

As an example of how the B-series depends on the A-series. Every event is past, present or future. If one event M is past and another event N is present, then we can derive their B-series positions as follows: M is earlier than N.

How then does the A-theorist respond to McTaggart's argument? They assert the primacy of the present in defining the positions in the A-series. McTaggart makes the mistake of treating event's possessions of properties at all times as equal. However, when it comes to describing how events occur in time, it is only true that events have their positions relative to one time. The time to which they have their A-properties is *the present* time.

And so, the A-theorist replies, there is no incompatible possession of properties. Events only really have the properties they have as defined according to *the present*.

For example, my last birthday is past; it is not present. It is also changing by becoming more past. According to my last birthday itself, it might be said to be present. This may seem to raise a contradiction: my birthday is present (to its own time) but also past (to when I write this)—indeed, it is changing by becoming more past. But there is no contradiction: it is not really present. It is only really past. (For defences of the A-theory, see, e.g. Loizou 1986; Lowe 2003).

5.4.1.2 B-Theory

According to the B-theory, events are really laid out in time by their locations in the B-series. If time is real, events really stand in the relations of earlier than/later than/simultaneous with each other. Their locations in the A-series, i.e. their being in the past, present and/or future, depends on their locations in the B-series, i.e. being earlier than/later than/simultaneous with one another. They agree with 2(ii) above but not with "1" or "2(i)".

As an example of this dependency: every event is later than, earlier than or simultaneous with every other event. If one event M is earlier than another event N, then we can derive their A-series positions as follows: (i) Relative to N: N is present and M is past. (ii) Relative to M: N is future and M is present.

If that is right, then there is no incompatible possession of A-properties. Events have each A-property relative to a different position in the B-series.

For example: My last birthday is past relative to a time later than it in the B-series (such as the time of writing). It is present relative to a time simultaneous with it (including, trivially, itself). It is future relative to a time earlier than it in the B-series. (For defences of the B-theory, see, e.g. Mellor 1998; Hoerl 1998; Sider 2001; Le Poidevin 2003; Callender 2008).

5.4.2 Theories of Time and Temporal Presence

A-theory and B-theory are active positions in contemporary debates about time. Neither theory's concept of presence is accepted by all sides in these debates. There is no room here to detail the various defences of each side's position on what is necessary or fundamental for time.[12] Let us move on to how each theory conceives of presence.

---

[12] This is due in part to implications from the physical concept of time found in relativity theory. Relevant to this paper is this: relativity further relativizes positions in the A-series to frames of reference defined by velocity (e.g. Lange 2002; Sider 2001; Power 2010).

5.4.2.1 A-Theory

For A-theory, real temporal presence defines real temporal locations in the A-series, locations events have without further qualification. And locations in the A-series define locations in time.

As such, this real presence is not relatively determined by its location in time. It is not like how spatial presence is determined. It comes before such indexing; the indexing of temporal events relative to locations in time is derived from locations in the A-series.

Note that this distinguishes temporal presence from spatial presence. Those who hold this view about presence in time do not hold it for presence in space. For example, Gale (Gale 1971b, 72)) writes that "[t]hose who have held that the present or now to be objective,[13] such as C.D. Broad and myself, have not been willing to countenance an equally objective here in nature, no less a here which "shifts" from place to place."

In A-theory, then, temporal presence is prior to any kind of "presence" relative to some time, such as the "presence" of my lecture at 2 pm. Such relative presence might also correspond to the real present, e.g. if my writing this is *really present*, then I can say that "present" as defined by when I am writing this paper (a relative "present") corresponds to the real present. However, these two kinds of "present" need not correspond. And for A-theory, real temporal presence is not merely presence relative to something. The real present divides the world in time in a fundamental way which is independent of any index or location. For this reason, it is a presence which is absolute.

Further, defenders of this view, like McTaggart, claim that this is the common sense conception of presence in time. Schlesinger (1991, 428) writes:

> [W]e feel strongly, there is a property which does confer upon m [some given moment] a unique privileged status, namely, its property of being in the present. [O]ur attitude to the present may be described as regarding it as distinct from every other temporal position, for while the future is yet to be born and the past is rapidly fading, the present is palpably real. This characteristic of m [...] is a transient feature of that moment; m goes bright, and come to life for an instant, after which its presence or immediacy is passed on to the next moment.

5.4.2.2 Presentism

This understanding of an absolute presence, and what is intuitive about it, is asserted by many of its defenders as going much further than a present which is a special moment. Those defenders hold the position of *presentism*, the view that only what is present is real. What is not present—what is past or future—is not real. Presentism

---

[13] Given the context of the entire quote, by "objective" Gale can be taken to mean a non-indexical, non-relative (and so absolute) present. See earlier in this paper for why I do not use this terminology.

is typically considered by its advocates as being more intuitive than any alternatives. For example, Bigelow (1996, 35) writes:

> I am a presentist: nothing exists which is not present. I say that this was believed by everyone, both the philosophers and the folk, until at least the nineteenth century; [...] Presentism was assumed by everyone everywhere, until a new conception of time began to trickle out of the high Newtonianism of the nineteenth century.

The following from Prior (2001, 290) is also often taken to express the presentist position: "The present simply is the real considered in relation to two particular species of unreality, namely the past and the future."

So, given presentism, the status of events at other times than the present is less significant than the present because they are not real. Reference to such events is more like reference to imaginary or merely possible things than, say, things at other spatial locations. This makes time, locations in time and temporal presence significantly different to space, locations in space and spatial presence. What is at a different place other than here is as real as what is here. But presentism denies this of what is at a different time than now; it is not as real as what is now. E.g. past dinosaurs and the future death of the Sun are closer to imaginary dragons than spatially distant things such as bacteria under the Antarctic ice. The bacteria are real (just at a different location). The dinosaurs and dragons are not real.

The opposing position to presentism is *eternalism*, the view that what is at any time is as real as anything at any other time; it does not matter which time is present to the different time's reality. Some A-theorists are eternalists (e.g. Loizou (1986) and Lowe (2003)), but a large fraction are presentists.

In any case, presentists or eternalists, one consequence of A-theory which is not so obvious is this: if there is an absolute temporal presence (as it must be if it is a fundamental feature of the world), then it is present for everything, including every event and time. This includes past times and events and future times and events. For those times and events, they are not happening in the real present. This is true no matter what one says about what is the case relative to them, including this: that something is present relative to them. This "relative" present cannot be the real present.

This may not be obvious but, given A-theory, it must be the case. If there is a present as asserted by A-theory—a fundamental, real and absolute present—then what is earlier than it, i.e. past, must be an absolute past. Now, one might still define a concept of "presence" relative to this past as follows: some event E is present relative to some past time by happening at that past time. However, by being past in the way required by A-theory, something is by definition not really present. Even if one defines an event as present relative to a past event, it is not really present. That an event is present relative to when the dinosaurs existed or when the sun will die—this does not make that event *really* present.

I bring this up to reinforce a point: even given A-theory, one might very well refer to relative presence as present relative to any event or time. Yet, this will not be the real present of A-theory. The real present is not defined by anything else. It is a fundamental feature of the world. For this reason, it is an absolute present.

We can see this more clearly if we consider presentism: There are no such past events *really*. There is no real past for anything to be relative to it. We may find it useful to speak as if there is—but such speech is like speech about relativity defined by imaginary things. E.g. the presence of events relative to when dinosaurs ruled the Earth is as real as the size of things relative to dragons.

5.4.2.3 B-Theory

For B-theory, temporal presence is defined by locations in the B-series. Locations in the B-series define locations in the A-series (past, present and future) as "here" and "there" are determined by non-indexical locations in space. Temporal presence, "now," is determined just like how "here" and spatial presence (and "there" and spatial absence) is determined.

So, temporal presence in B-theory is relative to a time—and these times are defined independent of the divisions between past, present and future. For example, I might define temporal presence as relative to when I am writing this paper. In that case, my writing is present trivially—it happens when it happens. However, so defined, your reading this may not be present. Still, your reading this can be present. The "present" needs only to be defined as what happens when you are reading this paper. In that case, your reading this is present and my writing this paper is not (because I have finished writing it before you read it).

For the B-theory, real temporal presence is a form of relative presence.

One consequence of this is B-theorists, if they ever make it explicit, subscribe to eternalism and not presentism (I can think of no exceptions). This is perhaps no surprise: distinctions in the A-series are merely relative for B-theory. Following presentism, one locates what is real only in the present, and not the past or future. For B-theory, this is to use what is merely relative to determine reality. It is at least intuitive to hold that what is relative does not determine what is real (I would expect significantly compelling reasons to deny this).

5.4.3 Temporal Presence and Duration

In this paper, duration is understood as being interchangeable with temporal extent. This makes it like the spatial extent as discussed in Sect. 5.3.4. As something has spatial extent by occupying multiple locations in space, so something has duration by occupying multiple locations in time.

In the section on spatial extent, it was argued that something could be spatially extended and spatially present. Indeed, they can do so within the same region of space. This is because the definition of spatial presence is relative to a location but the extent need not be defined relative to that location.

However, we are now considering temporal presence. Can a similar analysis be made of temporal presence? Can something be temporally present and have duration at the same time?

5.4.3.1   Relative Presence and Duration

The answer should be in the affirmative for relative temporal presence: the duration need not be defined relative to the same index as presence is defined. The multiple moments of a duration need not include the time which defines the present as merely a member. Just as with space, the time to which the present is defined can be all of those moments together.

Further, what determines duration is whatever defines a multiple of moments in time. In B-theory, this is the B-series. If there is a sequence of events as described by the B-series, the duration is simply the times required for that sequence of events. For example, if events x and y stand in the following relations in the B-series: x is earlier than y, then duration D is determined to be the period of time which occupies at least two moment—the moment of x and the moment of y.

Lastly, as with spatial presence and extent, one can generate a conflict between relative temporal presence and duration. One needs only hold that the same time which defines presence is also only *one* of the times which determine duration. In that case, the duration cannot be wholly at that time. It can only have it as one of its moments.

This should be enough for relative temporal presence, especially given the B-theory which motivates holding it. We return to it when we connect it to experienced change in the next section. Before doing that, we need to consider the alternative: absolute temporal presence.

5.4.3.2   Absolute Presence and Present Duration

The absolute present, especially as posited for A-theory, raises problems with holding that there is a present duration. As might be expected, these are problems which have no obvious spatial analogue.

The absolute is the same for all times and events. If I am writing in the absolute present, then that is the case whatever the position of earlier or later events. It would be a mistake to think that my writing is not really present because it is not "present" relative to your reading this. It is your reading this which is not really present; it is really future.

Of course, I might say that my writing this paper is past relative to your reading it and your reading it is future relative to my writing. However, given A-theory, my writing is only *really* past if your reading is *really* present. Your reading is only *really* future if my writing is *really* present.

Given A-theory, one cannot (a) temporally locate an event in some way independent of their locations in the A-series, such as their locations in the B-series and then (b) define the real past, present or future according to it. Events in times are really in the A-series. Then, one can determine their location in the B-series.

As with spatial extent, if the present does not have past or future moments in it, then any temporal extent, i.e. any *duration*, it has is present duration. Section 5.6

examines a possible way to have a durational absolute present, especially given a kind of presentism. However, I will argue here that this cannot be the absolute temporal presence of A-theory. A-theory presence cannot have real duration. The following is adapted but somewhat different to an argument by St. Augustine (1982; for discussion of St. Augustine's argument, McKinnon 2010, 307).

Assume that there is a present duration D and A-theory is true. Thus, there is an absolute present and it extends over multiple moments. Also assume the following (which should be agreeable to A-theorists): for any two moments[14] in that duration, they stand in temporal order to one another. That is, one precedes the other.

Let one moment in D be x. Let a later moment in D be y. Given how B-relations such as "later" and A-properties such as "past" and "future" are supposed to be related, we have this:

(i) Relative to x, y is future.
(ii) Relative to y, x is past.

Given B-theory, this relativity of pastness and futurity is all there is to pastness and futurity. If something is really past or future, then it is sufficiently so by being relative to some index (e.g. such as a moment in the B-series). Further, given B-theory, each event is also really present by being relatively present. This is how we get these properties—from positions in the B-series.

However, given A-theory, such relative pastness, futurity and presence is not how events really are in time. B-series positions are derived from A-series positions, not the other way around.

And so, since we are assuming that x and y together are moments in present duration D, x and y are really present in time. As such, neither x nor y are really past or future. Their A-properties in "i" and "ii" are not their real temporal properties.

The problem for A-theory at this point is this: Given A-theory, the properties of events in time depend on their positions in the A-series. This includes any B-series relations that might hold between them. Given A-theory, one gets B-series relations from A-properties. And so: x is really earlier than y is if x is really at a different location in the A-series to y, e.g. x is past and y is present; y is future and x is present.

However, we are assuming here that both x and y are really present—they are just both really present at different times. So it cannot be that one is past (or future) when the other is present.

A-theory and a present duration entail a contradiction. We have a duration in which either:

(i) x is past and present.
(ii) y is present and future.
(iii) Or both "i" and "ii": x and y are present, x is past and y is future.

Remember, these A-properties are the real A-properties of A-theory: they are absolute. They are not merely relative to some index or other; they are certainly not derivative.

---

[14] If one would rather avoid reference to "moments" here, one can substitute events. What is important, however, is whatever we refer to here—events or moments—that they are temporally ordered.

We can avoid the contradiction by removing one of the clashing A-series positions for *x* and *y*. We cannot remove the present because *x* and *y* are in present D. So, we must remove their pastness and futurity.

What we then get is this:

(a) *x* is earlier than *y*.
(b) *x* is not past.
(c) *y* is not future.
(d) *x* and *y* are present.

But again, for A-theory, this will not do. The B-series relations in "a" indicate different positions in the B-series. Points in the B-series are supposed to come from A-series positions. Yet, given "b", "c" and "d", there is no difference in the A-series positions of *x* and *y*. They are both present.

This means the following: the only way to get an absolute present which has duration—has multiple moments in it—is to hold that what differentiates times in that present is not defined by real positions in the A-series. This denies A-theory. It means that time is not fundamentally defined by the past, present and future; there are moments which stand in temporal relations—the moments in present duration—which confound such definition.

Still, perhaps we can deny A-theory and still hold to an absolute temporal presence with duration. We look at a way to do this later on in the paper.

5.5   Temporal Presence and Perceptually Experienced Change

The difference between absolute and relative temporal presence for time (and space) has been outlined. It has been explained why A-theory entails absolute temporal presence and B-theory relative temporal presence. In light of these differences, let us now return to the introductory issues around the perceptual experience of change and the specious present. It is argued here that the issues around the perceptual experience of change come from thinking of it in terms of absolute temporal presence. They do not occur given relative temporal presence.

5.5.1   Relative Temporal Presence and Perceptually Experienced Change

In his influential book *RealTimeII*, in the section on experience, the B-theorist Mellor (1998, 51) asks:

> [No-one] is impressed by the fact that all our experiences happen to us here, wherever here may be. [...] But why then, when so many are impressed by the temporal presence of experience, is no one impressed by its spatial presence? The answer, I am sure, is that no one thinks the latter to be any more than a trivial tautology. Because my experiences are wher-

ever I am, my belief that they are here must be made true by the fact that this token belief is where they are. Can we say something similar to the experience and temporal presence? That what we experience seems to be, or is, now is entirely up to what we define as now?

Given that temporal presence is relative, the answer to Mellor's question is "Yes". Experience is temporally present (or not) depending on

(a) What relatively defines temporal presence and
(b) How whatever defines presence in "a" temporally relates to the experience.

Although "a" determines if something is present or not, it does not mean that what we experience fails to be temporally present if "b" doesn't match it. It just means it fails to be temporally present relative to whatever index is used in "a". It can still be temporally present to something else. Given relative temporal presence, what is temporally present in "a" depends only on what time we pick, something arbitrary (within a range of times, anyhow). Indeed, if we use the experience itself to define the present, then it is a trivial matter that it is temporally present. Given relative presence, everything in time (including experience) is present to itself.

Let us return to the experienced change—the "specious" present. If experience occupies this specious present, then given relative presence, the specious present is present. It is present relative to the time of the experience. That is all that is required given relative presence.

There is a terminological consequence of this: if "specious" x implies "false" x, then this present is not specious at all. It satisfies the conditions of presence for relative presence—just as does *anything* that occurs in time.

What about Kelly's difficulty: that it is very strange to claim that we are in "direct perceptual contact" with the past or future?

Consider the spatial analogy to this claim: we are in "direct perceptual contact"[15] with spatially distant things—things that not *here* (analogous to not *now*) but are *there* (analogous to *then*). This is not obviously a strange claim—that we are in direct perceptual contact with spatially distant things. So, again, why is it strange to claim that we are in such contact with temporally distant things?

If temporal presence were relative, there should be no special problem here. Relative presence implies relative pastness and futurity (just as relative "here" implies relative "there"). What is past or future is only relatively so.

If we define what we are in "direct perceptual contact" as happening when experience happens, and then presence by the same index, then what we are in "direct perceptual contact" with is present. It is not past or future. But it is also not much of a problem. We do not need to have events at different times be defined as "past" or "future." It all depends on what index we are using.

---

[15] So far as I can tell, Kelly's reference to direct perceptual *contact* is a usage particular to him. "Contact" might suggest a kind of touching of the things perceived. If that is what's meant, then it does seem strange to say that we "touch" the past. Yet, it also strange to say that we touch, for example, planes that we are seeing lifting off from a runway, or many other kinds of seen change. That is one of Kelly's examples so I assume "contact" here is not to be taken to be a kind of touching.

From the position of B-theory and relative presence, Kelly's claim that there is a problem here because we are in perceptual contact with past things rests on a confusion of indices. Relative to one index, we are in perceptual contact with past things; but relative to another index, we are in perceptual contact with present things. There is no reason to prefer one over the other. If we do not like talking about experienced events as past, then just change the temporal location by which we define past, present and future.

Again, given relative presence, temporal presence should play no more a role than spatial presence. What does not matter to questions about experience in space is whether or not it is defined as *spatially present*. Presence depends on what index we pick to define it. In choosing something, we already have other spatial relationships between experience and the index. Those other relationships will determine whether or not the experience is located at the index (and so here) or has parts at it (here but also extended beyond it). Further ascriptions of spatial presence contribute nothing to this account. They come after the work is done.

Given relative presence in B-theory, it is perhaps unsurprising how easily B-theorists claim that the perceived present has a duration and also suggest accounts of experience that involve the experience being extended (as opposed to punctal). For example, going by his 2001, Dainton subscribes to B-theory. In his (2003, 1), he writes:

> [G]iven the fact that we directly apprehend change and persistence, albeit only over quite short intervals, the present of experience—the phenomenal or specious present—cannot straightforwardly be [...] durationless [...]. If change and persistence are directly experienced, the phenomenal present cannot be strictly instantaneous, it must—in some manner—have duration.

Still, Mellor's original quote asks why temporal presence is not being treated like spatial presence. Saying that relative presence does not struggle with it does not answer this. As we saw earlier with the specious present, there is a puzzle here for some theorists. Why is that?

I think it is because of absolute temporal presence—and through that, A-theory.

5.5.2 Absolute Temporal Presence and the "Specious Present"

Given A-theory, real presence is absolute presence. Temporal presence is not simply defined relative to some index. We are not free to pick just any time or event—e.g. as defined by the B-series or some event such as my writing this paper—and then hold that this time or event is *really* present. The present is independent of such indexing. If there is any relationship between the present and definition in time, the present defines.

This gives the following: If *experience* really has presence, then its presence is absolute. It is not just had because it is at any time (e.g. such as when you are reading this paper). It is because it is at the present; if your reading is past, then there present experience is not.

This makes a difference to thinking about the specious present. Given the absolute present, one can ask if we experience the specious present or the real present. As stated, the evidence of the phenomenology seems to point at the *specious* present: it is of perceived durational changes. Yet, some theorists have had problems with that view. These problems, I believe, come from A-theory and presentist intuitions, not from what is necessary about time.

Recall the consequences of absolute presence: First, what is present—what is *really* present—is independent of times as otherwise described. The real present is not defined relative to anything, e.g. such as the moment of dinosaurs' extinction. Second, given absolute presence, the real present is absolutely punctal. It has no duration.

In that case, a duration cannot be contained in the present. It must occupy the past and/or future. As such, by extending over a duration, the "specious" present is indeed specious. It is not the real present.

Of course, the real present can lie *within* the specious present. It can be a point in that duration. The duration contains the present. The present is a proper part of the duration.

Yet, that does not make the real present identical to the specious present. The real present is no more identical to the specious present than the location of a house is identical to the location of the whole street.

Here is one response for an A-theorist (or any advocate of an absolute present): We experience the whole duration, including the past and future. It is just that the whole duration *seems present*. This does give us an experience of change, a change which extends into either the past, the future or both.

Yet, it does so at a disadvantage to relative presence. This apparent presence is not real presence. Some of it is non-present—whatever happens to be past and/or future. Recall Kelly's difficulty: it is strange to hold that we are in direct perceptual contact with the past or future. This has force with absolute presence: these changes are in *the past* or *the future*. Unlike relative presence, one cannot say that this change can have all of these A-series positions, i.e. it is not present to some indexes; past and/or future to others.

A-theorists and advocates of absolute presence may bite the bullet on this and hold that, Kelly's reservations or not, we experience non-present things. Note, however, that there is a bullet to bite here. There is none for relative presence.

Recall also Kelly's expansion of the problem: the past is no longer happening. An A-theorist may hold that, still, it can be experienced. However, this is further complicated once we bring presentism into the picture.

Given presentism, there are no real past or future things. Experiencing past or future things is a form of experiencing unreal things.

As such, if an experience exists *at all*, it does so in the present. If we experience anything real—as we seem to do in perceptual experience—we must experience something present. If it is not present, it does not exist.

Temporal presence, in this case, plays a central role in accounts of perception. It determines what is real, and so also what is real *about* perception.

So, going back to the specious present, perhaps our experience seems to have duration. But if presentism is true, and our experience is real, then what we experience does not have duration. As such, its apparent duration is an illusion of some kind.

Even if A-theorists are eternalists, if they hold that our experience occurs only in the absolute present, they need to give some account of how an experience at a time can be of a time in which it is embedded. Saying it is extended does not pick out the absolute present. It just picks out the set of A-series positions in which the absolute present is located.

In conclusion, in order for A-theorists or anyone proposing experience happens at an absolute present, in order to explain our experience of time, one needs to turn to what happens to an experience at a punctal moment.

This discussion suggests that, for the perceptual experience of change at least, one should prefer relative temporal presence over absolute temporal presence. As such, one should also prefer the theories which commit one to relative temporal presence, e.g. B-theory, over the theories which commit one to absolute temporal presence, e.g. A-theory.

Given the overall metaphysics debate about time, this could be considered a good result. It brings descriptions of the experience of time closer to the dominant side in the metaphysical debate about time. The larger proportion of contemporary philosophers[16] of time subscribes to B-theory.[17] As mentioned throughout this paper, the standard interpretation of relativistic physics also more closely matches B-theory and eternalism than A-theory and presentism. And here we have a situation where a description of temporal experience, of perceptually experienced change, corresponds to the dominant theory of time and change.

Yet, unlike space and spatial presence, there are still sides on the debate about temporal presence. If taken as it stands, this is a loss for A-theory and presentism. A-theory and presentists argue that their positions are intuitive in a very important and fundamental way. One who is not focused on the debate about time, e.g. anyone who is not a philosopher of time, may share these intuitions; they may not want to give them up. So, in the final section, let us consider some objections to the claim that relative presence and B-theory is better for experienced present change than A-theory and presentism.

---

[16] In the analytic tradition, at least. I do not claim anything about other philosophical traditions—or even link the concepts employed here—which are predominantly analytic—to those other traditions.

[17] This is evident in a recent Philpapers survey, in which amongst philosophers familiar with the debate, B-theory was more accepted than A-theory. For summaries and discussion of this survey, see Chalmers and Bourget (2013). Specific discussion and summaries on A-theory/B-theory are scattered throughout the commentary and tables.

5.6 Durational Presentism

Two anonymous referees have raised the possibility that some recent work by a number of philosophers can have a theory of an extended (or durational) and absolute present. This account comes of holding the position that reality extends in time. There is more than one real time. This is very like the eternalism of B-theory. However, this temporal extension is contained only within *the present*. That is, this account is a form of presentism.

Dainton (2001), Hestevold (2008), and McKinnon (2010) have separately explored the idea of such a durational present—where it is understood that this is *the present* so important to A-theory (and thus a non-relative present, an absolute present). They all do so in the context not just of A-theory generally but of presentism in particular. Dainton refers to the position as *compound presentism* (Dainton 2001, 95), Hestevold refers to it as the *Thick Presentism* (Hestevold 2008, 330), McKinnon varies his terminology but at one point refers to it as *durational presentism* (McKinnon 2010, 317). The common idea is of a durational present under a concept of time acceptable to presentists. I will refer to all such views here under the umbrella of *durational presentism.*

Durational presentism, then, is the view that reality extends over a multiple of times but does not extend over all times. A multiple of times, lying within a particular duration, are real. The duration is *present* duration; these are all present times. It is still presentism because other non-present times are unreal, i.e. the past and the future which lie outside this privileged present group.

Here are the advantages of durational presentism over a durationless or *punctal* presentism.

First, it's clear that this kind of presentism allows there to be an experience of present change. Indeed, this is one reason why Hestevold suggests it over a durationless (or punctal) present (Hestevold 2008, 331).

Second, holding that we do have this experience does not force one also to hold that we experience anything past or future. As such, we can have the "specious present" of experience in the way relative presence and B-theory can have it, i.e. not "specious" at all.

Third, it satisfies most of the presentist intuitions, at least beyond the time of what we experience, i.e. that the past and future are unreal. Such durational presentism makes the following description true: The change that is my typing these words is a real change because it happens in the duration of the present. However, the change that is your processing of these words as you read them is not real—because it is in the future. If this is an important intuition, as presentists argue, then it is preserved by durational presentism.

However, as already discussed, present duration is not compatible with the tenets of A-theory. To reiterate: whatever defines the temporal relationships between the multiple moments of a present duration, where all moments in it are separated from each other in time, it is not their locations in the A-series; they all have the same location, the present. For A-theory, the fundamentality of the A-series to time is central to its arguments—as indeed is the primacy of the present in that series.

Still, perhaps this only means that we abandon A-theory while keeping durational presentism. If one is a durational presentist, then one should not be an A-theorist. One should be what might be called a *non-A-theorist presentist*.

5.6.1 Non-A-Theory Presentism

Positing a non-A-theory presentism is stepping outside current philosophical debate. So far as I know, no philosopher has developed such a position on time. It is not within the scope of this paper to develop it either. However, since there is interest in a durational presentism, and such presentism does not look as if it can be A-theory, it is worth spending a section on speculation within the concepts of time already introduced.

If one is to be a presentist without being an A-theorist, then what kind of presentist can one be? So far, we have discussed two opposing positions. So, perhaps one could commit to the alternative to A-theory—B-theory or some variant of it. I will argue that this has its own problems. These are not perhaps as great as with A-theory but they still involve denying B-theory as it stands. This is because one must deny the B-theory understanding of presence (and tense generally).

We may refer to the present as a sequence of real times (of experienced change). One assumes that these times stand in temporal relations: one is earlier than one other and later than yet another. If this were typical B-theory or A-theory, we could then say that, for the later events, the earlier events are past. For example, we could say that the first moment of that duration is past to all the others and the last one is future to those previous to it. Further, we can say that each time is present to itself and not present to other times.

Yet, if the present is what they all share, then we cannot say that these times are not present. Further, if past and future are defined in relation to this shared present, we cannot say these times are past and future. So what can we say that about the earlier and later times in relation to one another?

In B-theory, one derives A-series positions by relativizing them to B-series positions, e.g. if x is earlier than y, then (a) relative to y, x is past and not present and (b) relative to x, y is future and not present. These relativized A-series positions just are tenses in B-theory; they are as real as tenses get.

Yet, these are not tenses in durational presentism. x and y are both present—they are in the absolute present containing the real duration. The relativized tenses (including that x is present to itself) are not tenses. Yet, this is all that there is to tenses according to B-theory.

As such, whatever it is that is the case with (a) the temporal relationships in the real duration and (b) real A-series positions, it is not what is the case according to B-theory. This durational presentism cannot describe its real time in terms of B-theory.

As such, holding to an absolute presence which is extended, given presentism, is neither compatible with A-theory or B-theory. To summarise:

- A-theory cannot have such a solution because the solution involves a set of temporal relationships (those within the real duration) which are not based in the A-series. It is central to A-theory that times be based in the A-series.
- Real events in a much shortened B-series might be thought of as compatible with this solution. The temporal relationships are described as being in the B-series. However, one cannot appeal to B-theory in analysing these temporal relationships because the tenses cannot be understood as they are in B-theory, i.e. as relative.

As such, if one does wish to a durational presentist, it looks as if one ought to develop a concept of time that is neither A-theory nor B-theory.

5.7 Conclusion

In this paper, we have discussed the following:

- The relative and absolute concepts of presence as applied, first, to spatial presence and then to temporal presence.
- The relationship between absolute/relative presence and extent. Why absolute temporal presence must be absolutely punctal.
- Why some philosophers of time hold that there is absolute temporal presence and why others hold that there is relative temporal presence.
- The consequences for thinking in terms of either absolute or relative temporal presence for the experience of temporal presence. In particular, how thinking in terms of either affects one's understanding of perceived present change, i.e. the specious present.
- An objection to the foregoing arguments, and what that objection entails.

At this point, I should admit a bias. Of the concepts and theories of time detailed here, I prefer B-theory, eternalism and relative temporal presence. This is mainly because of physical theory: B-theory fits current models of relativistic physics better than A-theory. Central to its fitness is its approach to temporal presence. This is because, in relativistic physics, presence in time is relative—and indeed not even to times but, also in some cases, to inertial frames of reference (e.g. Mellor 1998; Sider 2001; Power 2010).

In practice, this means that when I consider the structure of experience in time, I do so from the B-theory. For example, questions of the presence, pastness or futurity of experience or its constituents are something I consider after the fact. I do not include these in the explanatory or descriptive elements of experience, nor see a need to do so. Or, at least, for experience, I see no need to do so any more than appeals to spatial presence or spatial absence are needed to explain the structure of experience.

To close, I give some recommendations about what to do given the difference between relative and absolute temporal presence.

1. Most philosophers of time subscribe to B-theory and eternalism. If one is motivated by philosophical thinking, then I recommend that one suspend thinking in terms of absolute temporal presence and A-theory. If one wishes to take a side on the philosophical debate, given current thinking, I recommend that one work with eternalism, B-theory and relative temporal presence (again, of course, I am biased). One can of course suspend thinking in terms of either, if that is possible.

    Although not examined in any detail here, it is also worth noting that the physical concept of time is in terms of B-theory, eternalism and relative presence (for discussion, Mellor 1998; Sider 2001; for explorations of ways to deny this way of understanding time given physics, see Lowe 2003; for discussions on experience and contemporary physics, see Power 2010).

2. A-theory, presentism and absolute temporal presence are said to have intuitive force. So, in thinking about the problems of temporal experience, it is also important to keep in mind that problems one has might be because one tends to think in terms of A-theory and presentism.

    This is evident in the analysis of the specious present, where the problems seem to be based on treating the present, past and future as fundamentally dividing how things are in the world.

As such, when confronted by a problem concerning time perception and temporal experience, I suggest asking oneself if either absolute or relative temporal presence is being assumed in thinking about the problem.

Finally, there are other discussions one might have around the topic of experience and presence which have been neglected here. One is the idea that the presence of experience is as described by A-theory, i.e. as something unique, as standing out, and perhaps absolute. If so, one might argue that relative presence does not work for experience. Another is this: how the plausibility of general models of consciousness, e.g. such as retentionalism and extensionalism (discussed in other papers in this volume; see Arstila (Chap. 9), Rashbrook-Cooper (Chap. 7), and Wittmann (Chap. 6)), might vary given different concepts of presence. Perhaps different models suit other different concepts of presence. If so, there is further interesting work to be done using this distinction.

Acknowledgments My thanks to two anonymous referees for their very thorough and insightful comments on an earlier draft of this paper. I am also grateful to Valtteri Arstila for additional thoughts and comments. This paper is based on two talks given at the source of this volume, a 2013 conference on presence at the University of Turku. My further thanks for the insights of participants at the workshop. Finally, of course, I am grateful for the invitation to present at the workshop (as well as the funding provided by the TIMELY research network to do this).

References

Bigelow, John. 1996. Presentism and properties. *Philosophical Perspectives (Metaphysics)* 10: 35–52.

Callender, Craig. 2008. The common now. *Philosophical Issues* 18: 339–361.

Chalmers, David, and David Bourget. 2013. What do philosophers believe? *Philosophical Studies* 170(3): 465–500.

Dainton, Barry. 2001. *Time and space*. Chesham: Acumen.

Dainton, Barry. (2003). Time in experience: Reply to Gallagher. *Psyche* 9(12).

Gale, Richard M. 1971a. Has the present any duration? *Noûs* 51: 39–47.

Gale, Richard M. 1971b. "Here" and "now". In *Basic issues in the philosophy of time*, ed. Eugene Freeman and Wilfrid Sellars. La Salle: The Open Court Publishing Co.

Grush, Rick. 2007. Time and experience. In *Philosophie der Zeit*, ed. Thomas Müller, 27–44. Frankfurt am Main: Klostermann.

Hestevold, H. Scott. 2008. Presentism: Through thick and thin. *Pacific Philosophical Quarterly* 89: 325–347.

Hoerl, Christoph. 1998. The perception of time and the notion of a point of view. *European Journal of Philosophy* 6(2): 156–177.

James, William. 1918. *The principles of psychology*. New York: Dover.

Kelly, Sean D. 2005. The puzzle of temporal experience. In *Cognition and the brain: The philosophy and neuroscience movement*, ed. Andrew Brook and Kathleen Akins, 208–240. Cambridge: Cambridge University Press.

Lakoff, George, and Mark Johnson. 1980. *Metaphors we live by*. Chicago: University of Chicago Press.

Lange, Marc. 2002. *An introduction to the philosophy of physics*. Oxford: Blackwell Publishing.

Le Poidevin, Robin. 2003. *Travels in four dimensions: The enigmas of space and time*. Oxford: Oxford University Press.

Loizou, Andros. 1986. *The reality of time*. Aldershot: Gower.

Lowe, Edward Jonathan. 2003. *A survey of metaphysics*. Oxford: Oxford University Press.

McGinn, Colin. 2004. *Mindsight: Image, dream, meaning*. Cambridge, MA: Harvard University Press.

McKinnon, Neil. 2010. Presentism and consciousness. *Australasian Journal of Philosophy* 81(3): 305–323.

McTaggart, J.M. Ellis. 1908. The unreality of time. *Mind* 17(68): 457–474.

Mellor, David Hugh. 1998. *Real time II*. London: Routledge.

Perrett, Roy W. 1999. Musical unity and sentential unity. *British Journal of Aesthetics* 39(2): 97–111.

Power, Sean Enda. 2010. Complex experience, relativity and abandoning simultaneity. *Journal of Consciousness Studies* 17(3–4): 231–256.

Power, Sean Enda. 2012. The metaphysics of the "specious" present. *Erkenntnis* 77(1): 121–132.

Prior, Arthur Norman. 2001. The notion of the present. In *Metaphysics: Contemporary readings*, ed. Michael J. Loux, 289–293. London: Routledge.

Sartre, Jean-Paul. 1986. *The imaginary: A phenomenological psychology of the imagination*. Trans. J. Webber. London: Routledge.

Schlesinger, George N. 1991. E pur si muove. *The Philosophical Quarterly* 41(165): 427–441.

Sider, Ted. 2001. *Four-dimensionalism: An ontology of persistence and time*. Oxford: Oxford University Press.

St Augustine. 1982. *Confessions*. Trans. E.B. Pusey. Harmondsworth: Penguin.

Wittmann, M. (2011). Moments in time. *Frontiers in Integrative Neuroscience*, 5(66). doi:10.3389/fnint.2011.00066

Chapter 6
The Duration of Presence

Marc Wittmann

Abstract Regarding the present experience in the here and now, the question arises as to what the temporal limits of conscious awareness are. At least three levels of temporal present pertaining to temporal integration with different duration can be discerned: (1) in the range of milliseconds, the *functional moment* defines whether events are perceived as simultaneous or as appearing temporally ordered; (2) in the range of up to 2 or 3 s, the *experienced moment* is related to temporal segmentation which enables the conscious awareness of the present moment; (3) in the range of multiple seconds, continuity of experience is formed by working memory processes leading to the sense of *mental presence*. Present experience is a single unitary state. Therefore, experiences on lower levels of temporal integration are embedded and discontinuously fused into the highest level of integration: mental presence. Events occurring within an experienced moment are phenomenally present and integrated into working memory-related mental presence.

6.1 The Dual Aspect of Time Consciousness

Conscious experience evolves over time. The sensed passage of time constitutes itself through the anticipation, subsequent experiencing, and eventual remembering of an event. Metaphorically, time is therefore often described as a stream or flow. The experience of a seemingly continuous stream of events over time contrasts with another temporal aspect of conscious awareness: the sense of living and experiencing in the present moment. Phenomenal consciousness is bound to present experience. The contents of consciousness are phenomenally present—now. Phenomenal analysis points to this dual aspect of experience: the passage of time and the feeling of a present moment (James 1890; Husserl 1928). Phenomenal consciousness thus consists of an island or window of presence in the continuous flow of experiences related to events happening right now (Metzinger 2004).

M. Wittmann (✉)
Institute for Frontier Areas of Psychology and Mental Health,
Wilhelmstr. 3a, 79098 Freiburg, Germany
e-mail: wittmann@igpp.de

© Springer International Publishing Switzerland 2016   101
B. Mölder et al. (eds.), *Philosophy and Psychology of Time*, Studies
in Brain and Mind 9, DOI 10.1007/978-3-319-22195-3_6

Experienced presence implies duration[1]: an experience within a window of presence has a beginning and an end; several events following each other in rapid succession may be experienced as temporally extended entity (Revonsuo 2006). Movement and change as part of individual perceptual experiences (a passing car, notes of the musical scale played, a tender stroke on my back) cover an interval of time in which distinct events can happen (Hoerl 2013). Extended experiences have temporal structure with parts that are temporally ordered but nevertheless are perceived as unity (Kiverstein and Arstila 2013). Present experience, thereafter, is not a durationless instant in time but is embedded in a temporal field. The content of present experience is always extended through time reaching into the past as memory of what has just happened and into the future as anticipation (Husserl 1928; Kiverstein 2009; Lloyd 2012).[2]

In philosophers' discussions at least two conceptualizations can be found regarding the question of how we can perceive transition and movement, or experience events as stretched over time (Dainton 2008; Hoerl 2009, 2013; Kiverstein 2010; Benovsky 2013): In the *intentionalist* (or *retentional*) *account* perceptual experience is intentionally directed to the past, present, or future. A just passed event is experienced through memory-type components (retentions) that are actualized next to momentary sensory experience, thus creating a sense of duration through simultaneous awareness of all components past and present. In the *extensionalist account* perceptual experience is extended over time just as the events in the world are. We perceive change and succession or persistence because individual experiences unfold over time as whole intervals, perceived events falling within the extended present experience. As I will attempt to show, these two accounts may be referring to different types of present experience with different duration (Wittmann 2011). Moreover, the *specious present*, which is conceptualized as having perceptually experienced duration (Power 2012, Chap. 5 of this volume), is estimated to range between milliseconds (Dainton 2008) and several dozen seconds (James 1890). This discrepancy can be explained by the fact that the selected examples of experience pertain to different types of present (Wittmann 2011).

Temporal experience according to the extensionalist account has a lower and an upper limit because phenomena are only perceivable or discriminable within these limits (Hoerl 2009, 2013). We can see the second hand of the clock moving but not the movement of the hour hand. Although the movement of the hour hand may be mechanically constructed in the way that it moves continuously, this movement cannot be perceived as it is too slow; it can only be inferred from memory that the

---

[1] See the distinction between absolute and relative presence by Sean Enda Power's Chap. 5 of this volume. My analysis is consistent with the concept of relative presence which can be durational. Absolute presence would be punctual.

[2] Dan Lloyd (2004, 2012) provides an intuitive example for a temporal field in music. Someone familiar with the Beatles song "Hey Jude," when she hears Paul McCartney start to sing the "Hey" accompanied with the well-known tune will automatically anticipate the "Jude." The "Jude" is somehow present but it actually physically not yet existent (the recording could suddenly be interrupted). When McCartney sings the "Jude" the "Hey" is still somehow present although no longer physically there. "Hey Jude" forms a unit of present experience.

position of the hour hand now is in a different position than some minutes before. This is due to an upper limit of temporal experience. The experienced present is temporally not extended enough for the slow changes of the hour hand. The lower limit of temporal perception on the other hand can be inferred from experimental designs probing for visual flicker and auditory flutter fusion thresholds. If the flickering of a light is physically too fast subjects do not perceive the flickering, the light seems to shine continuously (double pulse detection is possible with inter-stimulus intervals above ca. 40 ms; Poggel et al. 2012). A higher temporal resolution is found in the auditory domain where the flutter fusion threshold typically lies in the range of around 11 ms in young subjects (Rammsayer and Altenmüller 2006).

Independent of philosophical conceptions underlying the phenomenon of a sensed presence, we all seem to experience an extended period of time (Power 2012). That is, succession, rhythmic grouping and motion are directly perceived as constituents of present experience (Stern 1897). If the present experience of these features has extension one can then ask: what is the duration of the present? Edmund Husserl (1928/1991, 32) actually speaks of duration of the temporal field, "which is manifestly limited, precisely as in perception's case. Indeed, on the whole, one might dare to assert that the temporal field always has the same extensions."

Thresholds of temporal integration pertaining to the subjective grouping of separated events are identified and quantified in experimental psychology; in the neurosciences neural system states are assumed to be created by neural oscillations with defined duration of their periods. That is, perceptual information is processed in discrete units, i.e., at regular moments in time (van Wassenhove 2009). The first question is whether these functional states, as identified with methods of the cognitive sciences, can be related to the experience of presence. A second question follows, namely why despite the discrete nature of neural processes in perception and cognition experience seems to evolve continuously (VanRullen and Koch 2003). This question is mirrored by the puzzle formulated by philosophers concerning the dual aspect of consciousness. If there are temporal windows of a present spanning certain duration, how can it be that we experience a continuous flow of time (Dainton 2010)? The two questions might be answered when one studies the very nature of present experience and its underlying processes.

6.2 Discontinuous Processes in Perception

I am watching white clouds slowly and steadily pass the blue background of the patch of sky I am overlooking sitting in a café. As smooth as the movement of the clouds may appear, while I try to fixate a certain spot in the sky the "neural machinery" underlying perception is in fact processing information in discrete ways on several pre-conscious as well as consciously accessible levels ranging from some tens of milliseconds to several tens of seconds; several further discretely operating processing mechanisms come into play when considering the full spectrum of cognitive processes in human perception and interpersonal action such as when communicating with others (Trevarthen 1999, Table 1; Tschacher et al. 2013). Moreover, as an embodied perceiver, many rhythmic body processes as represented in the brain are implicated with cognitive processes; in fact, the feeling of presence is dependent upon the experience of an emotional and bodily self that is constantly changing (Craig 2009; Seth et al. 2012).

When focusing on the visual impression of the moving clouds, an active fixation system is necessary to keep the eyes on the chosen spot. Nevertheless very small involuntary movements, of which we are not aware, micro-saccades, drift and tremor with duration in the milliseconds range, occur for maintaining accurate vision (Martinez-Conde et al. 2009). After a short while of gazing at a certain area of an interestingly shaped cloud which I followed with smooth pursuit eye movements, I make a voluntary eye movement to another spot. This rapid ballistic eye movement of some tens of milliseconds duration is initiated to reach the target, where I fixate another spot in the sky for the duration somewhere between 150 ms and 2 s before another eye movement happens. We are mostly unaware that on average three saccades per second occur under normal circumstances. During a saccade, sensitivity of vision is strongly distorted, but this goes unnoticed by the observer who overall experiences visual stability due to compensation mechanisms (Ross et al. 2001). Potentially, a mechanism to temporally extend the target's percept fills in this perceptual "gap" during saccadic suppression, thus creating the experience of undistorted visual continuity (Yarrow et al. 2001). Moreover, we are mostly unaware that there are eye blinks, which—for a considerable duration of 200–250 ms (Caffier et al. 2003)—interrupt the stream of visual input. Potentially, an active top-down mechanism is responsible so that eye blinks go noticed (Bristow et al. 2005).[3]

As the phenomenal and functional analysis of visual perception shows, ongoing conscious experience seems continuous, or at least does not seem discontinuous, as there are no apparent gaps or temporal boundaries (Rashbrook 2013), despite the discontinuous processing of the underlying "machinery." My present experience can be summarized as "presently experiencing the clouds in motion and being aware that I have been watching them for a while." The question here can be formulated as: What is the upper limit of present and continuous experience? More generally stated, what are the temporal boundaries of experience that allow us to be in the here and now? The whole answer might require the differentiation between different types of present (Pöppel 1988, 2009; Varela 1999; Wittmann 2011, 2014; Montemayor 2013). Three different temporal integration levels on different time scales are presented, each level covering different aspects of conceptions and examples pertaining to present experience.

---

[3] Just as an aside: Why would the neural system create a mechanism which lets the eye blinks get unnoticed? Indeed the duration of 200–250 ms is quite long. But functionally, a top-down mechanism does not compensate for the loss of visual input. Is this mechanism for letting eye blinks go unnoticed an indication that phenomenal consciousness beyond functionality is an important feature for the conscious observer?

6.3   The Experience of the Living Self: Mental Presence

The following phenomenal description by Barry Dainton (2000, 117) combines the two aspects of time consciousness, the metaphoric stream and the temporally extended present:

> We have an immediate experience only of what is present, a present that is surrounded by the comparative darkness of the remembered past and the anticipated future; the experienced present is not momentary, we seem to be directly aware of intervals of time as wholes; within these wholes there is a continual flow of content, and each experienced whole seamlessly gives way to the next.

There are several aspects contained in this description. There is a remembered past which lies in darkness because it is not directly present. Only when I recall yesterday's important telephone call the contents of long-term memory are activated. The same applies to when I suddenly am reminded of tonight's plans. Only when I imagine what might happen in a few hours' time, the anticipated event steps out of darkness, while I am watching the clouds pass in the sky. This experience is not in the dark but is in the illuminated clearing (Martin Heidegger's *Lichtung* in *Sein und Zeit*; see Dainton 2008, 116) "within which the world presents itself and we live our lives." That is, this presence involves an extended window of experience of a perceiving and feeling agent ("the self"). The extended window of presence is manifest through the awareness of a flow of events (from the near-immediate future to the past of what has just-happened) forming a temporal platform within which also long-term memories or future plans can be actualized.

Regarding its temporal structure, phenomenological analysis implies that with the tripartite structure of a possible future (what is about to occur) and a history (of what has just happened) within present experience, self-reflective consciousness is enabled (Lloyd 2004; Kiverstein 2009): We become aware of what is happening to me through memory of what has happened to me and expectations of what might happen to me. Through this temporal structure of consciousness the realization of a self emerges. Being conscious implies "that there is something it is like to be that organism—something it is like for that organism" (Nagel 1974, 436). In other words, there is a first-person mode of givenness in conscious states; experience is inherently given to me; phenomenal experience is mine (Zahavi 2005; Metzinger 2004). This type of presence includes all our momentary perceptions, thoughts and feelings as they relate to me. Within this temporal horizon of mental presence (Wittmann 2011) the representation of a narrative (autobiographical) self emerges which enables personal identity and continuity over time (Gallagher 2000).

Mental presence refers to the representation of a unified experience based on a temporally limited platform created by working memory. As one of the renowned neuroscientists studying working memory function once phrased it: "Working memory provides a temporal bridge between events—both those that are internally generated and environmentally presented—thereby conferring a sense of unity and continuity to conscious experience" (Goldman-Rakic 1997). The capacity of short-term retention is dependent upon the gradual loss of memorized elements as time passes. Experimental studies show how the correct recall of items decreases with increasing interval length in the range of multiple seconds (Peterson and Peterson 1959; Rubin and Wenzel 1996). As a consequence of the gradual memory loss of just experienced events over time in the multiple modalities of mental presence, boundaries or discontinuities are not manifest in the phenomenology of experience. That is, the sliding window of mental presence co-occurs with the constant fading out of multiple memory contents while new experiences are constantly appearing (James 1890, chapter XIV). Mental presence thereafter is a temporal platform with no fixed duration but is a variable temporal interval spanning multiple seconds. In this type of working memory-related mental presence the content of experience is tensed. Concurring with the intentionalist account of presence, I am aware that there are earlier and later steps in the series of perceptual experiences, thoughts and imaginations.

6.4   The Present Now: An Experienced Moment

It is important to note that mental presence is related to a unified multimodal temporal platform of short-term memory content enabling the representation of a multidimensional self (Gallagher 2013). This type of presence has to be differentiated from a sensorimotor present, *the experienced moment* (Wittmann 2011, 2014), which is based on a temporal integration window containing sensory information but also enabling accurate motor behavior up to 2–3 s duration (Fraisse 1984; Pöppel 1988, 2009). The function of the present is not confined to the rather passive notion of pure perception, but to inter-subjective synchronization and communication through the means of a common inter-personal temporal platform (Wittmann and Pöppel 2000; Franck 2012; Tschacher et al. 2013). Findings of many qualitatively different experiments suggest temporal integration up to a few seconds in perception and action. In fact, segmental processing creating temporal windows of representation have been reported for an approximate time range between 300 and 3,000 ms. For example, when hearing a metronome "tick" at a moderate speed we automatically integrate and accentuate every n-th beat to form rhythmic units (1–2, 1–2, 1–2, or 1–2–3, 1–2–3, etc.). These unified percepts are constructs; physically speaking they do not exist (Pöppel 2009). There is a certain speed range within which one can perceive individual "ticks" as being part of such a temporal gestalt: when inter-beat intervals range between approximately 300 ms as a lower limit (fastest speed) and 2–3 s as an upper limit (slowest speed). If the metronome is too fast a train of "ticks" is perceived without the representation of discrete individuated elements within the train of events. If the metronome is too slow the integration capacity breaks down and a series of individually separated events is heard (Szelag et al. 1996; London 2002). When one is instructed to follow regular metronome beats with finger tapping by pressing a button each time a beat occurs (sensorimotor synchronization), the task can effortless and accurately be accomplished within the same time range of inter-beat intervals (Peters 1989; Mates et al. 1994). A tempo with inter-beat intervals smaller than roughly 250–300 ms is too fast to accurately follow; inter-beat intervals of 2–3 s duration lead to button presses that are too early or too late (reaction times) or the observer has to actively use a counting strategy to accurately keep up with this slow pace.

The behavioral indices to some extent also concur with subjective impressions of discontinuities around the discussed durations of an upper and lower limit in the perception of unfilled temporal intervals as collected by verbal reports of volunteers in systematic experimental studies (Benussi 1913; Nakajima et al. 1980). Two similar temporal borders are delineated when instructed to tap with the finger consecutively in a self-paced tempo. Typically subjects tap with a frequency of around 2–3 Hz (with inter-tap intervals somewhere between 300 and 600 ms). This is done in order to be able to sequentially distinguish and control every individual finger tap ("now–now–now…") which is only possible with inter-tap intervals longer than ca. 250 ms (Peters 1989; Wittmann et al. 2001). When subjects are instructed to tap in the slowest tempo for them possible but nevertheless maintain a smooth and regular rhythm, they do so with tapping intervals of 2–3 s (McAuley et al. 2006). These and other examples such as the switching rate of ambiguous figures (i.e., the Necker cube) suggest that individual events separated by duration between 300 ms and 3 s are integrated to form units of sensorimotor control, i.e., experienced moments in perception and action (Pöppel 1988; Atmanspacher et al. 2004; Wittmann 2011; Montemayor 2013).

6.5   Temporal Binding: Below and Above the Functional Moment

The analysis of the temporal constraints in sensorimotor processing actually provides us with yet another type of present, an experience of perceived events between which there is no before-after relation. "Tap a table with your fingers, at a regular intervals of about a second; after each new tap, ask yourself if you can still hear its immediate predecessors. If the span of your auditory specious present is anything like mine, the answer will be 'no'." Barry Dainton's (2008) example of the present thus relates to a temporally unified whole with duration defined by the experienced fusion of perceived elements—in his case a sequence of auditory events produced by finger taps. Following from the phenomenal and functional analysis above, Dainton's specious presence is a functional moment of event integration spanning duration of ca. 300 ms.

When we tap our fingers on a table with maximum speed, we do so with inter-tap intervals of around 150 ms. This repetitive movement is too fast, sensory resolution is too low, to be able to have a representation of individual finger taps in an ordered sequence. That is, no temporal ordering in the sense that we would perceive a sequence of finger taps as "now–now–now," such as during a self-paced tempo, is possible. Only when the movement slows down and inter-tap intervals have duration of at least 250–300 ms individuated finger taps as following each other in a temporal sequence are experienced (Peters 1989; Wittmann et al. 2001). This temporal resolution is not restricted to the sensorimotor modality. When four sounds are presented in sequence or four colored discs appearing at one location and following each other, a stimulus-onset asynchrony of at least 300 ms between consecutively presented stimuli has to exist (individual stimuli may be shorter with pause intervals in-between) before an individual can reliably indicate the correct temporal order (Warren and Obusek 1972; Ulbrich et al. 2009). Below this threshold of a functional moment perception of temporal order is not possible (Wittmann 2011).[4]

Temporal order thresholds of around 300 ms are typically detected for series of four stimuli. The temporal onset of two complex and meaningful stimuli such as for auditory-visual stimuli (lip movements and voice onset) leads to context-dependent thresholds between several tens of milliseconds and up to 200 ms (van Wassenhove et al. 2007; Vatakis and Spence 2007). Temporal order thresholds for two short events such as two light flashes, two tactile stimuli or two sounds lie at 20–60 ms, inter-model thresholds being slightly higher (Exner 1875; Hirsh and Sherrick 1961; Fink et al. 2006; Miyazaki et al. 2006). Related to these empirical findings, which could be indicative of discrete processing mechanisms in perception, it has been suggested that the brain creates a-temporal system states for the processing of incoming information. This mechanism would enable to bind intra- and intermodal information (Pöppel et al. 1990; Pöppel 2009). Related to a non-temporal domain in visual perception, discrete processing in the 7 Hz range as assessed with EEG has been shown to reflect discrete neural processing cycles (Busch and VanRullen 2014). Temporal binding of spatially separated features has been identified for the human visual system, two temporal mechanisms—a fast one with ca. 30 Hz and a slow one with ca. 3 Hz (Holcombe 2009)—which correspond to the two levels of the functional moment. These two time scales are most likely related to different levels of processing, the lower (or fast) level to coarse pre-attentive processing and the higher (slow) level to content-related processing that is attention driven (Fujisaki and Nishida 2010). The lower-level functional moment, which is related to temporal order perception of some tens of milliseconds, fuses two stimuli and thus is not experienced as having duration. The experience of duration necessitates a clearly demarcating onset A and offset B defining the interval with a temporal order A before B (Wackermann 2007). The temporal order threshold of 20–60 ms thus is the smallest interval where two events have a clear temporal relation. Below the temporal-order threshold, duration is not perceived.

Related to the task of recognizing the sequence of four acoustic stimuli in a given trial and presented once only—a buzz tone, a hissing sound, and two sinusoidal tones of 300 and 1,800 Hz with stimulus duration of 75 ms (Ulbrich et al. 2009)—subjects can reliably indicate the temporal order above a defined threshold with a stimulus-onset asynchrony of at least 300 ms (chance level is 25 %; the above chance

---

[4]This statement may only relate to the conscious awareness of temporal ordering. In certain visual temporal integration tasks it has been shown that some temporal-order processing may still happen on an unconscious level (Giersch et al. 2013; Pilz et al. 2013). Importantly, this unconscious coding of temporal ordering has been discussed creating a feeling for the continuous passage of time.

correct threshold was arbitrarily set to 50 %; 100 % correct detection would require even longer stimulus-onset asynchronies). That is, four of these sounds with duration of 300 ms form a unified stimulus with 1.2 s duration. Although participants, when stimulus-onset asynchronies are above threshold, can indicate the correct order above chance, they still make mistakes. As experience shows, these rapidly presented stimuli just above threshold are still too fast as to have them temporally represented in the sense that a listener can right away say: "a buzz, the low tone, the high tone, the hiss." This would require longer stimulus-onset asynchronies. What one actually does is try to keep the four acoustic events in working memory and to "play them back" repeatedly in mind and thus eventually generate an answer.[5] Perhaps this description comes close to what Christoph Hoerl (2009, 8) writes:

> I experience neither the whizz nor the bang as past (or future); my experience is rather as of each sound occurring in turn, and my experience's taking this course is what constitutes my being aware of the whizz being followed by the bang. In a nutshell, on a molecularist reading of the specious present, I am perceptually aware of the succession of sounds as and when it happens, because they fall within the scope of one temporally extended experience.

According to this extensionalist account (the term molecularist is used in the quote above), perceptual experience longer than the functional moment is extended over time, just as events in the world are.

6.6 Three Types of Present: One Experience

Despite the identification of at least three levels of temporal integration related to the concept of the present (the functional moment, the experienced moment, mental presence), experience is characterized as a continuous and unified whole which is tied to the temporal present as a single state (Franck 2012). According to the argument presented here, experienced continuity is established through working memory processes creating a platform of mental presence in the range of multiple seconds. This window of presence is characterized by the uniform transition from momentary appearances of thoughts and perceptual experiences to short-term memory related fading out of mental content (James 1890, chapter XIV). Depending on the type of present, different philosophical accounts apply: Mental presence as discussed here with an inherently tensed structure is reminiscent of the *intentionalist* (*retentionalist*) account of the present. Momentary experiences co-occur with just passed experiences in short-term memory. In contrast, an experienced moment is a

---

[5] The phenomenal experience related to listening to the sequence of these stimuli is not easy to capture. Although with a stimulus-onset asynchrony of 300 ms one has the clear impression of four consecutive sounds, one has yet to "replay" them several times in one's mind to come up with a definite answer. One could argue that temporal order is only inferred from a retrospective perspective after the perceptual gestalt has been perceived. The ordering of four cards representing the four different sounds makes the task definitely easier. I can provide probe stimuli in wav format with various stimulus-onset asynchronies for a personal listening experience.

perceptual whole from which temporal relations can be inferred, but to some extent only retrospectively after the perceptual *gestalt* has been perceived. The experienced moment is reminiscent of the *extentionalist* account for which perceived change and succession of elements unfolds over time as a whole and which correspond to the unfolding of events in the world. However, the philosophical models of time consciousness necessarily remain on the phenomenological-descriptive level; only psychological and neurophysiological work will reveal the underlying processing mechanisms (Mölder 2014). That is, the analysis of phenomenal consciousness has to be closely aligned with the output of theory-driven empirical and quantitative research in the cognitive sciences.

Within mental presence as the largest time scale of temporal integration of present experience several integration mechanisms on at least two smaller time scales are active, and which are related to the functional and the experienced moment (Wittmann 2011). Related to the functional moment, temporal integration enables the binding of spatially and temporally separated events to form unified percepts on a milliseconds level (van Wassenhove 2009). Related to the experienced moment, the temporal integration of sensory-motor elements leads to the temporal segmentation of meaningful perception and action units enabling inter-subjective synchronization and communication (Pöppel 2009).

The question was formulated at the beginning regarding the upper limit of present and continuous experience. Or, more generally stated, what are the temporal boundaries of experience for being consciously aware in the here and now? The answer to these questions, as I tried to elaborate, requires the differentiation between different types of present, namely in the range up to 300 ms (the perceptual moment), in a range of roughly between 300 and 3,000 ms (the experienced moment), and with intervals lasting several seconds (mental presence). However, I experience the present as a single unitary state. That is, multimodal perceptual experiences of lower levels of temporal integration must be embedded and discontinuously fused into the highest level of integration, mental presence. What I perceptually experience as occurring within an experienced moment is phenomenally present and integrated in mental presence, which itself is not phenomenally accessible but represented as transiently stored content of working memory. Only the experienced moment is phenomenally present: a perceptual now of up to a few seconds that is embedded within mental presence enabling the narrative self.

Acknowledgments Over the years the empirical and conceptual work presented here was supported by grants from the Bundesministerium für Bildung und Forschung (Berlin), the Max Kade Foundation (New York), the National Institute of Drug Abuse (Bethesda), the Kavli Institute for Brain and Mind (San Diego), the tri-national Neuroscience Network Neurex (Illkirch), and the Fundação Bial (Porto). The author was also supported by the European project COST ISCH Action TD0904 "Time In MEntaL activitY: theoretical, behavioral, bioimaging and clinical perspectives (TIMELY; www.timely-cost.eu)." Thanks go to Bruno Mölder, Valtteri Arstila, and Peter Øhrstrøm for hosting the Turku workshop on "The philosophy and psychology of time: continuity, presence and the timing of experience" (14.8.–15.8.2013). At this meeting I got the chance to exchange with many philosophers and scientists on the topic of temporal presence. From this meeting and from continuous exchange, inspiration came from several people, which I would like to mention, namely

Valtteri Arstila, Sean Power, Christoph Hoerl, Oliver Rashbrook, Ian Phillips. Finally, I want to mention the conceptual work on the topic presented here with Carlos Montemayor.

References

Atmanspacher, Harald, Thomas Filk, and Hartmann Römer. 2004. Quantum zeno features of bistable perception. *Biological Cybernetics* 90: 33–40.

Benovsky, Jiri. 2013. The present vs. the specious present. *Review of Philosophy and Psychology* 4: 193–203.

Benussi, Vittorio. 1913. *Psychologie der Zeitauffassung*. Heidelberg: Carl Winters Universitätsbuchhandlung.

Bristow, Davina, John-Dylan Haynes, Richard Sylvester, Christopher Frith, and Geraint Rees. 2005. Blinking suppresses the neural response to unchanging retinal stimulation. *Current Biology* 15: 1296–1300.

Busch, Niko, and Rufin VanRullen. 2014. Is visual perception like a continuous flow or a series of snapshots? In *Subjective time: The philosophy, psychology, and neuroscience of temporality*, ed. Valtteri Arstila and Dan Lloyd, 161–178. Cambridge: MIT Press.

Caffier, Philipp, Udo Erdmann, and Peter Ullsperger. 2003. Experimental evaluation of eye-blink parameters as a drowsiness measure. *European Journal of Applied Physiology* 89: 319–325.

Craig, (A.D.). Bud. 2009. How do you feel—now? The anterior insula and human awareness. *Nature Reviews Neuroscience* 10: 59–70.

Dainton, Barry. 2000. *Stream of consciousness*. Abindgon: Routledge.

Dainton, Barry. 2008. Sensing change. *Philosophical Issues* 18: 362–384.

Dainton, Barry. 2010. Temporal consciousness. In *The Stanford encyclopedia of philosophy*, ed. Edward N. Zalta. http://plato.stanford.edu/entries/consciousness-temporal/. Accessed 17 May 2014.

Exner, Sigmund. 1875. Experimentelle Untersuchung der einfachsten psychischen Processe. III. Abhandlung. *Pflügers Archiv für die Gesamte Physiologie* 11: 403–432.

Fink, Martina, Pamela Ulbrich, Jan Churan, and Marc Wittmann. 2006. Stimulus-dependent processing of temporal order. *Behavioral Processes* 71: 344–352.

Fraisse, Paul. 1984. Perception and estimation of time. *Annual Review in Psychology* 35: 1–36.

Franck, Georg. 2012. What kind of being is mental presence? Toward a novel analysis of the hard problem of consciousness. *Mind and Matter* 10: 9–24.

Fujisaki, Waka, and Shin'ya Nishida. 2010. A common perceptual temporal limit of binding synchronous inputs across different sensory attributes and modalities. *Philosophical Transactions of the Royal Society B* 277: 2281–2290.

Gallagher, Shaun. 2000. Philosophical conceptions of the self: Implications for cognitive science. *Trends in Cognitive Sciences* 4: 14–21.

Gallagher, Shaun. 2013. A pattern theory of self. *Frontiers in Human Neuroscience* 7(443).

Giersch, Anne, Laurence Lalanne, Mitsouko van Assche, and Mark A. Elliott. 2013. On disturbed time continuity in schizophrenia: an elementary impairment in visual perception? *Frontiers in Psychology* 4(281).

Goldman-Rakic, Patricia. 1997. Space and time in the mental universe. *Nature* 386: 559–560.

Hirsh, Ira J., and Carl E. Sherrick Jr. 1961. Perceived order in different sense modalities. *Journal of Experimental Psychology* 62: 423–432.

Hoerl, Christoph. 2009. Time and tense in perceptual experience. *Philosopher's Imprint* 9: 1–18.

Hoerl, Christoph. 2013. A succession of feelings, in and of itself, is not a feeling of succession. *Mind* 122: 373–417.

Holcombe, Alex O. 2009. Seeing slow and seeing fast: Two limits on perception. *Trends in Cognitive Sciences* 13: 216–221.

Husserl, Edmund. 1928. *Vorlesungen zur Phänomenologie des inneren Zeitbewußtseins*. Halle: Max Niemeyer Verlag. English edition: Husserl, Edmund. 1991. Lectures on the phenomenology of the consciousness of internal time. In *On the phenomenology of the consciousness of internal time (1893–1917)*. Dordrecht: Kluwer Academic Publishers.

James, William. 1890. *The principles of psychology*. London: Macmillan.

Kiverstein, Julian. 2009. The minimal sense of self, temporality and the brain. *Psyche* 15: 59–74.

Kiverstein, Julian. 2010. Making sense of phenomenal unity: An intentionalist account of temporal experience. *Royal Institute of Philosophy Supplement* 85: 155–181.

Kiverstein, Julian, and Valtteri Arstila. 2013. Time in mind. In *A companion to the philosophy of time*, ed. Heather Dyke and Adrian Bardon, 444–469. Chichester: Wiley.

Lloyd, Dan. 2004. *Radiant cool: A novel theory of consciousness*. Cambridge: MIT Press.

Lloyd, Dan. 2012. Neural correlates of temporality: Default mode variability and temporal awareness. *Consciousness and Cognition* 21: 695–703.

London, Justin. 2002. Cognitive constraints on metric systems: Some observations and hypotheses. *Music Perception* 19: 529–550.

Martinez-Conde, Susana, Stephen Macknik, Xoana Troncoso, and David Hubel. 2009. Microsaccades: A neurophysiological analysis. *Trends in Neurosciences* 32: 463–475.

Mates, Jiří, Ulrike Müller, Tomáš Radil, and Ernst Pöppel. 1994. Temporal in sensorimotor synchronization. *Journal Cognitive Neuroscience* 6: 332–340.

McAuley, J. Devin, Mari Riess Jones, Shayla Holub, Heather M. Johnston, and Nathaniel S. Miller. 2006. The time of our lives: Life span development of timing and event tracking. *Journal of Experimental Psychology: General* 135: 348–367.

Metzinger, Thomas. 2004. *Being no one: The self-model theory of subjectivity*. Cambridge: MIT Press.

Miyazaki, Makoto, Shinya Yamamoto, Sunao Uchida, and Shigeru Kitazawa. 2006. Bayesian calibration of simultaneity in tactile temporal order judgment. *Nature Neuroscience* 9: 875–877.

Mölder, Bruno. 2014. How philosophical models explain time consciousness. *Procedia—Social and Behavioral Sciences* 126: 48–57.

Montemayor, Carlos. 2013. *Minding time: A theoretical and philosophical approach to the psychology of time*. Leiden: Brill.

Nagel, Thomas. 1974. What it is like to be a bat? *The Philosophical Review* 83: 435–450.

Nakajima, Yoshitaka, Shinsuku Shimojo, and Yoichi Sugita. 1980. On the perception of two successive sound bursts. *Psychological Research* 41: 335–344.

Peters, Michael. 1989. The relationship between variability of intertap intervals and interval duration. *Psychological Research* 51: 38–42.

Peterson, Lloyd R., and Margaret J. Peterson. 1959. Short-term retention of individual verbal items. *Journal of Experimental Psychology* 58: 193–198.

Pilz, Karin S., Christina Zimmermann, Janine Scholz, and Michael H. Herzog. 2013. Long-lasting visual integration of form, motion, and color as revealed by visual masking. *Journal of Vision* 13: 1–11.

Poggel, Dorothe, Bernhard Treutwein, Claudia Calmanti, and Hans Strasburger. 2012. The Tölz temporal topography study: Mapping the visual field across the life span. Part I: The topography of light detection and temporal-information processing. *Attention, Perception, & Psychophysics* 74: 1114–1132.

Pöppel, Ernst. 1988. *Mindworks: Time and conscious experience*. New York: Harcourt Brace Jovanovich.

Pöppel, Ernst. 2009. Pre-semantically defined window for cognitive processing. *Philosophical Transactions of the Royal Society B* 364: 1887–1896.

Pöppel, Ernst, Kerstin Schill, and Nicole von Steinbüchel. 1990. Sensory integration within temporally neutral system states: A hypothesis. *Naturwissenschaften* 77: 89–91.

Power, Sean. 2012. The metaphysics of the "specious" present. *Erkenntnis* 77: 121–132.

Rammsayer, Thomas, and Eckart Altenmüller. 2006. Temporal information processing in musicians and nonmusicians. *Music Perception* 24: 37–48.

Rashbrook, Oliver. 2013. The continuity of consciousness. *European Journal of Philosophy* 21: 611–640.

Revonsuo, Antti. 2006. *Inner presence: Consciousness as a biological phenomenon.* Cambridge: MIT Press.

Ross, John, M. Concetta Morrone, Michael E. Goldberg, and David C. Burr. 2001. Changes in visual perception at the time of saccades. *Trends in Neuroscience* 24: 113–121.

Rubin, David C., and Amy E. Wenzel. 1996. One hundred years of forgetting: A quantitative description of retention. *Psychological Review* 103: 734–760.

Seth, Anil K., Keisuke Suzuki, and Hugo D. Critchley. 2012. An interoceptive predictive coding model of conscious presence. *Frontiers in Psychology* 2: 395.

Stern, William. 1897. Psychische Präsenzzeit. *Zeitschrift für Psychologie und die Physiologie der Sinnesorgane* 13: 325–349.

Szelag, Elzbieta, Nicole von Steinbüchel, Mathias Reiser, Ernst Gilles de Langen, and Ernst Pöppel. 1996. Temporal constraints in processing of nonverbal rhythmic patterns. *Acta Neurobiologiae Experimentalis* 56: 215–225.

Trevarthen, Colwyn. 1999. Musicality and the intrinsic motive pulse: Evidence from human psychobiology and infant communication. *Musicae Scientiae* (Special Issue 1999–2000) 3: 155–215.

Tschacher, Wolfgang, Fabian Ramseyer, and Claudia Bergomi. 2013. The subjective present and its modulation in clinical contexts. *Timing & Time Perception* 1: 239–259.

Ulbrich, Pamela, Jan Churan, Martina Fink, and Marc Wittmann. 2009. Perception of temporal order: The effects of age, sex, and cognitive factors. *Aging, Neuropsychology, and Cognition* 16: 183–202.

van Wassenhove, Virginie. 2009. Minding time—An amodel representational space for time perception. *Philosophical Transactions of the Royal Society B* 364: 1815–1830.

van Wassenhove, Virginie, Ken W. Grant, and David Poeppel. 2007. Temporal window of integration in auditory-visual speech perception. *Neuropsychologia* 45: 598–607.

Vanrullen, Rufin, and Christof Koch. 2003. Is perception discrete or continuous? *Trends in Cognitive Sciences* 7: 207–213.

Varela, Francisco J. 1999. Present-time consciousness. *Journal of Consciousness Studies* 6: 111–140.

Vatakis, Argiro, and Charles Spence. 2007. Crossmodal binding: Evaluating the "unity assumption" using audiovisual speech stimuli. *Perception & Psychophysics* 69: 744–756.

Wackermann, Jiří. 2007. Inner and outer horizons of time experience. *Spanish Journal of Psychology* 10: 20–32.

Warren, Richard M., and Charles J. Obusek. 1972. Identification of temporal order within auditory sequences. *Perception & Psychophysics* 12: 86–90.

Wittmann, Marc. 2014. Embodied time: The experience of time, the body, and the self. In *Subjective time: The philosophy, psychology, and neuroscience of temporality*, ed. Valtteri Arstila and Dan Lloyd, 507–523. Cambridge: MIT Press.

Wittmann, Marc, Ernst Pöppel. 2000. Temporal mechanisms of the brain as fundamentals of communication—with special reference to music perception and performance. *Musicae Scientiae* (Special Issue 1999–2000) 3: 13–28.

Wittmann, Marc. 2011. Moments in time. *Frontiers in Integrative Neuroscience* 5(66).

Wittmann, Marc, Nicole von Steinbüchel, and Elzbieta Szelag. 2001. Hemispheric specialisation for self-paced motor sequences. *Cognitive Brain Research* 10: 341–344.

Yarrow, Kielan, Patrick Haggard, Ron Heal, Peter Brown, and John C. Rothwell. 2001. Illusory perceptions of space and time preserve cross-saccadic perceptual continuity. *Nature* 414: 302–305.

Zahavi, Dan. 2005. *Subjectivity and selfhood: Investigating the first-person perspective.* Cambridge: MIT Press.

Part III
Continuity and Flow of Time in Mind

Chapter 7
The Stream of Consciousness: A Philosophical Account

Oliver Rashbrook-Cooper

Abstract In this chapter I provide characterisation and explanation of what the "streamlikeness" of consciousness consists in. I distinguish two elements of stream-likeness—Phenomenal Flow, and Phenomenal Continuity. I then show how these elements of the phenomenology can be explained within an Extensionalist account of temporal experience. I also provide criticism of attempts to conceive of the streamlikeness of consciousness in terms of the absence of "gaps" in conscious experience. The "gapless" conception of streamlikeness generates a worry about the stream of consciousness potentially being illusory, as psychological research reveals the processes underlying consciousness to be gappy. The account of streamlikeness I provide generates no such worry, and thus provides a way to reconcile phenomenological and psychological research into the stream of consciousness.

7.1 Introduction

In this chapter I explore what it means to say that consciousness is "continuous" *and* "flowing." I will mostly be concerned with characterising how consciousness seems from the perspective of the experiencing subject. The notions of "continuity" and "flow" tend to be appealed to when theorists attempt to cash out the metaphor of the "stream of consciousness"—the thought being that "continuity" or "flow" provide a theoretical account of what "streamlikeness" merely gestures toward.

In this paper I begin by giving a characterisation of what it is in the phenomenology of experience that we are picking out by saying that consciousness is "stream-like." I take the "streamlikeness" of consciousness to be picking out a ubiquitous feature of consciousness—something that it is present *whenever* a subject of experience is phenomenally conscious. I argue that the attempt to characterise "stream-likeness" in terms of gaps—a dominant theme in the literature on the stream of consciousness—fails to adequately capture the phenomenology.

O. Rashbrook-Cooper (✉)
Christ Church, St. Aldates, Oxford OX1 1DP, UK
e-mail: o.w.rashbrook@gmail.com

© Springer International Publishing Switzerland 2016
B. Mölder et al. (eds.), *Philosophy and Psychology of Time*, Studies in Brain and Mind 9, DOI 10.1007/978-3-319-22195-3_7

I provide a positive account of the "streamlikeness" of consciousness in terms of the notions of "Phenomenal Flow"—consciousness strikes us as an "ongoing" or "occurrent" phenomenon—and "Phenomenal Continuity"—the temporal limits of experience aren't directly manifest in the phenomenology. Having established that these features of consciousness are what jointly render it streamlike, I provide an explanation of how consciousness could possess such features. I suggest that these features are well accommodated by an "Extensionalist" account of temporal experience, and that "Phenomenal Continuity" may present a challenge for "Atomist" accounts. This chapter aims to account for aspects of the phenomenal character of the conscious experience of time. In taking this approach, it is to be contrasted with the approach of its companion piece, which is only interested in consciousness as functionally characterised.

7.2 Aspects of Consciousness

In order to coherently discuss the continuity and flow of conscious experience, we need to first distinguish between some different "aspects" of consciousness. In particular, we need to distinguish between the *state* of consciousness, the *stream* of conscious experience, and the items *represented* in conscious experience (hereafter the *contents* of consciousness).

(a) The State of Consciousness
(b) The Stream of Consciousness
(c) The Contents of Consciousness

By "The State of Consciousness," I just mean that state of being awake as opposed to asleep or in a vegetative state. In this paper, I shall not be talking about the state of consciousness, but mention it only to put it to one side. The crucial components of the above list are the *stream* of consciousness and the *contents* of consciousness.

If we distinguish between conscious experience and the objects represented in conscious experience (the contents of consciousness), then questions arise about their respective natures. In order to understand consciousness we need to understand all of its many and varied features, and one important feature of consciousness is that it fills time in a distinctive way. It is this distinctive way of filling time that philosophers and psychologists have in mind when they claim that consciousness is "streamlike," "continuous," or "flowing." Wakeful consciousness involves my having an uninterrupted series of phenomenally conscious experiences. It is these uninterrupted stretches of experience that are sometimes described as "streamlike," "continuous," or "flowing." These properties aren't, in this instance, being ascribed to the objects of experience, but to experience *itself*.

The third item on the list, the contents of consciousness, refers to the items "represented" in the stream of consciousness. The distinction between the two levels is the distinction between my experiences, and what my experiences are experiences of. Over the course of the unfolding of an uninterrupted stream of consciousness, a large variety of items are typically represented. We need to distinguish between the stream of consciousness and its contents when discussing the concepts of "continuity" and "streamlikeness" in order that it is clear whether these concepts are intended to apply to the stream, its contents, or both.

In theorising about the continuity and flow of conscious experience, it isn't enough to merely distinguish between the stream and its contents. We also need to consider how the stream and its contents are related to one another. In particular, how it is that we are able to make claims about the character of the stream of experience given the "transparency of experience"? The contemporary notion of "transparent" experience is that, when we attempt to introspectively attend to the properties of the stream of experience, we fail. Introspective attention can only be focussed on the contents of consciousness, not the stream itself.

Michael Tye (2003, 96) articulates the tension between claiming that experience is "streamlike"/"continuous"/"flowing" and Transparency as follows:

> When we introspect, we are not aware of our experiences at all . . . So, we are not aware of our experiences as unified or as continuing through time or as succeeding one another.

However, even if we were to agree with Tye about the transparency claim, such agreement doesn't preclude our being able to learn about the structure of temporal experience by attending to the contents of consciousness. In the next section, I shall give an example of how this might work.

7.3   Temporal Limits

Temporal experience has temporal limits. By this I mean that there are limited intervals into which changes in objects in the subject's environment must fall in order for them to be perceived as changing. This point is illustrated nicely in C.D. Broad's (1923, 351) example of the hour-hand and the second-hand of a clock:

> To see a second-hand moving is a quite different thing from "seeing" that an hour-hand has moved. In the one case we are concerned with something that happens within a single sensible field; in the other we are concerned with a comparison between the contents of two different sensible fields.

In order to perceive a hand of the clock as moving, rather than merely becoming aware that the hand has moved, the changes in the hand's position must fall within a limited interval of time. It is in this sense that temporal experience has temporal limits (or, in Broad's terms, temporal experience a "sensible field").

This case thus gives us an example of how, even if we agree with the transparency claim, we can learn something about the properties of the stream of experience. By reflecting upon the contents of experience—the difference between the hour- and the second-hand cases—we learn that temporal experience has temporal limits. Later in this paper I will argue that capturing the "Flow" of consciousness requires us to abandon the transparency claim.

For the moment, the important thing to stress is that we only come to realise that experience has temporal limits indirectly—by reflecting upon the contents of consciousness. This tells something about the status of temporal limits: while they are certainly an important aspect of consciousness, they are not directly manifest in the phenomenology. Over the next three sections I shall argue that it is only by saying more about the failure of temporal limits to be directly manifest in the phenomenology that we can give an adequate account of the continuity of consciousness.

Before moving on to a positive characterisation of the continuity of consciousness, I will first criticise existing attempts to capture continuity. These attempts don't appeal to the notion of temporal limits, but rather to the idea that consciousness" temporal profile can be characterised in terms of the "strict" or "mathematical" notion of continuity. I first discuss the view that the temporal profile of consciousness is best characterised by the way it exhibits strict continuity. I then discuss a contrasting view (that of Strawson) according to which we ought to characterise consciousness as positively seeming to lack strict continuity. I argue that neither kind of appeal to strict continuity is adequate.

7.4 "Strict" Continuity

According to the "Strict" sense of continuity, consciousness is continuous if and only if it doesn't have any gaps. On this reading, if we were to somehow keep "dividing," and examining smaller and smaller sections of an extended period of conscious experience, we would never find any portions of time not filled with conscious experience. This notion of continuity as "Strict" as it has its origins in mathematics. One reason that we ought to be suspicious about characterising how consciousness experientially seems in terms of Strict Continuity is that the notion is often introduced by contrasting it with "Density":

> The rational numbers (the positive and negative fractions) under the standard ordering are dense. If time is dense there will be another instant between any pair of distinct instants... While the rationals are dense there is a sense in which there are "gaps" in the rational numbers. There is, for example, no rational number whose square is 2. In order to fills these "gaps" we add to the rationals the irrationals... The resulting system is the real number system whose salient characteristic is that it is not only dense but lacks "gaps"—an idea we express by saying that the real number system is *continuous*. (Newton-Smith 1984, 113)

If consciousness is to seem either dense or continuous, it needs to be characterisable in terms of instants. There will either seem to be an instant of conscious experience between any two other instants (density), or seem to be no gaps between any two instants of conscious experience (continuity). The problem with trying to characterise consciousness in this way is that consciousness is fundamentally to be characterised in terms of intervals of time, not instants. I shall illustrate this point by first discussing the *contents* of consciousness, and then suggesting that the same considerations can be ascribed to the stream.

There is a temptation, when characterising how experience seems from the perspective of the experiencer, to provide a description that builds in an unrealistic degree of richness into the phenomenology. This can sometimes be seen in discussions of the experience of "continuous" or "homogeneous" expanses of colour. Consider the following quotation from Sellars (1963, 26):

> The manifest ice cube presents itself to us as something which is pink through and through, as a pink continuum, all the regions of which, however small, are pink. It presents itself to us as ultimately homogeneous; and an ice cube variegated in colour is, though not homogeneous in its specific colour, "ultimately homogeneous," in the sense to which I am calling attention, with respect to the generic trait of being coloured.

The suggestion is that to describe an object of experience as "continuously" or "homogeneously" pink is to say that, in Clark's (1989, 280) words, "…between any two pink points on the cube there is a third pink point." The difficulty with this account of how the experience of the homogeneously pink ice cube strikes us is that it requires us to be able to discern *points* in experience. The reason for resisting this attempt at characterizing experience is simple—we cannot discern such point-like entities in experience. When presented with a uniformly coloured region we are unable to attend to point-like regions, and this fact is explained by the absence of point-like regions in experience.

I suggest that we ought to think of the experience of "continuous" or "homogeneous" pink differently. Rather than thinking of the experience as one in which "it seems as if between any two pink points there is a third pink point," we should think of it as one in which "it doesn't seem as if between any two pink points there is a non-pink point." For an object to appear as continuously pink, it is enough that we (a) represent the object as pink, and (b) don't represent the object as having any non-pink areas.

If this proposal is to work, then we need to think of experience as having a "top-down" structure. The representation of an object as continuously pink isn't composed out of a collection of representations of the state of the object at points. Rather, the basic experiential element just is a representation of the spatially extended object as pink—and for it to seem continuously or homogeneously pink is just for us not to represent it as having any non-pink areas.

If we accept that the experience of the pink ice cube should not be characterised in terms of the experience of points because we can't discern points in the phenomenology, then similar considerations apply to the temporal case. Just as, in the representation of a spatial expanse, we can't discern points, neither can we discern instants in the representation of a temporal expanse. The starting point of an account of the representation of a uniform expanse of colour needs to be the experience of a spatial expanse, rather than of an extensionless point. Likewise, in the temporal case, the starting point needs to be the experience of a temporal extent.

The potential mistake of conceiving of a temporal stretch of experience as consisting of a series of instants is also supported by research showing that experience has a finite simultaneity threshold. Items are experienced as simultaneous despite occurring at distinct times—in audition, for example, the relevant threshold is 2–3 ms. Given that our perceptual sensitivity isn't best explained in terms of sensitivity to instants, it would be highly unusual if experience seemed phenomenologically to involve such sensitivity.

Finally, turning to the stream of consciousness, just as it doesn't seem as if we represent temporal or spatial expanses as consisting of a collection of points or instants, neither does it seem to us that we can discern instants when we try and attend to the character of the stream of consciousness. The general lesson of this section is that a characterization of how consciousness experientially seems ought not to mention instants or points. This, in turn, means that a characterization of consciousness in terms of either strict continuity or density is not appropriate.

The failure of the "Strict" notion of continuity to capture how consciousness seems from the perspective of the experiencer has the consequence that the stream-likeness of consciousness should not be considered a "Grand Illusion." The idea of a "Grand Illusion" is that we could be "misled as to the true nature of consciousness" or that "we are mistaken in our assessment of how things seem to us to be" (Noë 2002, 202). The purported "Grand Illusion" stems from the thought that while the stream of consciousness seems subjectively to be strictly continuous, objective empirical investigation provides substantial evidence suggesting that it is discontinuous (see, for instance, Dehaene 1993; VanRullen and Koch 2003; van Wassenhove 2009).

This Grand Illusionist line of thought (found in James 1890, 130–1; Dennett 1993, 356; Blackmore 2002, 17) rests upon the claim that consciousness seems subjectively to be strictly continuous—a claim that I have just argued against. This claim is also assumed in section one of this chapter's companion piece—see Madl et al. Sect. 8.1. On the position I am adopting, it isn't that consciousness (or indeed the contents of consciousness) *seems* to have *no* gaps, but rather that it *doesn't* seem to *have* gaps. Just as in the spatial case of experiencing a uniform expanse of the colour pink (best characterised in terms of an absence of non-pink regions), temporal consciousness is best characterised in terms of absence of awareness of gaps, rather than awareness of an absence of gaps.

If this is right, then there is no longer a clash between how experience seems, and how it really is. All we have is a failure of an objective property of consciousness to be manifest in subjective experience. The list of such properties (e.g. being dependent upon an organ housed in a skull, involving 40 Hz oscillations) is long and untroubling. Consciousness doesn't *phenomenologically seem* to involve 40 Hz oscillations, but no-one thinks this presents a deep paradox about conscious experience.

I have argued elsewhere in more detail that the claim required by the Grand Illusionists (that consciousness positively seems to have no gaps) isn't warranted by the phenomenal character of subjective experience (Rashbrook 2013b). In this paper, I want to simply challenge the Grand Illusionists to provide evidence that supports the claim that experience positively seems to have no gaps, rather than the weaker claim that it doesn't seem to have gaps. To give an example, a subject's *lack* of awareness of *petit mal* seizures is clearly not evidence that consciousness seems to have no gaps, but merely that it doesn't seem to have gaps.

7.5   Strawson's Gappy Stream

One diagnosis of the failure of the "Strict" notion to capture what we mean by the "Continuity" of consciousness might be to suggest that this is because consciousness *does* seem to have gaps. This claim can be found in Galen Strawson's recent discussion. In fact, Strawson goes beyond using his discussion to object to the idea that consciousness is strictly continuous—he claims that the *gappiness* of conscious is "a constant feature of normal consciousness" (Strawson 2009, 240).

Strawson distinguishes between three different kinds of "break" or "gap" that one could find in the stream of consciousness: Content Breaks; Flow Breaks; and Temporal Breaks. Each of these kinds of break are to be defined in terms of how things strike the experiencing subject, rather than how things are objectively with the subject.

A content break is a radical change in the content of consciousness. Strawson (2009, 234) suggests:

> Trains of thought are constantly broken by detours—by-blows—fissures—white noise. This is especially so (in my experience) when one is just sitting and thinking.

The claim is that, when embarked upon a train of thought about one topic, we will typically find ourselves thinking about an entirely unrelated topic. Strawson (2009, 235) suggests that this is something "true to a greater or lesser extent of all thought." However, these kinds of "Content Break" don't show that consciousness lacks continuity, for, as suggested earlier, the kind of continuity we are interested in is a property of the *stream* of consciousness, not a property of the content of consciousness. As James (1890, 240) notes when considering a similar issue:

> The transition between the thought of one object and the thought of another is no more a break in the *thought* than a joint in a bamboo is a break in the wood. It is a part of the *consciousness* as much as the joint is a part of the *bamboo*.

More startling than Content Breaks, are Strawson's notions of "Flow" and "Temporal" Breaks. In introducing these ideas, Strawson's intention is to argue that, even if the content of consciousness was held fixed, it would still be possible to discern discontinuities in the stream of consciousness. The idea of a "Flow Break" is set out in the following:

> When I'm alone and thinking, I find that my fundamental experience of consciousness is one of *repeated launches of consciousness as if from nothing*, where "as if from nothing" isn't meant to indicate any sort of positive sense of a preceding temporally extended period of *non*-consciousness (although it isn't meant to rule it out either) but just—a sense of complete beginning. (Strawson 2009, 238)

For Strawson, a crucial part of the phenomenology of conscious experience is that of an episode of consciousness having just begun. We sometimes notice that our attentional focus has shifted from the external world (e.g. looking at an object, listening for a sound), to the internal (e.g. reflecting upon a philosophical problem, daydreaming about some flight of fancy). At points it looks as though this is the phenomenon Strawson (2009, 239) has in mind[1]:

> There's a familiar and distinctive experience of realizing retrospectively that one has in fact been briefly absent as one tries to maintain continuous visual attention.

While it is certainly true that our attentional focus can shift between different kinds of conscious activity—and that one might sometimes only come to realise that such a shift has occurred retrospectively—it isn't persuasive that we ought to characterise this as a "break," rather than a shift of attentional focus within an ongoing stream.

Now on to the final type of "break"—a "Temporal Break"—"An experience that has the character of there having just been a complete absence, however brief, of consciousness." (Strawson 2009, 241) One example that appears to possess this characteristic is that of waking. When we wake from dreamless sleep, it seems to us that we are typically noninferentially aware that we have just awoken from unconsciousness. It certainly doesn't seem as if I have to *infer* whenever I wake up that I have just been asleep—but nevertheless I am aware of having been unconscious. However, this kind of experience marks the beginning of a stream of consciousness, rather than a "gap" within a single stream, and thus doesn't bear on the issue of what a single stream's "streamlikeness" consists in.

Another case that appears to satisfy Strawson's characterisation of a "Temporal Break" is supplied by Armstrong (1993, 93):

> One can "come to" at some point and realize that one has driven many miles without consciousness of the driving, or, perhaps, anything else. One has kept the car on the road, changed gears, even, or used the brake, but all in a state of "automatism."

The relevant version of Armstrong's description, for Strawson's purposes, is one in which one suddenly realises that one previously lacked consciousness of the driving or of anything else. This kind of case is extremely difficult to interpret, as it is a controversial matter as to whether it involves absence of attention, phenomenal consciousness, or self-consciousness. One difficulty with treating the case as one in which the stream of consciousness involves a "Temporal Break" is Dennett's (1993, 137) plausible suggestion that "if you had been probed about what you had just seen…you would have had at least some sketchy details to report."

It is thus far from clear that there really is a "temporal break" in this situation, rather than "a case of rolling consciousness with swift memory loss." (Dennett 1993, 137) Even if Dennett's suggested interpretation here was shown to be inadequate, there is a further problem for Strawson as this kind of experience is not ubiquitous. Rather, it typically occurs in monotonous conditions and so doesn't support Strawson's positive thesis that "gappiness" is a distinctive aspect of consciousness.

---

[1] Note that this is my interpretation of Strawson, which I have supplied as he is slightly unhelpful in making these notions precise: "The notions of a content break and a flow break are not sharp…" (Strawson 2009, 241).

7.6 Streamlikeness: Phenomenal Continuity

We have already seen that attempts to characterise the streamlikeness of consciousness by using the notion of "Strict" continuity fail. Neither the claim that consciousness seems gapless, nor Strawson's claim that consciousness is fundamentally gappy, get at features of the phenomenology that are suitably extant or ubiquitous. With this failure in mind, I want to propose an alternative approach to thinking about the "streamlikeness" of consciousness.

My alternative proposal is that we ought to characterise streamlikeness in terms of two properties: Phenomenal Continuity, and Phenomenal Flow. In this section I set out what is meant by Phenomenal Continuity. My claim is that consciousness is Phenomenally Continuous, where this means that the boundaries of temporal experience fail to be manifest in the phenomenology of experience.

As established earlier, there is a clear sense in which temporal experience possesses boundaries—it is these temporal boundaries that explain the difference between the second hand and hour hand cases. However, we can discern these boundaries only indirectly—and this provides a marked contrast to visual spatial experience. This contrast is key to spelling out the notion of Phenomenal Continuity.

In visual experience, there is a spatial region into which objects must fall if they are to be visually perceived. There is thus a parallel between visual spatial experience—in which perception is *spatially* limited—and temporal experience—in which, as we have already seen, there are temporal limits. While both varieties of experience have limits as a matter of fact, things are very different when we compare their respective phenomenology.

Crucially, in spatial visual experience, the boundaries of the spatial extent into which objects must fall, if they are to be perceived, are themselves manifest in the phenomenal character of experience. Richardson (2009, 233) sets out the notion of the spatial visual field as follows:

> [To say that there is a visual field is] to say that the boundaries or limitations of the cone, the apex of which is the point of origin for visual experience, are present in visual experience.

Further, Richardson (2009, 239) claims that:

> My awareness of there being more space than currently falls within my sensory limitations, and more things to be seen there than I can see without changing those limits is … what gives visual experience its field-type character.

It is distinctive of spatial visual experience in that, when having such experience, we are aware that there is a distinction to be drawn between the space falling within the visual field—space for current possible objects of visual perception—and the space outside the visual field.

One simple way to spell out what it means to say that the boundaries of spatial visual experience are "manifest" is to consider a situation in which you are being questioned about which items in an unchanging array you can currently see. It is reasonably clear to us which items we need to mention in order to give a complete characterisation of our visual experience—and this is to be explained by the fact that the experience involves a manifestation of our visual limitations.

However, when a temporal element is introduced into the above kind of scenario—for instance, if you were to be asked which of the notes in an A-minor scale played at three notes per second you are currently hearing, it is not clear where we ought to draw the boundary between those notes we are currently hearing, and those notes we merely retain in some form of short term memory. The key contrast being that it is *not* reasonably clear to us in the temporal case which items we need to mention in order to characterise what we are currently experiencing, whereas it was in the case of spatial visual experience.

Another way of spelling out this distinction between the two cases is suggested by Soteriou, who proposes that the absence of manifest boundaries in the temporal case may explain why, in contrast to a case in which an object is too large to fall within our spatial visual field, we don't feel as if our inability to hear a "whole one hour symphony from one's current temporal location is something that is to be explained by one's sensory limitations" (Soteriou 2013, 114). We don't feel like this, because our sensory limitations are not manifest in temporal experience in the same way as they are in spatial experience.

It is this disanalogy between spatial vision and temporal experience that renders determining the extent of the "specious present" (if we take "specious present" to denote *the* limits of temporal experience that explain why we can experience the second hand of a clock as moving, but not the hour hand) so intractable. We don't take there to be a fascinating and deep problem about investigating the spatial limits of visual experience, whereas we do take there to be one in the temporal case.

The lack of manifestation of the limits of temporal experience in the phenomenology is what forces investigation of the limits of the specious present to proceed via indirect means. While we can determine the spatial limits of spatial visual experience regardless of the content of the visual experience in question, we can only investigate the temporal limits of temporal experience via isolation of various salient aspects of the contents of consciousness. It is this kind of indirect strategy that is being employed in using a comparison of the experience of seeing the second hand *moving*, and merely seeing that the hour hand *has moved*, in order to establish that temporal experience has limits.

We can see this indirect strategy being employed in the various psychological studies of the limits of temporal experience cited in (Wittmann 2011, 4). In attempting to determine the limits of the specious present only via appeal to salient aspects of the contents of consciousness, these studies implicitly acknowledge that the temporal limits of experience are not manifest in the phenomenology in the way that they are in vision.

For example, we see attempts to glean the duration of our temporal limits that appeal to various perceptual gestalts—the grouping of metronome beats into perceptual units:

> When listening to a metronome at moderate speed, we do not hear a train of individual beats, but automatically form perceptual gestalts as an accent is perceived on every nth beat (1–2, 1–2, or 1–2–3, 1–2–3). (Wittmann 2011, 4)

How these beats are grouped appears to depend upon their falling within a 2–3 s window. We might then be inclined to claim that the "specious present" has a temporal extent of 2–3 s. Regardless of whether we are willing to accept this claim, the crucial point for our purposes is that this figure is arrived at via noting salient features of the contents of consciousness. The temporal limits of temporal experience are thus not themselves directly manifest in the phenomenology, unlike the spatial limits of spatial visual experience.

The notion of Phenomenal Continuity provides a way of understanding James' initial introduction of the metaphor of the stream of consciousness:

> Consciousness, then, does not appear to itself chopped up in bits. Such words as "chain" or "train" do not describe it fitly as it presents itself in the first instance. It is nothing jointed; it flows. A "river" or a "stream" are the metaphors by which it is most naturally described. (James 1890, 239)

On the "Phenomenal Continuity" understanding of the continuity of consciousness, the claim that consciousness doesn't appear "chopped up" or "jointed" expresses the fact that the temporal limits of experience are not manifest in the phenomenology.

7.7 Streamlikeness: Phenomenal Flow

"Phenomenal Continuity" captures something of what it means to say that consciousness is streamlike. However, it is only part of the story. An additional respect in which consciousness is "streamlike" is that it exhibits "Phenomenal Flow." To introduce this notion, consider the following from Dainton (2006, 180):

> Whenever we choose to inspect our streams of consciousness, our attentive gaze will itself always possess some temporal duration; throughout this duration we will be aware of content which is continually flowing. Or as Bradley put it: "in the ceaseless process of change in time you may narrow your scrutiny to the smallest focus, but you will find no rest."

One problem with Dainton's conception of the "flow" of consciousness is that he characterises it in terms of the *content* of consciousness. This means if there were to be an absence of flowing content (for example, imagine a case in which you are staring at an unchanging expanse of painted wall, with no other ongoing change in the contents or consciousness, nor any attentional shifts), consciousness will no longer exhibit "flow"—a bad result if we are looking for an account of a ubiquitous feature of conscious life.[2] I thus suggest we follow O'Shaughnessy's account of

---

[2] Whether such a situation could ever take place in the stream of consciousness is an interesting question. It may be that there is some principled reason to think that the contents of consciousness must exhibit continual change. If this is the case, then Dainton and O'Shaughnessy's accounts could be reconciled with one another. My thanks to an anonymous referee for pointing out that finding actual cases where the stream of consciousness exhibits no change in its contents is not straightforward.

flow, where this property is attributed to the stream of consciousness itself, rather than merely to its contents:

> Even when experience is not changing in type or content, it still changes in another respect: it is constantly *renewed*, a new sector of itself is there and then *taking place*. (O'Shaughnessy 2003, 42)

Even when presented with an unchanging environment, our stream of consciousness will nevertheless strike us as "going on" or "unfolding." Our awareness that, even in the absence of changes in the contents of consciousness, a new portion of conscious experience is taking place, constitutes the "Phenomenal Flow" of temporal experience. Note that this aspect of the phenomenology constitutes a counterexample to Tye's transparency claim (that we only find the contents of consciousness when we introspect), as here we are able to discern a phenomenological feature that isn't explained by the character of the contents of consciousness.

O'Shaughnessy's thoughts on this experiential "flow" or "flux" are importantly connected to his insightful discussion of our experience of the present (O'Shaughnessy 2003, 51). He suggests that an experiencing subject differs from a non-experiencing subject—for instance a dreamlessly sleeping subject—in a crucial regard. Only the experiencing subject can pick out the moment that is in fact present *as* the present.[3] The experiencing subject's experience unfolds in such a way that it makes available to him a new moment of time that he can pick out as "the present moment."

So, even if experience is not "changing in type or content," it will still strike us as "streamlike" in that it is constantly making available to us a new moment of time to pick out as "the present moment." To say that consciousness exhibits Phenomenal Flow is thus to say that consciousness is constantly making a new moment of time available to be picked out as "the present moment." This feature of consciousness is, like Phenomenal Continuity, suitably ubiquitous to count as picking out a key component of what renders consciousness "streamlike."

Both of the features of consciousness that render it streamlike—Phenomenal Flow and Phenomenal Continuity—share the feature of being aspects of consciousness that are present regardless of the contents of consciousness. They are what we might call "structural features" of consciousness—features of our awareness of consciousness" contents, rather than features of the contents themselves (for detailed discussion of "structural features" see Richardson 2009; Soteriou 2013). It is this that makes them suitable for a characterisation of "streamlikeness"—for "streamlikeness" is a feature that is ubiquitous in all conscious experience.

Because the characterisation of "streamlikeness" that I have suggested is not given in terms of "gaps," it isn't vulnerable to Strawson's suggestion that consciousness doesn't really seem continuous—for Strawson is conceiving of continuity in terms of gaplessness. The notion of "streamlikeness" proposed here is also not threatened by empirical work which shows that the physical basis of consciousness

---

[3] I use the term "moment" here in order to leave it deliberately open whether experience allows us to pick out strictly instantaneous points in time, or only minimal intervals.

may exhibit gappiness. The claim that consciousness exhibits Phenomenal Flow and Phenomenal Continuity is neutral as regards the gappiness of its underlying physical processes.

Having provided a phenomenological characterisation of the stream of consciousness, I shall spend the remainder of the paper setting out how a certain kind of extensionalist model of temporal experience can accommodate this phenomenology.

7.8   Extensionalism and Streamlikeness

The minimal commitment for an Extensionalist account of temporal experience is the idea that the temporal properties of experience have an explanatory role to plan in an account of our perception of temporal properties. It is thus to be contrasted with Atomist accounts, according to which the temporal properties of experience have no such explanatory role to play.

Atomists typically think of the stream of consciousness as consisting of a series of instantaneous (or as close to instantaneous as their background theoretical commitments allow) experiences. Each experience in the series represents a temporally extended interval. By contrast, Extensionalists hold that for a fundamental set of temporal properties (for instance succession and duration) the portion of experience in which those properties feature must itself bear those properties. For the Extensionalist, a representation of succession requires a succession of representations, and a representation of duration must itself possess duration (for further discussion of this point see Hoerl 2013; Rashbrook 2013a).

These two accounts of temporal experience thus differ in what they take to be explanatory of our perception of temporally extended phenomena. The atomist takes the character of the stream of consciousness to be explained independently of its temporal properties. The extensionalist, upon the other hand, holds that—at least for certain durations of time—the temporal properties of the stream of consciousness are explanatorily crucial. For the extensionalist, the character of experience *at* a particular time is always to be explained in terms of the character of experience *over an interval* of time (for more discussion of this kind of view, see Rashbrook 2013a; Soteriou 2013; Phillips 2014).

If it is crucial for the extensionalist that experiences are always extended in time, then this can be put to work in explaining the feature of consciousness I have labelled "Phenomenal Flow." An extensionalist account can explain Phenomenal Flow by saying something about *how* temporally extended experiences fill time. This is the move made by O'Shaughnessy. Immediately after introducing the idea that experience is always in flux, he (2003, 42) proposes that this is because…

> … the domain of experience is essentially a domain of occurrences, of processes and events. In this regard we should contrast the domain of experience with the other great half of the mind: the non-experiential half.

We can distinguish between experiential items that are fundamentally occurrent or ongoing (events and processes), and items that merely obtain (states). O'Shaughnessy suggests that we can draw this distinction by appealing to a thought experiment—namely, by thinking about which phenomena could exist at 0° Absolute.

His proposal is that while we can conceive a subject frozen at 0° absolute retaining all of their beliefs, we cannot conceive of them having experiences. This is due to the fact that experiences are fundamentally occurrent items (events and processes), while beliefs are not (they are states). The claim that experiences are of the kind "event" or "process" thus serves to provide an explanation of the phenomenological datum picked out by Phenomenal Flow. Experience exhibits Phenomenal Flow because it is, in fact, fundamentally an occurrent phenomenon.

One component of "Streamlikeness" is thus explained by the proposal that temporally extended experiences are occurrent items. In order to see how an extensionalist can explain Phenomenal Continuity, we need to take a more detailed look at how extensionalist accounts function. I shall do this by focussing upon Dainton's extensionalist account, taking it to exemplify a number of key extensionalist claims.

Dainton suggests that the stream of consciousness consists of a series of experiences—and that these experiences overlap with one another by sharing common parts. He proposes that the temporal limits of experience are to be explained by appeal to the primitive relation of "co-consciousness."[4] Co-consciousness binds together experiences over brief intervals of time. Co-consciousness is a "primitive" relation in the sense that it is *unanalysable*—Dainton proposes that we can't say anything more about than that it is responsible for explaining our temporal limits.

While its unanalysability renders this relation ultimately unsatisfying as part of an account of temporal experience, it has the advantage of simplicity. I will thus talk in terms of the relation "co-consciousness" for the remainder of the paper, though I take "co-consciousness" to label something in need of a deeper explanation in terms of a developed account of the metaphysics of experience (for such an attempt, see Soteriou (2013)).

Figure 7.1 (taken from Dainton 2008, 65) gives an example of how Dainton's account is to work. The stream of consciousness consists of a series of experiences of different tones, and these tones are bound together by the relation of co-consciousness to form the series of overall experiences P_1, P_X, P_2, P_Y, and P_3. In the first such experience (P_1) the tones "do" and "re" features, in the second (P_X) "re" and "mi" feature, and so on. Each overall experience in the series shares a temporal part with at least one another.

That the diagram has slightly counterintuitive labelling, as regards spelling out the order in which the experiences occur, is because Dainton is using the diagram to illustrate a point about the character of temporal experience. If consciousness in this instance was structured in such a way that it only consisted of experiences P_1, P_2, and P_3, then "re" would be experienced as following on from "do," and "fa" would

---

[4]While Dainton doesn't frame his account in terms of explicit talk of "temporal limits," they are certainly part of what his appeal to "co-consciousness" is intended to explain.

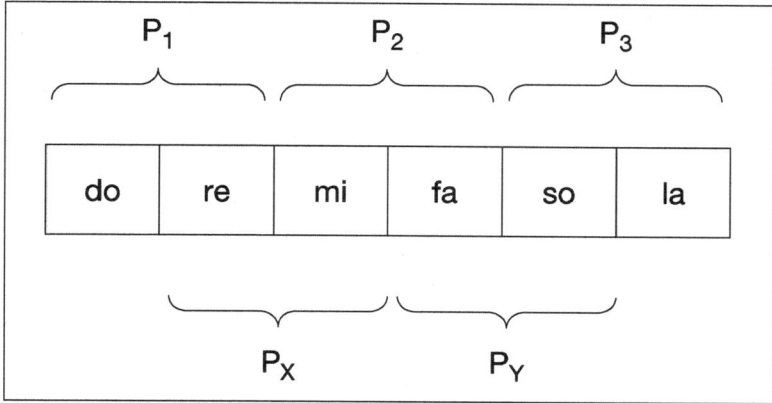

Fig. 7.1 Dainton's extensionalist model (Diagram reproduced by permission of Oxford University Press)

be experienced as following on from "mi." However, "mi" would not be experienced as following on from "re," as "mi" and "re" feature in distinct experiences, and items can only be experienced as following on from one another if they feature together in the same experience.

This hypothetical structure, while it might map out a *possible* way in which temporal experience could be structured, doesn't map out the way in which *our* temporal experience is structured. When we experience the scale, every tone is experienced as following on from the tone that preceded it.

If this characterisation of the experience of a scale is typical, then it can't be the case that temporal experience consists of a non-overlapping series of distinct conscious experiences (as it is when modelled as consisting just of experiences P_1, P_2, and P_3). In order to give the right account of our temporal experience, we need to either deny that there are distinct conscious experiences (e.g. Tye 2003) or that distinct conscious experiences are non-overlapping (e.g. Broad 1923; Husserl 1991). This experiential overlap is what Dainton is proposing in introducing experiences P_X and P_Y into the model.

With this account of temporal experience in mind, we can now set about the business of explaining how it can account for Phenomenal Continuity—the failure of the temporal boundaries of experience to be manifest in the phenomenology. Note that when inspecting Dainton's diagram, one question we might want to ask concerns what the subject is experiencing at the instant when the note "re" is exactly midway through sounding. After all, the experience of "re" is part of two overall experiences—P_1 and P_X—and it is thus unclear which of these provides the correct characterisation of what the subject is experiencing at that time.

The right extensionalist response to this question, however, is to note that they are claiming that facts about what the subject is experiencing *at* a time are determined by how experience unfolds *over* time. If this is so, then it is a mistake to think that the question about how things stand with the subject at an instant can be answered independently of specifying some interval within which that instant falls. There are thus a number of different, non-rivalrous, candidate intervals that we might appeal to in specifying what the subject is experiencing midway through the sounding of "re."

It is this feature of Extensionalism that serves to explain Phenomenal Continuity. Even if we accept that there are set limits (perhaps the earlier-suggested duration of 3 s) to what can be encompassed by co-consciousness, for any given time at which we ask the subject "what are you experiencing *now*?" there are a number of candidate intervals of the relevant duration, all of which supply a correct answer. Dainton's diagram contains only two such intervals, though this is a deliberate oversimplification, in reality there will be many more intervals, all of which can provide satisfactory accounts of "what it is like" for the subject at the time in question.

On the Extensionalist model there is only a fact of the matter about the character of experience with respect to an interval. So in order to enquire about the character of experience we need to first specify the interval of time that we have in mind, and then ask about the character of experience over that interval. Given this, we ought not to find it surprising that the temporal boundaries of experience are not manifest in the phenomenology—for there is no such thing as *the* bounded interval whose boundaries could be manifest. It isn't as if the subject is, at an instant, having just one experience or representation, the boundaries of which they might be able to discern (as they would be on an atomist view).

Rather, there are any number of candidate temporal regions in terms of which their experience at that instant could be specified. It is this indeterminacy of which temporal region it is that we are experiencing at a time that renders it impossible for us to discern experience's temporal *limits* or *boundaries*. In enquiring about the character of experience, we need to specify the interval about which we are enquiring. Thus there could be no prospect of finding out the temporal limits of the unique interval that the subject is perceptually aware of at a particular moment in time.

In this sense, the Extensionalist account of how we go about determining the character of a subject's experience at a time bears a resemblance to Dennett's (1993, 138) claim that "there are no fixed facts about the stream of consciousness independent of particular probes."[5] On the Extensionalist account, we need to specify the interval we have in mind before asking for an introspective report of the character of the stream of consciousness, and so there is no prospect of ascertaining the temporal boundaries of experience by just attempting to determine what the subject is experiencing "now." In order for us to ensure that we are "probing" the interval that corresponds to the temporal limits of experience we would need to have some prior grasp of what those temporal limits are. Any attempt to try and determine the temporal limits of experience by simply enquiring about what the subject is experiencing "now" must therefore presuppose knowledge of the limits that it is attempting to determine.

---

[5] This point of comparison is made in Soteriou (2013).

Implicit in the above is a challenge for the atomist approach to temporal experience. The atomist holds that, at a time, there *is* some fact of the matter about which temporal region the subject is experiencing. There is, at a time, an interval-independent fact of the matter about what the subject is aware of. If this is their view, then one potential difficulty for them concerns how they can explain the trouble we have discerning the duration of the "specious present." If we are, as the atomist proposes, presented, at an instant, with a temporal region with a given duration, then it looks mysterious that we cannot simply solve the problem of the duration of the specious present via introspection. Accounting for Phenomenal Continuity thus presents a problem for Atomism, and I leave it to the defenders of this approach to show how they might solve it.

7.9 Conclusion

In this paper I have provided criticism of a tendency to characterise the streamlikeness of consciousness in terms of gaps, by arguing that such attempts fail to pick out a suitably ubiquitous or extant feature of consciousness. I have provided a positive alternative account of what "streamlikeness" consists in: Phenomenal Flow and Phenomenal Continuity. Consciousness strikes us as a fundamentally ongoing phenomenon (it "flows"), and also fails to include phenomenological manifestation of its temporal boundaries (continuity). Both of these features are ubiquitous, and thus serve to adequately characterise what it means to say that consciousness is "streamlike."

These features are well accounted for by an Extensionalist theory of temporal experience. Extensionalism lends itself to a simple account of Phenomenal Flow, whereby consciousness seems fundamentally to be an ongoing phenomenon precisely because it *is* one. It also lends itself to a more complex account of Phenomenal Continuity, whereby the boundaries of temporal experience fail to be phenomenologically manifest due to there being a number of candidate intervals in terms of which a subject's experience might be characterised at any given instant.

In conclusion, the account of "Streamlikeness" provided in this paper helps to avoid some illusory puzzles (the so-called "Grand Illusion" of streamlikeness) that can be caused by an incorrect approach to thinking about streamlikeness. It has also provided some insight into why certain kinds of psychological investigation proceed in the way that they do—for instance, attempts to discern the limits of temporal experience proceed in the way they do because those limits fail to be manifest phenomenologically in the way that spatial limits are manifest in vision. This conception of what it means to say that consciousness is like a stream thus provides a way to reconcile philosophical and psychological approaches to investigating the stream of consciousness.

References

Armstrong, David. 1993. *A materialist theory of the mind*. London: Routledge.
Blackmore, Susan. 2002. There is no stream of consciousness. *Journal of Consciousness Studies* 9: 17–28.
Broad, Charles Dunbar. 1923. *Scientific thought*. London: Routledge & Kegan Paul.
Clark, Austen. 1989. The particulate instantiation of homogeneous pink. *Synthese* 80: 277–304.
Dainton, Barry. 2006. *Stream of consciousness: Unity and continuity in conscious experience*, 2nd ed. London: Routledge.
Dainton, Barry. 2008. *The phenomenal self*. Oxford: Oxford University Press.
Dehaene, Stanislas. 1993. Temporal oscillations in human perception. *Psychological Science* 4: 264–270.
Dennett, Daniel. 1993. *Consciousness explained*. London: Penguin.
Hoerl, Christoph. 2013. A succession of feelings, in and of itself, is not a feeling of succession. *Mind* 122: 373–417.
Husserl, Edmund. 1991. *On the phenomenology of the consciousness of internal time (1893–1917)*. Dordrecht: Kluwer Academic Publishers.
James, William. 1890. *The principles of psychology*. New York: Henry Holt and Company.
Newton-Smith, William. 1984. *The structure of time*. London: Routledge & Kegan Paul.
Noë, Alva. 2002. Is the visual world a grand illusion? *Journal of Consciousness Studies* 9: 1–12.
O'Shaughnessy, Brian. 2003. *Consciousness and the world*. Oxford: Oxford University Press.
Phillips, Ian. 2014. The temporal structure of experience. In *Subjective time: The philosophy, psychology, and neuroscience of temporality*, ed. Valtteri Arstila and Dan Lloyd, 139–159. Cambridge: The MIT Press.
Rashbrook, Oliver. 2013a. An appearance of succession requires a succession of appearances. *Philosophy and Phenomenological Research* 87: 584–610.
Rashbrook, Oliver. 2013b. The continuity of consciousness. *European Journal of Philosophy* 21: 611–640.
Richardson, Louise. 2009. Seeing empty space. *European Journal of Philosophy* 18: 227–243.
Sellars, Wilfrid. 1963. Philosophy and the scientific image of man. In *Science, perception and reality*, 60–105. London: Routledge & Kegan Paul.
Soteriou, Matthew. 2013. *The mind's construction: The ontology of mind and mental action*. Oxford: Oxford University Press.
Strawson, Galen. 2009. *Selves: An essay in revisionary metaphysics*. Oxford: Oxford University Press.
Tye, Michael. 2003. *Consciousness and persons*. Cambridge: The MIT Press.
van Wassenhove, Virginie. 2009. Minding time in an amodal representational space. *Philosophical Transactions of the Royal Society, B: Biological Sciences* 364: 1815–1830.
VanRullen, Rufin, and Christoph Koch. 2003. Is perception discrete or continuous? *Trends in Cognitive Sciences* 7: 207–213.
Wittmann, Marc. 2011. Moments in time. *Frontiers in Intergrative Neuroscience* 5(66): 1–11. doi:10.3389/fnint.2011.00066.

Chapter 8
Continuity and the Flow of Time: A Cognitive Science Perspective

Tamas Madl, Stan Franklin, Javier Snaider, and Usef Faghihi

Abstract Modern tools and methods of cognitive science, such as brain imaging or computational modeling, can provide new insights for age-old philosophical questions regarding the nature of temporal experience. This chapter aims to provide an overview of functional consciousness and time perception in brains and minds (Sect. 8.2), and to describe a computational cognitive architecture partially implementing these phenomena (Sects. 8.3, 8.4 and 8.5), and its comparison with data from human behavioral experiments (Sect. 8.6).

8.1 Introduction

The life (existence?) of each of us as human being consists introspectively of a continual flow of conscious or consciously-mediated experience over time. This assertion seems to raise all sorts of questions. What, if anything, "out there" is being experienced? Is this continual flow "really" continual, or do we create the illusion of continuity from a rapid sequence of frames? For the latter, what can we say about the structure of one of these frames? And, what is meant by "over time"? In this chapter we propose possible answers to these questions derived from cognitive neuroscience with the help of an integrated, systems-level cognitive model of how minds work. We hypothesize an answer to the first question above by assuming the existence of a real, physical world that can only be known to us in part through our various senses. We assume that when a tree falls in the forest there are vibrations of the air, but sound would exist only in the mind of some organism (or artificial

T. Madl (✉)
School of Computer Science, University of Manchester, Manchester, UK

Austrian Research Institute for Artificial Intelligence, Vienna, Austria
e-mail: thomas.madl@gmail.com

S. Franklin • J. Snaider
Institute for Intelligent Systems, University of Memphis, Memphis, TN, USA

U. Faghihi
Department of Computing and Technology, Cameron University, Lawton, OK, USA

© Springer International Publishing Switzerland 2016
B. Mölder et al. (eds.), *Philosophy and Psychology of Time*, Studies in Brain and Mind 9, DOI 10.1007/978-3-319-22195-3_8

agent?) equipped with an appropriate auditory sense and concomitant cognitive abilities with which to represent and perhaps understand the sound. This process can be thought of as the organism (agent) cognitively modeling its world, at least in part. We say "in part" since the frequency range of the auditory sensory apparatus is typically limited. This view leads us to hypothesize perception as a creative cognitive process at least partially dependent on our senses.

In contrast to other modes of perception, such as taste, color or sound, there is no specific physical sense for time. However, we perceive information from the other senses *over* time; we perceive time in response to change in our sensations. Thus time is viewed here as being fundamental to our cognitive processes. Instead of asking "How can time be perceived?" we will consider "How can a sense of time be produced by a cognitive system?" We hypothesize that our perception of time is constructed by cognitive processes of an organism or other agent. In this chapter we propose to explore possible such processes for producing a sense of time.

Philosophers have proposed that our phenomenal flow of consciousness over time as composed of individual frames (episodes of experiencing), and have given three different accounts of their structure. One of them refers to these three as the cinematic, retentional, and extensional models (Dainton 2010). The *cinematic model* views our introspective flow of time as consisting of a continuous succession of very brief, motion-free frames lacking any (or significant) extension. The *retentional model* takes an entirely similar view, except that the content of each frame is allowed to refer to frames representing intervals. Thus, these contents can represent, though not constitute, temporally extended time intervals. As the name would suggest, the *extensional model* considers each frame to have a brief temporal extension, to comprise an interval of time.

According to our first hypothesis above, what we know of the presumed outside physical world is constructed by us from our conscious perception. Our introspection tells us that this ongoing stream of conscious perception is continuous, extended over time, without gaps, other than those produced by deep sleep. Each of the three models discussed above assume the continuity of our perception of time, But, might it be that our introspection has deceived us as it does when we perceive a sufficiently rapid sequence of still frames in a movie theater as continuous in time? Here we will argue that this is precisely the case, that we in fact construct our apparently continuous flow of conscious perception from a rapid (5–10 Hz) sequence of discrete frames of conscious content (Madl et al. 2011). This view is consistent with recent neuroscience results, which suggest conscious access to arise from periodic phases of information integration (Baars et al. 2013; Dehaene et al. 2014; VanRullen et al. 2014). We will also argue that events in the same frame are consciously perceived as simultaneous (Snaider et al. 2012), and that each frame allows some small amount of motion[1] within its duration (VanRullen and Koch 2003).

---

[1] Confirmed by Christof Koch in personal communication with one of the authors.

8.2 The Cognitive Neuroscience of Consciousness and Time Perception: A Brief Introduction

The cognitive neuroscience of conscious perception is concerned with trying to find minimal neuronal mechanisms which distinguish "conscious" mental states from unconscious ones, as reported by experimental subjects (Crick and Koch 1990; Koch 2004). Consciousness is a difficult phenomenon to study, due to its intrinsically introspective nature; and its experimental investigation is further complicated by some ambiguity as to what exactly is meant by the term.

In this chapter, we will talk only about the functionally relevant aspects of consciousness—"functional consciousness" or "access consciousness" in neuroscience (Block 1995; Baars 2005; Dehaene and Changeux 2003). We will neglect phenomenal consciousness or "qualia" (e.g., what experiences might feel like—such as taste qualia in the case of wine) (Dennett 1988), since it is notoriously difficult to study in a formal, systematic setting. In contrast, functional or access consciousness are described in terms of the availability of mental states to higher-level cognitive processes. If a state or percept enters an agent's (biological or artificial) functional consciousness, it can influence decision making (for example, when a subject correctly presses a button in response to a stimulus, or verbally reports his perception of that stimulus).

The brain mechanisms underlying functional consciousness can be studied in paradigms contrasting conscious and non-conscious brain states. Example conditions in which visual stimuli can be presented such that they cannot be consciously perceived include visual illusions (Kim and Blake 2005), masking[2] (Kouider and Dehaene 2007), or binocular rivalry[3] (Doesburg et al. 2009; Pitts and Britz 2011). Such paradigms help investigate the "neural correlates of consciousness" by identifying which parts of brain activity patterns might correspond to conscious percepts, as opposed to unconscious percepts. Apart from sensory areas such as the visual cortex, brain imaging experiments have indicated that prefrontal and posterior parietal networks exhibit activation strongly correlated with visual awareness (Rees et al. 2002). Unfortunately, there does not seem to be a set of brain areas exclusively involved with conscious processing (Dehaene et al. 2014), casting in doubt the idea of a specific cognitive processor being responsible for consciousness. For example, even areas associated with high-level cognition such as task switching in the prefrontal cortex can be triggered non-consciously (Lau and Passingham 2007; Reuss et al. 2011).

It has been suggested that the difference between conscious and non-conscious processing might be due to differences in temporal coherence or synchronization of

---

[2] Masking involves the elimination of the visibility of one briefly presented stimulus by the presentation of a second brief stimulus (the "mask").

[3] Binocular rivalry involves presentation of different visual stimuli to the left and right eyes of subjects. In this paradigm, conscious perception alternates between the two stimuli—see also Fig. 8.1.

neural activity in the same anatomical substrate (Melloni et al. 2007; VanRullen et al. 2014; Singer 2011) (Fig. 8.1 top). Unlike unconscious perception, which involves local coordination and propagation of sensory information to progressively higher-level representations, conscious perception might require global coordination of widely distributed neurons. This global coordination might be facilitated by long-distance synchronization (Dehaene et al. 2006, 2014), which can temporarily integrate neurons into coherent assemblies and facilitate long-range communication between distant brain areas. There is a large amount of empirical support for this idea—for example, cortical and thalamic neurons discharge synchronously during wakefulness (Steriade 2006) and synchrony is enhanced for consciously perceived

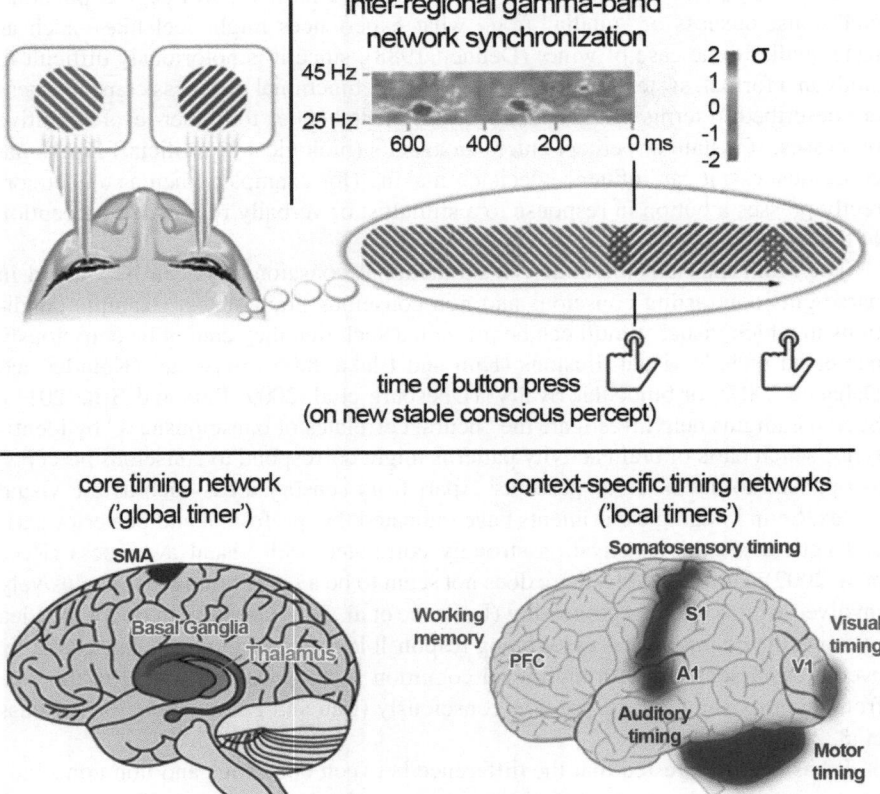

Fig. 8.1 Oscillatory synchrony, and major neural correlates of time perception. *Top*: schematic of a binocular rivalry experiment, and periods of synchrony dissolving and re-forming at each conscious episode (Synchrony data from Doesburg et al. (2009), head image from Dieter and Tadin (2011)). *Bottom*: Neural bases of the core timing network (thalamus, cortex, *BG* basal ganglia and *SMA* supplementary motor area), and example context-specific timing networks including the visual cortex (*V1*), and auditory and somatosensory cortex (*A1* and *S1*), and the cerebellum (Brain images modified from Wikimedia 2009, 2010, based on Merchant et al. 2013)

stimuli (Palva et al. 2005). In masking paradigms, increased gamma frequency band synchrony is induced only by words reported as perceived by subjects (Melloni et al. 2007). Furthermore, in the binocular rivalry paradigm, gamma-synchronous activity locked to an ongoing theta rhythm precedes perceptual switching (as indicated by subjects pressing a button when the stimulus which they are conscious of changes (Doesburg et al. 2009)). Finally, neural activity is globally disintegrated and fragmented in time in unconscious subjects, e.g., those undergoing anesthesia (Lewis et al. 2012), and awake vs. unconscious states can be reliably separated using a measure of the amount of information shared by distant cortical sites (Casali et al. 2013). See Singer (2011) for further evidence.

More recent theories of consciousness are consistent with such empirical results, suggesting consciousness to be a process involving large-scale brain activity, instead of attempting to confine it to one or few brain areas. Prevalent examples include the Global Workspace Theory (which proposes that consciousness is facilitated by a fleeting memory capacity enabling access between spatially separate brain functions (Baars 2005; Dehaene and Changeux 2003; Baars et al. 2013)), and Neural Darwinism (which proposes that conscious experience arises from reentrant neural activity in the thalamocortical system (Edelman and Tononi 2000)).

If periodic large-scale integration via oscillatory synchrony is indeed necessary for conscious processing, then this would have important implications for the structure of experience. Most importantly for the present topic, it would imply that consciousness is a discrete mechanism, since large-scale synchrony in brains is not continuously present, but has been observed to arise and dissolve periodically several times per second (VanRullen et al. 2014; Doesburg et al. 2009; Singer 2011; Madl et al. 2011). Although there is no definite answer to whether consciousness is discrete or continuous, there is substantial neuroscientific (see above) as well as psychophysical evidence supporting the discrete hypothesis, such as the wagon wheel illusion, in which a turning wheel is perceived to rotate in the wrong direction, presumably due to discrete sampling (see VanRullen and Koch (2003) for a review of psychophysical evidence of discrete perception).

Another important consequence of a periodic mechanism facilitating consciousness is that such a mechanism can be used to estimate the durations of events by counting the occurrences of cycles, similarly to pulse accumulator models of time perception in psychology (Grondin 2010). Large-scale oscillatory activity in a cortico-thalamic-basal ganglia circuit has been described as the "core timer" of the brain (Merchant et al. 2013) (the cortico-thalamic system has also been suggested to be involved with conscious perception (Edelman et al. 2011; Steriade 2006)). There is also substantial psychophysical evidence for the existence of a global timing mechanism, e.g., the observation that the variability of interval timing is proportional to the duration of the interval across a large number of tasks, sensory modalities, and species (Gibbon et al. 1997; Buhusi and Meck 2005).

Apart from such a central timing mechanism, there is evidence for "local timers," brain areas with neurons able to measure temporal intervals: see Fig. 8.1 bottom (it has even been argued that timing is a ubiquitous ability of cortical networks, and that a central clock might not be needed (Karmarkar and Buonomano 2007)). Cells associated with temporal processing in the medial premotor cortex are one example confirmed by recording studies (different neurons in this area react most strongly to different time intervals preceding an action such as a button press (Zarco et al. 2009)). Local timers also include several sensorimotor areas with their own local oscillatory cycles, such as the visual, auditory and somatosensory cortices for timing stimuli perceived in these modalities, or the cerebellum for motor timing (see Merchant et al. 2013 for a more comprehensive discussion).

8.3   Models of Time Perception

Here, we will focus on three main aspects of time from the point of view of cognition, namely succession, duration, and temporal perspective (Block 2014). Succession refers to a sequence of events which can be used to perceive temporal order and successiveness. Duration denotes a length of time during which an event might persist, or between events. Temporal perspective in turn addresses the separation of events into past, present, and future. Below, we will discuss a model of time perception focusing on succession and duration, which accounts for these concepts, as well as others including continuity, the duration of the immediate present, perceived length of time.

Many perspectives model time perception. At the end of nineteenth century, William James (1890) developed one of the first, which is relevant to this work. However, most cognitive models that try to explain time are only focus on one or two aspects of it. For example, Michon (1990) studied duration of events, and Block (2014) the sequence of events. Well-known psychological models focusing on duration include the scalar expectancy theory and the pulse accumulator model (Gibbon et al. 1984; Buhusi and Meck 2005). These models use a pacemaker, generating pulses at regular intervals, and a pulse accumulator to estimate event durations. The accumulator facilitates the estimation of event durations by storing the pulses generated by the pacemaker, and comparing them to pulses in a reference memory. Other authors, including Boltz (1995), Grondin (2010), Zakay and Block (1996, 1997), Zakay (1992), and Zakay et al. (1994) describe how the structure of an event influences our perception of its duration. In particular, they consider how the event structure and its complexity affects the accuracy of duration judgments. Most prior work studies event duration perception on the order of magnitude of dozens of seconds or more, whereas this chapter focuses on shorter durations.

In neuroscience literature, time perception is most commonly used to refer to the perception of event duration (Ivry and Schlerf 2008), although some authors including Eagleman (2008) adopt a more general perspective, accounting for duration as well as perception time scale and sequence. Studies on the perception of time abound in both the neuroscience and the behavioral literature. Some are related to memory processes, the order of events as we experience them. They distinguish recalling when an experienced event happened from estimating its duration. Others are related to consciousness, the awareness of subjective time. Still others are concerned with time in relation to sensory processing, for example the processing of speech, music and successive visual images. Grondin (2010) offers pointers to the literatures of each of these, as well as many others. Ivry and Schlerf (2008) contribute a review of dedicated and intrinsic models of time perception.

8.4 Global Workspace Theory and the LIDA Cognitive Architecture

In contrast to most previous models of time, which are limited to one or few cognitive phenomena, our model is based on a general model of cognition: LIDA (Learning Intelligent Distribution Agent), a conceptual and computational cognitive architecture partially implementing and fleshing out the Global Workspace Theory (GWT) of consciousness and a number of other prevalent cognitive science and neuroscience theories, including Anderson (2003), Glenberg and Robertson (2000), Varela et al. (1991), perceptual symbol systems (Barsalou 1999), working memory (Baddeley and Hitch 1974), memory by affordances (Glenberg 1997), long-term working memory (Ericsson and Kintsch 1995), transient episodic memory (Conway 2002), and Sloman's H-CogAff cognitive architecture (Sloman 1999).

8.4.1 Global Workspace Theory

Among different theories of cognition, we choose to work from Baars' (1997) GWT, a prevalent psychological and neurobiological theory of consciousness. According to the GWT, the nervous system is a distributed parallel system incorporating many specialized processes. Various coalitions of these specialized processes facilitate making sense of sensory data currently coming in from the environment. Other coalitions sort through the results of this initial processing and pick out items requiring further attention. In the competition for attention a winner emerges, and occupies the global workspace, the winning contents of which are presumed to be at least functionally conscious. The presence of a predator, enemy, or imminent danger should be expected, for example, to win the competition for attention. However, an unexpected loud noise might well usurp consciousness momentarily even in one of these situations. The global workspace contents are broadcast to processes throughout the nervous system in order to recruit an action or response to this salient aspect of the current situation. The contents of this global broadcast also enable many modes of learning, which explains why it needs to be global. This broadcast provides large-scale integration via access consciousness as discussed in Sects. 8.1 and 8.2 above. We hypothesize that it is accomplished through oscillatory synchrony (Baars et al. 2013). We will argue that Learning Intelligent Distribution Agent (LIDA), which implements Baars' GWT, may be suitable as an underlying cognitive architecture with which to explicate and investigate ideas and hypothesis regarding time.

8.4.2   The LIDA Cognitive Architecture

Autonomous agents (including humans, animals and artificial agents) have to frequently sample (sense) their environments and choose appropriate responses
(actions). Agent's "lives" can be thought of as consisting of a sequences of such
cycles, which we call cognitive cycles. Each such cycle consists of units of sensing,
attending and acting. Cognitive cycles can be thought of as moments of cognition,
cognitive "atoms," and are similar to action-perception cycles in neuroscience
(Fuster 2002; Freeman 2002). Based on evidence from empirical neuroscience, and
consistent with psychophysical paradigms measuring reaction time, we have estimated the duration of cognitive cycles to be approximately 200–500 ms (Madl et al.
2011). However, these cycles can partially overlap (Fig. 8.5b), leading to a rate of
5–10 cycles per second (Baars et al. 2013; Franklin et al. 2013). The LIDA cognitive
cycle is not built into the model, but rather, emerges from it. Almost all of the modules as seen in Fig. 8.2, run continuously and asynchronously in parallel.

There are three phases in each cycle: the understanding phase, the attending
phase, and the action selection and learning phase. In the understanding phase, the
agent tries to make sense of its situation by updating its representations of external
entities (perceived through the senses), as well as internally generated features. In
the attending phase, the agent selects the most salient, important or urgent part of
the constructed representation—the part that needs to be attended to. This part is
sent to the rest of the system as the conscious broadcast (and thus becomes the current content of consciousness). In the third phase, internal resources are recruited
based on this content—potential actions for the action selection mechanism to
choose from. Furthermore, the conscious contents facilitate and modulate learning
into multiple different memories. Figure 8.2 shows this process, starting in the
upper left and proceeding roughly clockwise. Although the descriptions will be in
terms of modules and processes, LIDA makes no commitment regarding whether
the neural structure in humans is modular or localized. However, it is possible to
tentatively assign neuronal correlates to LIDA's modules based on functional correspondence (Franklin et al. 2013), which we will briefly mention below.

The understanding phase starts with incoming sensory stimuli from the external
and internal environments activating low-level feature detectors in Sensory Memory
where they are partially interpreted by short term (tens of milliseconds) processes
(sensory memory corresponds to sensory brain areas, such as the visual and auditory cortices). Results thereof proceed to LIDA's Perceptual Associative Memory
(PAM) (long term associative recognition memory) to be processed by higher-level
feature detectors, which can activate more abstract representations, e.g. objects, categories, actions, events, etc., as well as to the preconscious Workspace (a
preconscious working memory with duration in tens of seconds). LIDA uses graphical representation,[4] nodes and links, in PAM and in the Workspace to represent

---

[4] More specifically, LIDA often uses directed graphs composed of nodes and links to represent
items (nodes) and relationships between them (links).

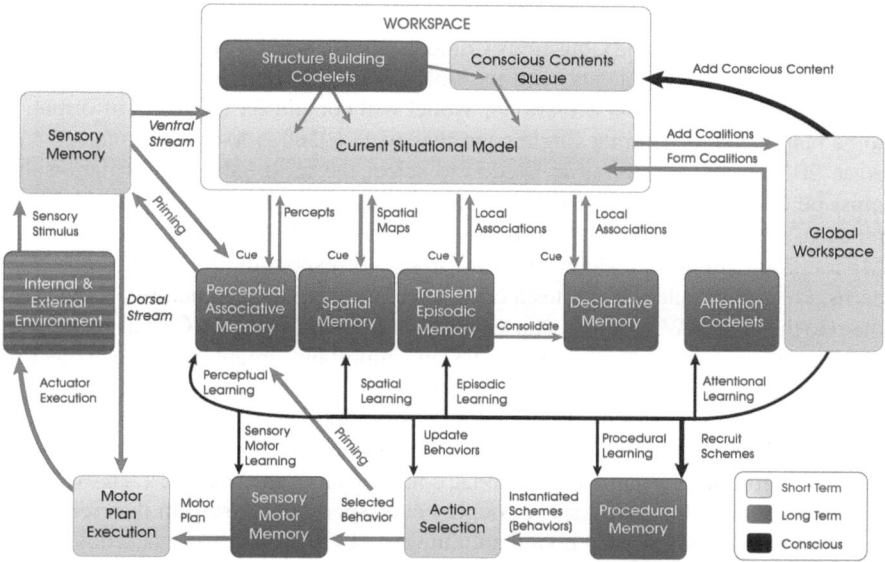

Fig. 8.2 The LIDA cognitive cycle

features, objects, categories, actions, feelings, events, etc. Localizing brain areas functionally corresponding to representations in PAM is not straightforward, as they are distributed and multimodal (Fuster 2004, 2006; Barsalou 2008; Fuster and Bressler 2012). Some areas involved in such representations include the perirhinal and orbito-frontal cortices and the amygdala.

Contents of the Workspace continually cue PAM, Spatial Memory (long term), Transient Episodic Memory (lasting a few hours or a day), and Declarative Memory (long term). Local associations recalled from the cueing of these various memories return to (or perhaps only point to) the Workspace. Neural correlates corresponding to these long-term memories include the hippocampus. On the other hand, the Workspace might correspond to temporo-parietal and frontal lobes and the entorhinal cortex (Franklin et al. 2013).

Workspace contents are operated upon by structure building codelets,[5] with the results being used to update the agent's preconscious Current Situational Model (CSM) within its Workspace. The agent's understanding of events occurring right now (i.e. within the last few cognitive cycles) is represented in this Current Situational Model (Snaider et al. 2012). The understanding phase is concerned with updating this CSM within the Workspace. Representations within the workspace

---

[5]A codelet is a small, single purpose, independently running piece of code, corresponding to a process in Baars' Global Workspace Theory. Structure building codelets build structures of nodes and links.

may persist in subsequent cognitive cycles, until they decay away. Another submodule of the Workspace, the Conscious Contents Queue will be discussed in Sect. 8.5.

For many complex agents with multiple senses, "living" in complex, dynamic environments, the Current Situational Model will contain far too much information to be responded to within a single cognitive cycle (~200–500 ms in humans (Madl et al. 2011)). Some filtering is needed to select the most salient information that must be attended to. In LIDA, attention codelets begin this filtering, or attention, phase of the cognitive cycle by creating coalitions of parts of the Current Situational Model. Each attention codelet looks for items corresponding to its particular concerns. On finding such, it creates a coalition containing their structures, and moves them to the Global Workspace. Subsequently, a competition in the Global Workspace chooses the most salient (the most relevant, important, urgent, novel, unexpected, loud, bright, moving, etc.) coalition, which then becomes the content of consciousness, and is broadcast globally to facilitate action selection and multiple modes of learning (implementing the large-scale integration and broadcasting mechanism suggested to underlie human functional consciousness in Sect. 8.2). The winning contents of the Global Workspace roughly correspond to neurons in different brain areas which are temporarily bound and integrated via oscillatory synchrony (Baars et al. 2013). This broadcast completes the attention phase of LIDA's cognitive cycle.

The third and final phase of the cognitive cycle is concerned with learning in several modes, and with action selection and its execution. Since these modules and processes play little significant roles in the perception of time, and have been described in detail elsewhere, we will describe them only briefly here. Based on Drescher's (1991) schema mechanism, data structures in Procedural Memory are called schemes. Each scheme consists of a context, an action, a result, and a base-level activation which measures the likelihood of the result happening should the action be taken in the scheme's context. Each of the first three components are structures of nodes and links. Schemes whose context and/or results intersect the current conscious broadcast are instantiated as behaviors and passed to the Action Selection mechanism, where one is chosen and sent along for execution. Procedural Memory might correspond to the striatum and anterior cingulate, whereas the action selection mechanism might be grounded in the basal ganglia in brains (Franklin et al. 2013). Learning in the different modes takes place concurrently, completing the final phase of the LIDA cognitive cycle.

LIDA advocates a discrete view of consciousness, in accordance with GWT and the converging neuroscience evidence outlined in Sect. 8.2. As we shall see in the descriptions of the computational LIDA agents reproducing psychological experiments in Sect. 8.6, this view is also consistent with multiple behavioral paradigms investigating consciousness and attention, among others Allport's (1968) experiments on perceptual simultaneity (which have traditionally been interpreted to require consciousness to be continuous). This view is also consistent with the philosophical conception of streamlikeness provided by Rashbrook-Cooper in this book, which allows for subjective continuity despite of gaps in consciousness. Just like his conception, LIDA's view can be seen as an extensionalist account of temporal experience.

8.5 A LIDA-Based Model of Time Perception and Production

8.5.1 The Immediate Present Train Model

The LIDA model for time perception and representation is based on ideas from William James (1890). He discussed the "specious present," a term originally coined by E.R. Clay (1882). It has been called "specious" (plausible but wrong) since the present experienced by the human mind, instead of being a duration-less instant, is taken to comprise an interval.

Here we summarize the ideas introduced in Snaider et al. (2012). To describe our model we need first to briefly discuss some basic attributes of time: duration of short events, time duration scale, and succession. These attributes are fundamental for time flow perception, time concepts representation, and for defining what we call the Immediate Present Train (IPT), a more concrete instantiation for James's specious present.

Duration is probably the most well studied property of time. Saint Augustine (Warner 1963) discussed this issue, and argued that because the present is just an instant without duration, memory is required to measure an event's duration. We propose that without *any* memory it is not possible to have any notion of the concepts of past or of event duration. Notice that it is critical what we meant by "any memory" in the previous sentence. Using the LIDA concepts, this refers to the absence of transitive episodic and declarative memories. The workspace would only retain the present percept elements, but no past content is cued. Even in this reduced context, it is still possible to have some functionality, such as reacting to the present perception. However, memory is required to interpret the idea of something past. To evaluate the duration of an interval, some memory for the event (or events) is necessary, or at least some memory of their temporal properties (e.g., its starting time, or an accumulator that counts pulses). If we relax this idea, and we allow some memories of the past few seconds (probably in the preconscious workspace), it would be possible to model the concepts of past and duration.

The relative arrangement of events over time is an acknowledged property required for time awareness. Consider the events perceived by a subject. The arrangement that these events have is, in many cases, a piece of information as important as the events themselves. Processes such as detecting cause and effect situations, planning, and learning a path are possible when the perception and modeling of the sequence of the participant events are available. A nice metaphor for this is a family photo album. The photographs' order is telling us a story. A different arrangement may tell us a completely different one. If instead of an album we have a pile of pictures in no particular order, even the concept of story disappears.

When the intervals and durations of situations are relatively large (i.e. durations of some minutes, hours, days or even longer ones), we assume that an episodic memory module (as described in LIDA) participates in the process of maintaining the chronology. However, when events have durations between a fraction of second to a small number seconds, another entirely different mechanism is required. We consider the sequence of conscious broadcasts as the genesis of the stream of consciousness. Since humans are capable of *perceiving* this succession of broadcasts (Franklin et al. 2013), we introduce a structure that maintains this sequence, called the Conscious Contents Queue (CCQ) (Snaider et al. 2012), in LIDA's Workspace (see Fig. 8.2). James (1890, 606–607) clearly expressed this idea:

> If the present thought is of A B C D E F G, the next one will be of B C D E F G H, and the one after that of C D E F G H I—the lingerings of the past dropping successively away, and the incomings of the future making up the loss. These lingerings of old objects, these incomings of new, are the germs of memory and expectation, the retrospective and the prospective sense of time. They give that continuity to consciousness without which it could not be called a stream.

Although the CCQ name implies a queue's functionality (and in part this is true) it also resembles the behavior of a buffer. Its structure enables random access to its elements, while preserving their order. This allows several time related perceptual operations, such as the measure of an event's duration, or the detection of repeated event sequences.

For James, the specious present was "the prototype of all conceived times... the short duration of which we are immediately and incessantly sensible." James seems to imply a temporal interval, a "short duration," within which perceptions can be viewed to be in the present. For James, this duration could extend up to about 12 s. Latter users of the term "specious present" take it to mean "the (maximal) window through which we are directly aware of change and persistence..."

With this usage, there is considerable controversy as to the timespan of the specious present, but all current contenders are substantially less that James' 12 s (Dainton 2010). Wittmann (2011) reviews evidence suggesting the extent of an experienced moment to span a handful of seconds (mostly suggesting ~3 s; see also the chapter by Wittmann in this book). Block also estimates this duration at about 5 s (Block 2014).

Note that distinguishing events that last less than the estimated timespan of the specious present, and in some degree, modeling their chronology, are still possible. However, as in the case of timespans larger than the specious present that cannot be distinguished directly, some events are too excessively short to be identified as individual events. Images in TV screens are the prototypical example. Although we perceive them as moving images, they are actually static images presented in rapid succession. It is impossible to humans to perceive them as separated events. These ideas suggest that there is a range of event durations that humans can perceive directly, which we name perceptual time-range. Events with durations below this range are represented as a combination (e.g., the frames in the TV screen), or they may not be perceived at all. Events with durations above this time-range are still discernible using other cognitive processes such as episodic memory functionality or reasoning, but direct perception is not possible.

We hypothesize that these limits are not strict or fixed, and vary according to the nature and salience of the events. For example, when riding a rollercoaster, we would perceive a fast succession of stimuli, which may lead to a more than normal fine grain distinction of events.

In LIDA—although we often borrow from and build upon the ideas of James—we do not conceive of the "immediate present" as a fixed, absolute duration. Rather, we define it in terms of the events an agent (biological or artificial) might be currently conscious of. Every event that is broadcast consciously, and can be acted upon (or reported, or introspected upon), becomes a component of the "immediate present" when it is broadcast; and is included into the subjective past only once replaced by different events of a subsequent conscious broadcast.

Keep in mind that motion may be perceived in a single conscious broadcast, making us "directly aware of change and persistence." Thus with the previous estimation of specious present duration, we might claim it is comparable with that of a cognitive cycle, roughly 200–500 ms (Madl et al. 2011). Though motion can be directly perceived, events being perceived within the same cognitive cycle and becoming elements of the same conscious broadcast will be experienced as being simultaneous (e.g., flashes of light separated by a small distance and a few milliseconds), that is, as being a single event.

Snaider and colleagues (2012) combined these attributes of time into their Immediate Present Train where the (specious) present is modeled by a train, in which its extent corresponds to the timespan of the specious present. The cars in the train denote an ordered sequence of time steps, which contains the last few conscious events. In LIDA terms, the cars keep the elements of the recent broadcasts. Notice that the size of a car represents the extent of the shortest interval that can be distinguished directly. In other words, the train models the scope of events' durations discussed previously. We hypothesize that the train receives new conscious content every a few hundreds of milliseconds (for humans), and a car is appended to the front of the train with this content. Correspondingly, cars at the end are removed from the train. The train representation comprises several instants, which allows representing non-simultaneous events as components of the immediate present. In effect, events may be in different cars but still belong to the same train.

Note that an event shorter than the timespan equivalent to a single car may be represented directly as a change event. For example, the movement of a ball can be modeled as a moving-ball event, instead of a sequence of ball-position events.

Although this model may suggest that the duration represented by each car and the number of cars in the train are fixed magnitudes, the model actually allows variations in them. The interval comprising two consecutive conscious events may vary, thus affecting the duration represented by each car. Also, the elements in a few of the cars may decay away, removing these cars from the train (which changes the total duration represented by the train).

As we mentioned previously, the Conscious Contents Queue (CCQ), a submodule in the LIDA's workspace, is a more concrete instantiation of the Immediate Present Train and the specious present (Snaider et al. 2012). It is a combination of a queue and a buffer. It comprises a variable number of cells, similar to the cars described above. CCQ resembles a queue, since it has a head and an end, and the content of one cell is pushed back to the following cell when fresh element arrives

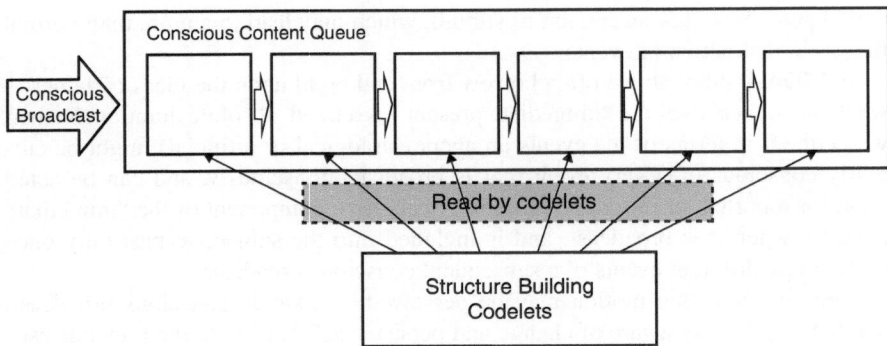

Fig. 8.3 The Conscious Contents Queue (From Snaider et al. (2012) with permission from Elsevier)

(Fig. 8.3). However, unlike queues, the cells can be accessed directly, allowing other process (particularly structure-building codelets) to read several cells simultaneously. With each broadcast, the CCQ receives new content, which is inserted in the head of the CCQ, while the old content is shifted towards the end. As other representations in LIDA, elements in the cells have activation, which decay over time. Some elements loss all their activation and are removed from the queue. Eventually, all elements in one cell may decay away, and then the cell itself is removed from the queue (even if it is not at the end). The frequency of the conscious broadcasts may vary over time as a factor of the many triggers of the Global Workspace module. As a consequence, the size of the CCQ (and the time that it ultimately represents) is not fixed. In other words, the time required to complete LIDA Cognitive Cycle phase defines the frequency of the broadcast and the duration determined by a cell in the CCQ. For humans each phase takes approximately 100 ms (for simple tasks, the understanding phase is estimated to take 80–100 ms, the attending phase an additional 120–180 ms, and the action selection phase 60–110 ms (Madl et al. 2011)). This duration determines the lower limit of the perceptual time-range, and the count of cells in the CCQ defines its maximum.

Structure-building codelets can approximately calculate the duration of short events by simply counting the number of cells that that event spans (see Fig. 8.4). Several factors affect the precision of these calculations. One of these factors is the activation decay of the elements in the CCQ described above. Another factor is the frequency of the conscious broadcast, which can vary. In general, when more stimuli are present, the frequency of conscious broadcast is higher (as in our example of a rollercoaster ride). This has the effect of filling the cells faster, and the 100 ms estimation for the duration represented by each cell becomes inexact. Actually, structure building codelets that inspect the CCQ elements will erroneously consider that each cell still represents 100 ms, and the event duration perception will be distorted producing the effect that the duration is longer than it really is, or in other words, producing the sensation that time flows more slowly, an effect that was reported in several experiments, e.g., Eagleman (2008).

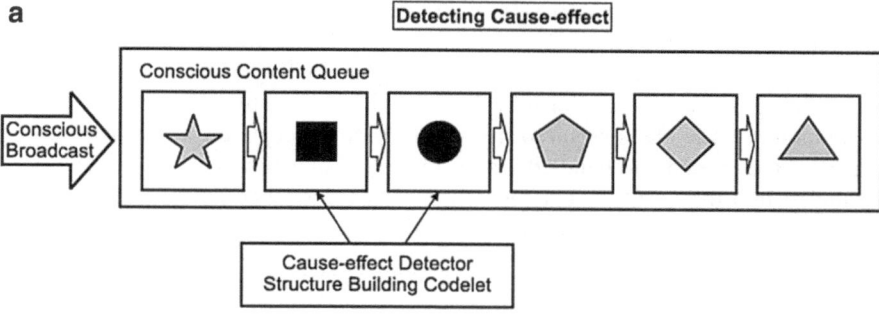

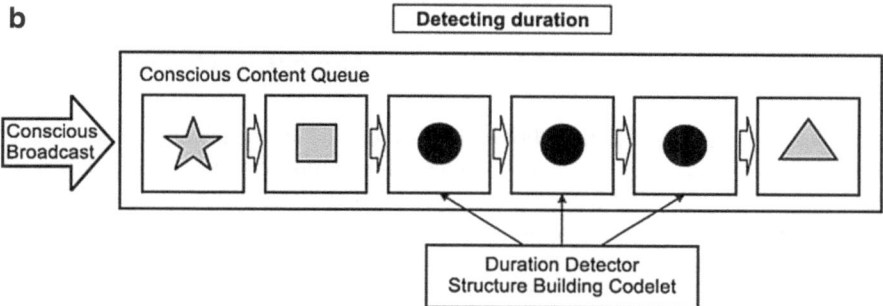

Fig. 8.4 Detecting causes and effects (**a**) and determining durations (**b**) using the Conscious Contents Queue. (**a**) A Cause-effect Detector Structure Building Codelet detects that the "*circle*" content precedes the "*square*" content in the CCQ, and would create a "*circle* before *square*" representation in the Workspace. (**b**) A Duration Detector Structure Building Codelet can select repetitions from the CCQ, and count the number of occurrences. The Codelet can then create a representation of the duration of the selected content in the Workspace, based on this number and on the duration of the cells in the CCQ (From Snaider et al. (2012) with permission from Elsevier)

Codelets may also perform other time related operations, such as determining cause-effect situations (for brief events) using the CCQ, thanks to its sequential order. If one event is present in a cell closer to the head than other event, a codelet could use this a signal for creating a cause-effect relationship between these two events (see Fig. 8.4). Another operation may be the detection of simultaneous or quasi-simultaneous events, depending on the number of cells considered for the tasks.

The current main representation in the LIDA architecture comprises nodes that represent concepts, and links, which denote relationships between these concepts. In the general case, nodes are *grounded* in sensor and motor memories. For example, the node representing the color red is ultimately rooted in the light sensors sensible to that color. However, in LIDA, time-based nodes, such as duration nodes, are grounded by the CCQ. Short duration nodes are instantiated, by codelets when they detect these intervals as we described previously. Other nodes for concepts such as fast and slow, can be derived from these duration nodes. The abstract notions of "duration" and "flow of time" can be created as categorizations of simpler nodes. To explain the creation (and perception) of nodes for larger spans, such as nodes representing minutes, hours, or even longer intervals, we hypothesize that an episodic memory module is required. However, these concepts for longer periods are correctly interpreted and handled thanks to their connection with the simpler ones grounded into the CCQ. In our view, the CCQ mechanism provides the seminal concepts for interpreting and working with time related concepts in LIDA.

8.6   Computational Reproductions of Experiments Involving Time

8.6.1   Consciousness and Continuity: The LIDA Allport Agent

The idea of consciousness possibly being discrete has been strongly criticized and in some cases even outright rejected based on those empirical results in the phenomenal simultaneity paradigm which seem to contradict discrete (e.g., cinematic) models. A number of frequently cited experiments were conducted by Allport (1968), who aimed to compare two prevalent competing theories of consciousness at that time, Stroud's (1967) Discrete Moment Hypothesis (DMH) and the Continuous Moment Hypothesis. The former states that consciousness comprises distinct and not overlapping conscious "moments," within which time-order information is lost, whereas the latter views conscious "moments" as corresponding to continuously moving segments of incoming sensory information. Allport's empirical results contradict the DMH, leading him to reject Stroud's discrete model.

However, although the LIDA model—like Stroud's—also proposes consciousness to be discrete, it can still account for this empirical evidence. To show this consistency, as well as to strengthen the claim that LIDA's GWT-based consciousness mechanism can model human functional consciousness, we have replicated Allport's experiment computationally with a LIDA-based cognitive software agent (Madl et al. 2011).

In Allport's (1968) experiment, participants faced a screen displaying a horizontal line in 1 of 12 possible positions on this screen (see Fig. 8.5a), and rapidly changing position moving upward. Each time the line reached the top position, the screen was first left blank for the amount of time it took for the line to traverse the screen, and then the line reappeared in the bottom position, moving upward. These cycles of the screen alternating between showing the moving line and being blank were repeated. Participants could control the cycle time (τ).

When cycle times were set to be large, participants were able to see the line jumping from one position to the next. When they reduced τ, participants saw multiple lines, moving together. However, at and below a small cycle time S, they reported perceiving an unmoving array of 12 lines which flickered in synchrony, instead of individual lines.

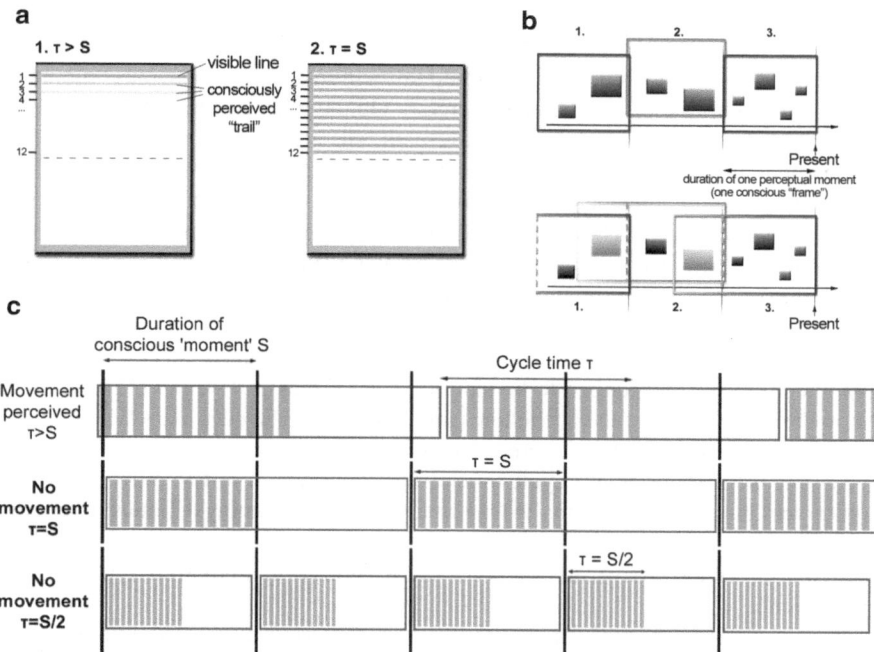

Fig. 8.5 Allport's experiment (**a**), and a comparison of the refuted DMH (**b**, *top*, and **c**) and LIDA's discrete consciousness mechanism (**b**, *bottom*). (**a**) The screen in Allport's (1968) experiment. A visible line was shown in 1 of 12 possible positions, moving upwards. Whenever it reached the top, the *line* vanished for the amount of time it took to reach the top. The cycle time is denoted by τ. When τ >S, participants could see movement (*left panel*). At τ=S, participants perceived all lines at the same time, and saw no movement (*right panel*). (**b**) Schematic comparison of the DMH (*top*) and LIDA's discrete consciousness hypothesis (*bottom*). The frames represent the temporal constraints of a perceptual moment or conscious "frame," and the solid rectangles symbolize incoming percepts. In LIDA, important percepts from previous conscious "frames" can remain conscious (*rectangles left* of the *dashed lines* in the frames in the *bottom* picture). (**c**) Predictions of the DMH. If conscious moments were discrete and distinct, there would be two cycle times at which subjects would perceive no movement (τ=S and τ=S/2). Instead, Allport (1968) reports only one cycle time (From Madl et al. 2011 with permission)

The task of the participants was to keep changing τ until they arrived at cycle time S and stopped perceiving moving lines. They were asked to do this in two types of trials, in which their cycle times were recorded. In the first type, they had to decrease the cycle time from a high value towards S (accelerating the cycles until they reached a cycle time τ₁ at which they saw stationary lines). In the second type, they increased cycle time from a low value towards S (slowing the cycles until they started seeing movement at cycle time τ₂)—see Fig. 8.5a.

The abovementioned hypotheses regarding consciousness make different predictions regarding the cycle times participants should arrive at in these two trial types. The Discrete Moment Hypothesis would predict that they should be different—

there should be two cycle times, τ_1 and τ_2, at which the 12 lines can be seen on the screen without movement. At τ_1=S, subjects would not perceive movement because everything happening on the screen should fall within a single conscious "moment" (all line positions as well as the blank screen). On the other hand, at a time τ_2 which equals S/2 there should also be no movement, since at this cycle time conscious "moments" might alternate between containing all line positions (taking S/2) and between containing a blank screen (also taking S/2)—thus, no moving lines should be seen, only flickering. Therefore, the DMH would predict that in the above experimental setting, subjects will arrive at two distinct cycle times in the two task types, τ_1=S when cycle times are decreased, and τ_2=S/2 when cycle times are increased.

The Continuous Moment Hypothesis, in contrast, would predict only a single cycle time τ_1=τ_2=S at which no movement can be perceived. According to this hypothesis, events are judged to be simultaneous if they fall within one conscious "moment." In this experiment, the lines are perceived to be stationary when all line positions as well as the blank screen fall within a conscious "moment," when the cycle time is S. However, at a cycle time of S/2, there would still be movement—the conscious "moment's" contents would change from containing 12 lines, over containing fewer and fewer lines, to finally only containing the blank screen. Thus, participants should arrive at the same cycle time S in both trial types in this experiment.

Allport (1968) reports that the cycle times in the two trial types were not significantly different. Based on this result, he argued for the implausibility of the Discrete Moment Hypothesis. However, despite LIDA's consciousness mechanism being discrete, we have reproduced Allport's result with a LIDA-based computational cognitive agent.

To simulate Allport's experiment, the LIDA Allport agent used the cognitive cycles outlined in Sect. 8.4. The Allport agent had a pre-defined PAM to model the experimental stimuli, containing a PAM node for each of the line positions on the screen, and feature detectors corresponding to each line passing activation to the respective node corresponding to the currently visible line. The agent also had a pre-defined Procedural Memory (PM) containing two behavior schemes, for the "movement perceived" and "no movement perceived" buttons. The former was activated when the agent perceived no line movement (i.e. when all 12 line positions were present in the conscious broadcast), whereas the latter was pressed by the agent whenever it perceived movement. Cycle times (τ) were adjusted gradually in the environment, and the agent only had to react to whether or not it could perceive movement (this was computationally easier to implement than letting the agent decide the cycle time, but did not make any difference in the implications and predictions of the discrete consciousness mechanism).

The environment first successively decreased the cycle times from a high value, and then successively increased it from a slow value, similarly to the two trial types of the Allport experiment. The button responses of the agent were recorded, and the cycle times at which the agent pressed the "no movement perceived" button compared between the two trial types. The agent pressed this button at the same cycle time in both conditions—at 96 ms (Madl et al. 2011), which matches the results of the human participants described above (in contrast to the predictions of the DMH), and suggests that the durations of conscious "moments" in LIDA approximately match those of humans.

In LIDA, conscious episodes are discrete, but contrary to the DMG as argued by Stroud (1967), not always distinct. Subsequent conscious "moments" might contain percepts from prior moments (symbolized by the rectangles left of the dashed lines in Fig. 8.5b). The duration of older percepts persisting in consciousness is influenced by multiple factors, including when (how long in the past) it was perceived, and on attentional modulation. Here we have an example of a systems level, computational cognitive model providing deeper understanding of an experimental result.

8.6.2 Attention

Two cognitive software agents were developed to reproduce experiments related to attention: the LIDA Attentional Blink (Madl and Franklin 2012) and the LIDA Attention agent (Faghihi et al. 2012).

The first LIDA agent accounted for the attentional blink (AB) (Madl and Franklin 2012), i.e. the observed phenomenon that subjects are frequently unable to report the second of two targets shown within 200–500 ms after the first, within a sequence of target and distractor stimuli (see Fig. 8.6). The AB has a number of observed properties. The second target (T2) can be consciously perceived and reported if it is presented after the first target (T1) but with no distractor in between ("lag-1 sparing"), but not if there are distractors between the targets. Furthermore, the AB effect can be reduced—the likelihood of the second target correctly being reported increased—by increasing its salience (Martens and Wyble 2010) or emotional arousal (Anderson 2005).

Many AB models have been proposed; however, most current models cannot account for all findings and properties in AB experiments (see Dux and Marois 2009 or Martens and Wyble 2010 for reviews). Furthermore, many of these models are specific to the AB, instead of being implemented within a general cognitive architecture.

We have developed a LIDA-based model of the AB (Madl and Franklin 2012) to computationally model the visual attentional blink experiment (Potter et al. 2010), reproducing human behavior data, and to conceptually account for a large number of phenomena. In LIDA, the attentional blink is mainly caused by a temporarily depleted attentional resource (which fully regenerates after ~500 ms), making attending to the second target difficult if it is presented very shortly after the first. Lag-1 sparing arises from both targets entering the same coalition and within the same cognitive cycle, and thus both coming to consciousness.

The second attention agent was based on a modified version of the experiment by Van Bockstaele et al. (2010). Its environment was composed of a screen with two white squares on both sides of a fixation cross (see Fig. 8.6). After a brief delay (fixation period in the original experiment), a colored cue randomly appeared in

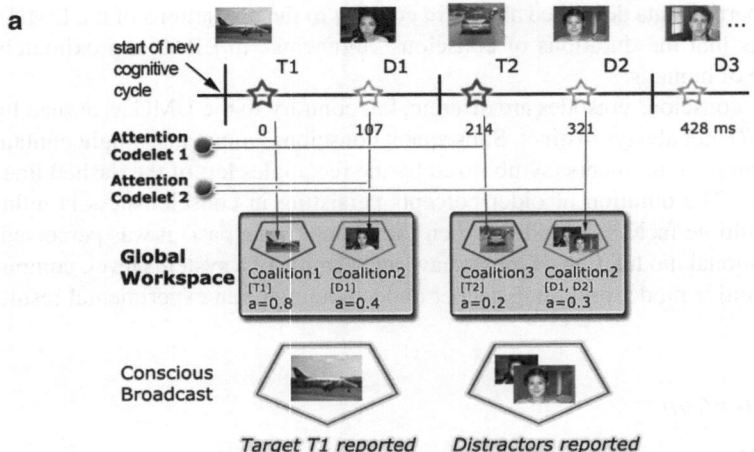

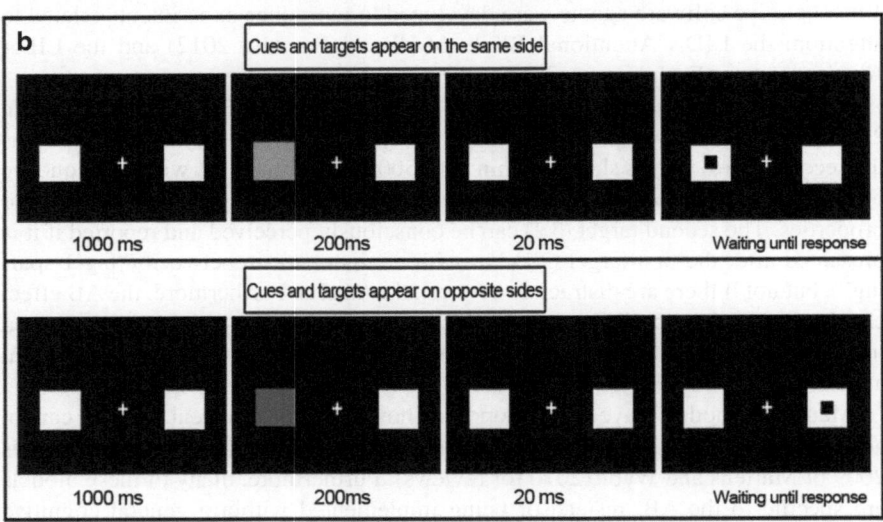

Fig. 8.6 Experimental paradigms reproduced by the LIDA Attentional Blink and LIDA Attention agents. (**a**) Sequentially presented images in the Attentional Blink paradigm. Two targets (vehicles; *T1* and *T2*) are presented with one distractor between them (*D1*) and several distractors after them (faces). In the figure, *T2* cannot be consciously perceived and reported, because in the second cognitive cycle the distractors win the competition for consciousness (starting times of cognitive cycles are marked by *bold vertical lines* on the timeline). If *T1* and *T2* were presented subsequently, they would both be bound within the same coalition, and perceived consciously. (**b**) Timeline of displayed cues in Van Bockstaele's experiment replicated by the LIDA Attention agent, with both target and cue being displayed on the same side in congruent trials (*top*), and on opposite sides in incongruent trials (*bottom*) (Fig. 8.6b from Faghihi et al. (2012) with permission from Elsevier)

either the left or the right square, for 200 ms, followed by the two empty white squares for 20 ms. Subsequently, a small black rectangle (the target) was presented in either the left or the right square, again at random. The agent (just like the participant in the original experiment) had to respond to the target, i.e. press the right one of two buttons, as fast as possible; response times were measured. The experience showed that both participants (Van Bockstaele et al. 2010) and the LIDA Attention agent (Faghihi et al. 2012) were faster in reaction by 20 ms on trials in which the cue and the target were shown on the same side (congruent trials), compared to trials where they appeared on opposite sides (incongruent trials)—average response times were 360 and 380 ms. The reason for this difference in the LIDA agent was the instantiation of the correct behavior scheme. That is, by the time the target arrives to consciousness the cue almost primes a behavior by sending more activation to it. In contrast, in trials with cue and target on opposite sides, different schemes from Procedural Memory needed to be instantiated and then a behavior will be selected and executed. The extra scheme instantiation cost to the Attention agent an additional 20 ms (Faghihi et al. 2012).

8.7 Conclusion

As we argued in the Introduction, the study of mind in all of its aspects, including the perception of time, is best approached from different perspectives. As we have seen throughout this work, it has proved useful to study such difficult questions as the seemingly continual flow of time using the various tools of each of the relevant disciplines, the introspection of the philosopher of mind (e.g., Block 1995; Dainton 2010; James 1890), the behavioral observation of the experimental psychologist (e.g., Buhusi and Meck 2005; James 1890; Michon 1990; Zakay et al. 1994), the brain imaging of the cognitive neuroscientist (e.g., Eagleman 2008; Ivry and Schlerf 2008), and the computational simulation of the cognitive modeler, a computer scientist (e.g., Buhusi and Meck 2005; Madl et al. 2011; Michon 1990; Snaider et al. 2012).

We have reviewed recent neuroscience evidence concerning large-scale integration by oscillatory synchrony as a possible mechanisms underlying functional consciousness, suggesting it to be discrete. We have also briefly reviewed recently suggested neural correlates of global and local time perception mechanisms in brains. After outlining Global Workspace Theory, a prominent theory accounting for functional consciousness, we have described a conceptual and partially computational model of cognition based on GWT—the LIDA cognitive architecture—and argued that it can account for time perception in a cognitively plausible fashion (substantiated by reproduced psychological experiments), and generate concepts such as continuity, immediate present duration, and perceived length of time.

Some of the modelers of time are only concerned with modeling time itself, or even one aspect of it, for example duration (Zakay et al. 1994). Here we have argued for the need to study time in the context of the study of mind, using a broad, systems-level cognitive model such as our LIDA (Franklin et al. 2013). We are not alone. Such arguments have been made earlier by a number of other researchers from disparate fields. Here we support our arguments by quotes from four such. From social psychology, Kurt Lewin says it quite concisely. "There is nothing so practical as a good theory" (1951, 169). A broad, systems level cognitive model is a theory of mind. From computer science, AI pioneer Allen Newell argues against the reliance on modeling individual laboratory tasks saying "You can't play 20 questions with nature and win" (Newell 1973). Making the same point for his field, psychological memory researcher Douglas Hintzman (2011) writes, "Theories that parsimoniously explain data from single tasks will never generalize to memory as a whole…" Hintzman's arguments rest precisely on the need for the type of cognitive models that we advocate, and apply broadly beyond memory research. Langley et al. (2008) wrote a review article entitled "Cognitive architectures: Research issues and challenges." In it they argue for the use of systems-level cognitive architectures such as our LIDA model, asserting that "Instead of carrying out micro-studies that address only one issue at a time, we should attempt to unify many findings into a single theoretical framework, then proceed to test and refine that theory." Several of the "open problems" described in their review have since been partially or fully solved by our LIDA. The reinterpretation of the Allport experiment provided by the LIDA Allport agent is one example of the value of such an approach. In a table allowing ready comparison of properties of some 26 "biologically inspired cognitive architectures" (Samsonovich 2010), LIDA compares rather well in terms of modeling a complete cognitive system, and also in terms of being truly biologically inspired.

We contend that such systems-level, conceptual and computational modeling can, if it is biologically plausible, integrate findings from the several disciplines, and produce hypotheses that will serve to guide further research.

References

Allport, David A. 1968. Phenomenal simultaneity and the perceptual moment hypothesis. *British Journal of Psychology* 59(4): 395–406.

Anderson, Michael L. 2003. Embodied cognition: A field guide. *Artificial Intelligence* 149(1): 91–130.

Anderson, Adam K. 2005. Affective influences on the attentional dynamics supporting awareness. *Journal of Experimental Psychology: General* 134(2): 258–281.

Baars, Bernard J. 1997. *In the theater of consciousness: The workspace of the mind*. Oxford: Oxford University Press.

Baars, Bernard J. 2005. Global workspace theory of consciousness: toward a cognitive neuroscience of human experience. *Progress in Brain Research* 150: 45–53.

Baars, Bernard J., Stan Franklin, and Thomas Zoega Ramsoy. 2013. Global workspace dynamics: Cortical "binding and propagation" enables conscious contents. *Frontiers in Psychology* 4: 200.

Baddeley, Alan D., and Graham J. Hitch. 1974. Working memory. *Psychology of Learning and Motivation* 8: 47–89.

Barsalou, Lawrence W. 1999. Perceptions of perceptual symbols. *Behavioral and Brain Sciences* 22(4): 637–660.

Barsalou, Lawrence W. 2008. Grounded cognition. *Annual Review of Psychology* 59: 617–645.

Block, Ned. 1995. How many concepts of consciousness? *Behavioral and Brain Sciences* 18(2): 272–287.

Block, Richard A. (ed.). 2014. *Cognitive models of psychological time*. New York: Psychology Press.

Boltz, Marilyn G. 1995. Effects of event structure on retrospective duration judgments. *Perception & Psychophysics* 57(7): 1080–1096.

Buhusi, Catalin V., and Warren H. Meck. 2005. What makes us tick? Functional and neural mechanisms of interval timing. *Nature Reviews Neuroscience* 6(10): 755–765.

Casali, Adenauer G., Olivia Gosseries, Mario Rosanova, Mélanie Boly, Simone Sarasso, Karina R. Casali, Silvia Casarotto, Marie-Aurélie Bruno, Steven Laureys, and Giulio Tononi. 2013. A theoretically based index of consciousness independent of sensory processing and behavior. *Science Translational Medicine* 5(198): 198ra105.

Clay, E.R. 1882. *The alternative: A study in psychology*. London: Macmillan.

Conway, Martin A. 2002. Sensory-perceptual episodic memory and its context: Autobiographical memory. In *Episodic memory: New directions of research*, ed. Alan Baddeley, Martin Conway, and John Aggleton. Oxford: Oxford University Press.

Crick, Francis, and Christof Koch. 1990. Towards a neurobiological theory of consciousness. *Seminars in the Neurosciences* 2: 263–275.

Dainton, Barry. 2010. Temporal consciousness. In The Stanford encyclopedia of philosophy. Stanford: The Metaphysics Research Lab, Center for the Study of Language and Information, Stanford University.

Dehaene, Stanislas, and Jean-Pierre Changeux. 2003. Neural mechanisms for access to consciousness. In *The cognitive neurosciences III*, ed. Michael Gazzaniga, 1145–1157. Cambridge, MA: MIT Press.

Dehaene, Stanislas, Jean-Pierre Changeux, Lionel Naccache, Jérôme Sackur, and Claire Sergent. 2006. Conscious, preconscious, and subliminal processing: A testable taxonomy. *Trends in Cognitive Sciences* 10(5): 204–211.

Dehaene, Stanislas, Lucie Charles, Jean-Rémi King, and Sébastien Marti. 2014. Toward a computational theory of conscious processing. *Current Opinion in Neurobiology* 25: 76–84.

Dennett, Daniel C. 1988. Quining qualia. In *Consciousness in contemporary science*, ed. A. Marcel and E. Bisiach, 42–77. Oxford: Oxford University Press.

Dieter, Kevin C., and Duje Tadin. 2011. Understanding attentional modulation of binocular rivalry: A framework based on biased competition. *Frontiers in Human Neuroscience* 5: 155. doi:10.3389/fnhum.2011.00155.

Doesburg, Sam M., Jessica J. Green, John J. McDonald, and Lawrence M. Ward. 2009. Rhythms of consciousness: Binocular rivalry reveals large-scale oscillatory network dynamics mediating visual perception. *PloS One* 4(7): e6142.

Drescher, Gary L. 1991. *Made-up minds: A constructivist approach to artificial intelligence*. Cambridge, MA: MIT Press.

Dux, Paul E., and René Marois. 2009. The attentional blink: A review of data and theory. *Attention, Perception, & Psychophysics* 71(8): 1683–1700.

Eagleman, David M. 2008. Human time perception and its illusions. *Current Opinion in Neurobiology* 18(2): 131–136.

Edelman, Gerald M., and Giulio Tononi. 2000. *A universe of consciousness: How matter becomes imagination*. New York: Basic Books.

Edelman, Gerald M., Joseph A. Gally, and Bernard J. Baars. 2011. Biology of consciousness. *Frontiers in Psychology* 2: 4.

Ericsson, K. Anders, and Walter Kintsch. 1995. Long-term working memory. *Psychological Review* 102(2): 211–245.

Faghihi, Usef, Ryan McCall, and Stan Franklin. 2012. A computational model for attentional learning in cognitive agents. *Biologically Inspired Cognitive Architectures* 2: 25–36.

Franklin, Stan, Tamas Madl, Sidney D'Mello, and Javier Snaider. 2013. LIDA: A systems-level architecture for cognition, emotion, and learning. *IEEE Transactions on Autonomous Mental Development.* doi:10.1109/TAMD.2013.2277589.

Freeman, Walter J. 2002. The limbic action-perception cycle controlling goal-directed animal behavior. *Neural Networks* 3: 2249–2254.

Fuster, Joaquín M. 2002. Physiology of executive functions: The perception-action cycle. In *Principles of frontal lobe function*, ed. Donald T. Stuss and Robert T. Knight, 96–108. New York: Oxford University Press.

Fuster, Joaquín M. 2004. Upper processing stages of the perception-action cycle. *Trends in Cognitive Sciences* 8: 143–145.

Fuster, Joaquín M. 2006. The cognit: A network model of cortical representation. *International Journal of Psychophysiology* 60(2): 125–132.

Fuster, Joaquín M., and Steven L. Bressler. 2012. Cognit activation: A mechanism enabling temporal integration in working memory. *Trends in Cognitive Sciences* 16(4): 207–218.

Gibbon, John, Russell M. Church, and Warren H. Meck. 1984. Scalar timing in memory. *Annals of the New York Academy of Sciences* 423(1): 52–77.

Gibbon, John, Chara Malapani, Corby L. Dale, and C.R. Gallistel. 1997. Toward a neurobiology of temporal cognition: Advances and challenges. *Current Opinion in Neurobiology* 7(2): 170–184.

Glenberg, Arthur M. 1997. What memory is for: Creating meaning in the service of action. *Behavioral and Brain Sciences* 20(1): 41–50.

Glenberg, Arthur M., and David A. Robertson. 2000. Symbol grounding and meaning: A comparison of high-dimensional and embodied theories of meaning. *Journal of Memory and Language* 43(3): 379–401.

Grondin, Simon. 2010. Timing and time perception: A review of recent behavioral and neuroscience findings and theoretical directions. *Attention, Perception, & Psychophysics* 72(3): 561–582.

Hintzman, Douglas L. 2011. Research strategy in the study of memory: Fads, fallacies, and the search for the coordinates of truth. *Perspectives on Psychological Science* 6(3): 253–271.

Ivry, Richard B., and John E. Schlerf. 2008. Dedicated and intrinsic models of time perception. *Trends in Cognitive Sciences* 12(7): 273–280.

James, William. 1890. *The principles of psychology*. Cambridge, MA: Harvard University Press.

Karmarkar, Uma R., and Dean V. Buonomano. 2007. Timing in the absence of clocks: Encoding time in neural network states. *Neuron* 53(3): 427–438.

Kim, Chai-Youn, and Randolph Blake. 2005. Watercolor illusion induced by synesthetic colors. *Perception* 34(12): 1501.

Koch, Christof. 2004. *The quest for consciousness*. Englewood: Roberts & Company.

Kouider, Sid, and Stanislas Dehaene. 2007. Levels of processing during non-conscious perception: A critical review of visual masking. *Philosophical Transactions of the Royal Society, B: Biological Sciences* 362(1481): 857–875.

Langley, Pat, John E. Laird, and Seth Rogers. 2008. Cognitive architectures: Research issues and challenges. *Cognitive Systems Research* 10: 141–160.

Lau, Hakwan C., and Richard E. Passingham. 2007. Unconscious activation of the cognitive control system in the human prefrontal cortex. *The Journal of Neuroscience* 27(21): 5805–5811.

Lewin, Kurt. 1951. In *Field theory in social science: Selected theoretical papers*, ed. D. Cartwright. New York: Harper & Row.

Lewis, Laura D., Veronica S. Weiner, Eran A. Mukamel, Jacob A. Donoghue, Emad N. Eskandar, Joseph R. Madsen, William S. Anderson, Leigh R. Hochberg, Sydney S. Cash, and Emery N. Brown. 2012. Rapid fragmentation of neuronal networks at the onset of propofol-induced unconsciousness. *Proceedings of the National Academy of Sciences* 109(49): E3377–E3386.

Madl, Tamas, and Stan Franklin. 2012. A LIDA-based model of the attentional blink. In *ICCM 2012 proceedings*, 283–288.

Madl, Tamas, Bernard J. Baars, and Stan Franklin. 2011. The timing of the cognitive cycle. *PloS One* 6(4): e14803.

Martens, Sander, and Brad Wyble. 2010. The attentional blink: Past, present, and future of a blind spot in perceptual awareness. *Neuroscience & Biobehavioral Reviews* 34(6): 947–957.

Melloni, Lucia, Carlos Molina, Marcela Pena, David Torres, Wolf Singer, and Eugenio Rodriguez. 2007. Synchronization of neural activity across cortical areas correlates with conscious perception. *The Journal of Neuroscience* 27(11): 2858–2865.

Merchant, Hugo, Deborah L. Harrington, and Warren H. Meck. 2013. Neural basis of the perception and estimation of time. *Annual Review of Neuroscience* 36: 313–336.

Michon, John A. 1990. Implicit and explicit representations of time. In *Cognitive models of psychological time*, ed. Richard A. Block, 37–58. Hillsdale: Lawrence Erlbaum Associates.

Newell, Allen. 1973. You can't play 20 questions with nature and win: Projective comments on the papers of this symposium. In *Visual information processing*, ed. W.G. Chase, 283–308. New York: Academic.

Palva, Satu, Klaus Linkenkaer-Hansen, Risto Näätänen, and J. Matias Palva. 2005. Early neural correlates of conscious somatosensory perception. *The Journal of Neuroscience* 25(21): 5248–5258.

Pitts, Michael A., and Juliane Britz. 2011. Insights from intermittent binocular rivalry and EEG. *Frontiers in Human Neuroscience* 5: 107.

Potter, Mary C., Brad Wyble, Rijuta Pandav, and Jennifer Olejarczyk. 2010. Picture detection in rapid serial visual presentation: Features or identity? *Journal of Experimental Psychology: Human Perception and Performance* 36(6): 1486–1494.

Rees, Geraint, Gabriel Kreiman, and Christof Koch. 2002. Neural correlates of consciousness in humans. *Nature Reviews Neuroscience* 3(4): 261–270.

Reuss, Heiko, Andrea Kiesel, Wilfried Kunde, and Bernhard Hommel. 2011. Unconscious activation of task sets. *Consciousness and Cognition* 20(3): 556–567.

Samsonovich, A.V. 2010. Toward a unified catalog of implemented cognitive architectures. In Proceedings of the 2010 conference on biologically inspired cognitive architectures, 195–244. Amsterdam: IOS Press. ISBN: 978-1-60750-660-7

Singer, Wolf. 2011. Consciousness and neuronal synchronization. In *The neurology of consciousness: Cognitive neuroscience and neuropathology*, ed. Steven Laureys and Giulio Tononi, 43–52. London: Academic.

Sloman, Aaron. 1999. What sort of architecture is required for a human-like agent? In *Foundations of rational agency*, ed. Michael Wooldridge and Anand Rao, 35–52. Dordrecht: Kluwer Academic Publishers.

Snaider, Javier, Ryan McCall, and Stan Franklin. 2012. Time production and representation in a conceptual and computational cognitive model. *Cognitive Systems Research* 13(1): 59–71.

Steriade, M. 2006. Grouping of brain rhythms in corticothalamic systems. *Neuroscience* 137(4): 1087–1106.

Stroud, John M. 1967. The fine structure of psychological time. *Annals of the New York Academy of Sciences* 138(2): 623–631.

Van Bockstaele, Bram, Bruno Verschuere, Jan De Houwer, and Geert Crombez. 2010. On the costs and benefits of directing attention towards or away from threat-related stimuli: A classical conditioning experiment. *Behaviour Research and Therapy* 48: 692–697.

VanRullen, Rufin, and Christof Koch. 2003. Is perception discrete or continuous? *Trends in Cognitive Sciences* 7(5): 207–213.

VanRullen, Rufin, Benedikt Zoefel, and Barkin Ilhan. 2014. On the cyclic nature of perception in vision versus audition. *Philosophical Transactions of the Royal Society, B: Biological Sciences* 369(1641): 20130214.

Varela, Francisco J., Evan Thompson, and Eleanor Rosch. 1991. *The embodied mind: Cognitive science and human experience*. Cambridge, MA: MIT Press.

Warner, Rex. 1963. trans. *The Confessions of St. Augustine*. New York: A Mentor Book.

Wikimedia. 2009. *Structures of brain: Wikimedia commons*. http://commons.wikimedia.org/wiki/
    File: Human–brain.SVG. Accessed 17 Apr 2014.
Wikimedia. 2010. *Diagram of human brain: Wikimedia commons*. http://en.wikipedia.org/wiki/
    File:BrainCaudatePutamen.svg. Accessed 17 Apr 2014.
Wittmann, Marc. 2011. Moments in time. *Frontiers in Integrative Neuroscience* 5: 66.
Zakay, Dan. 1992. The role of attention in children's time perception. *Journal of Experimental
    Child Psychology* 54(3): 355–371.
Zakay, Dan, and Richard A. Block. 1996. The role of attention in time estimation processes.
    Advances in Psychology 115: 143–164.
Zakay, Dan, and Richard A. Block. 1997. Temporal cognition. *Current Directions in Psychological
    Science* 6: 12–16.
Zakay, Dan, Yehoshua Tsal, Masha Moses, and Itzhak Shahar. 1994. The role of segmentation in
    prospective and retrospective time estimation processes. *Memory & Cognition* 22(3):
    344–351.
Zarco, Wilbert, Hugo Merchant, Luis Prado, and Juan Carlos Mendez. 2009. Subsecond timing in
    primates: Comparison of interval production between human subjects and rhesus monkeys.
    Journal of Neurophysiology 102(6): 3191–3202.

Part IV
The Timing of Experiences

Chapter 9
The Time of Experience and the Experience of Time

Valtteri Arstila

Abstract Philosophers have usually approached the concept of timing of experiences by addressing the question how the experiences of temporal phenomena can be explained. As a result, the issue of timing has been addressed in two different ways. The first, similar to the questions posed in sciences, concerns the relationship between the experienced time of events and the objective time of events. The second approach is more specific to philosophers' debates, and concerns the phenomenology of experiences: how is the apparent temporal structure of experiences constituted? In regard to both questions, this article shows why and how philosophers' views differ from those held by most scientists. To conclude, I present a combination of views that is not only compatible with that of scientists, but also addresses the problems that engage philosophers.

9.1 Introduction

This chapter focuses on the timing of experiences as it has figured in philosophy. More precisely, the topic of interest concerns the general principles that, considered from both an objective and subjective point of view, determine the moment when some experiential content is experienced.[1] Due to the nature of philosophical investigations, this topic has been approached through the general principles related to timing and philosophers' views have been strongly shaped by the other debates in which they are engaged. For example, philosophers are more concerned with the phenomenology and the metaphysics of time than scientists who often focus on performance in particular time-order tasks and measure the timing of experiences in milliseconds (Arstila 2011). Accordingly, philosophers' views are best understood

---

[1] By experience I mean the whole phenomenology of one's subjectively experienced moment regardless of whether that moment is subjectively speaking temporally extended or not. If not otherwise mentioned, (experiential) content refers to a conscious inner occurrence that is an individual element of an experience.

V. Arstila (✉)
Department of Behavioral Sciences and Philosophy, University of Turku, Turku, Finland
e-mail: valtteri.arstila@utu.fi

© Springer International Publishing Switzerland 2016
B. Mölder et al. (eds.), *Philosophy and Psychology of Time*, Studies in Brain and Mind 9, DOI 10.1007/978-3-319-22195-3_9

by examining the positions they oppose and problems they try to address. Let us therefore begin our consideration by explicating a view that, while simple and initially plausible, is endorsed by no philosopher.

Presumably, the simplest view on the temporal properties of experiences is the following: The presentation of a stimulus is first registered by sensory receptors, after which information about the stimulus is transmitted to the cortex. The main processing takes place in the cortex and—assuming that we experience the stimulus in the first place—we experience the stimulus as soon as the processing is completed, and it is experienced to occur at this time. Moreover, the experience ends when the neural processing is no longer sustained. In other words, our experience is an "online ... phenomenon, coming about as soon as a stimulus reaches its 'perceptual end point'" (Eagleman and Sejnowski 2000, 2036). Thus, for example, if a stimulus reaches this hypothetical perceptual end point before another stimulus reaches the end point the two are experienced to occur in this order.

This view is defined by three theses about the temporal properties of our experiences. The first, *the thesis of minimal delay*, concerns the temporal relationship between the objective time of occurrence of experiences and the events our experiences are about. It states that our experiences of external events are only delayed by the time it takes for light and sound to reach our sensory receptors and for our neural mechanisms to process the stimuli. To put this somewhat differently, because we experience events as soon as the processing is completed, that which we experience always occurred a bit in the past.

According to the second thesis, which I call *the thesis of temporal isomorphism* (see Mölder 2014a), the time when something is experienced to occur is isomorphic to the time of the neural processes realizing the experiences. This thesis thus concerns the relationship between the apparent or subjective time of an experience and the objective time when its neural correlates take place. Because the thesis claims that the time of the neural correlates of experiences matches the apparent time of experiences, the relationship is the simplest one possible. For example, because the apparent temporal order of experience simply mirrors the temporal order of neural events that underlie the experiences, we experience that A occurred before B because this is the order in which the cortical analyses are completed. This means that the temporal properties such as time-order do not need to be represented separately in experiences. Thus, this position has been referred to as the *time as its own representation view* (Kiverstein and Arstila 2013). Other expressions, e.g., *the braintime view* (Johnston and Nishida 2001) and *the brain time account* (Yarrow and Arnold's Chap. 10 of this volume), emphasize how experienced temporal properties are determined by the temporal properties of neural events.

Finally, if experience is an online phenomenon in which the experiential contents reflect what is processed at the perceptual end point, then without additional arguments, this view also suggests that the experiential contents are confined in moments. After all, once something is not processed at the perceptual end point, it is not part of our experience anymore. Thus, all that we experience we experience as occurring now. This leads to the last thesis, *the thesis of instantaneous contents*, according to which the contents of our experiences are confined in an instant.

These three theses make the approaches to the timing of experiences simplest possible one and I will hence refer to them collectively as *the simple view on the temporal properties of experiences* (in short, the simple view).[2] Despite its tempting simplicity, this view is almost unanimously rejected by philosophers. The reason for this will be discussed in the next section. It begins by discussing the thesis of instantaneous contents, which has drawn the most attention from philosophers. The remaining two theses, which relate more directly to the issue of the timing of experiences as the issue is often considered, will be discussed afterwards. If the thesis of instantaneous contents is rejected, then our experiences appear to us as temporally extended. How this apparent temporal structure of experiences is explained is the topic of Sect. 9.3. The simple view, and how it can meet the objections raised against it, will be revisited in the final section.

9.2 Three Theses of the Simple View and Philosophical Theories of Time Consciousness

9.2.1 Instantaneous Contents of Experiences

Philosophical theories of time consciousness, which aim to account for how time and temporal properties figure in our consciousness and as contents of phenomenal states, can be classified into roughly three groups. The first, called *the snapshot view*, is similar to the simple view. Both are committed to the thesis of instantaneous contents, which is the idea that our experiences are both objectively and subjectively confined to practically momentary points in time—to snapshots. Many scientific theories concerning the timing of experiences concur with this thesis as well, even if they reject the other theses of the simple view (e.g., Eagleman and Sejnowski 2000, 2007).

The snapshot view is rejected by all but a few philosophers because accounting for temporal experiences has proven difficult within this framework. Temporal experiences are those that imply the passing of time. Husserl's favorite example was the experience of hearing a melody. More recently, philosophers have focused on experiences of motion, succession and persistence. Thus consider, for example, an experience of motion. If we only experience what is taking place on a snapshot, then

---

[2] Rick Grush (2008) refers to the similar view as *the standard view*. There are, however, two differences between the views. First, Grush is "not concerned with" the small processing delays and thus do not differentiate between minimal delay and extra delay positions as regards the thesis of minimal delay. Second, whereas the thesis of temporal isomorphism is understood here as a claim that concerns the experienced temporal order of events and the temporal order of neural processes realizing these experiences, Grush makes this an issue of passive registration versus active construction of experiences. Given that one endorsing the simple view can hold both active and passive views on perceptual experiences, Grush's claim is an additional issue within the thesis of temporal isomorphism.

our experience of a moving object consists of the object in only one of its just-past (or predicted) positions—the experience of movement is missing. Likewise, while the snapshot view allows for a succession of experiences, this does not yet amount to the experience of succession. If succession is something that we can experience, then it seems that the snapshot view cannot account for it. Similarly, we could never really experience melody if our experience only consisted of the notes being currently played.

This line of reasoning has led philosophers to associate the snapshot view with the idea that, strictly speaking, we do not experience temporally extended events. Instead, motion, for instance, is merely inferred based on our memories of the previous positions of a stimulus and our perception of its current location. Hence, usually the snapshot view "reduces" to the view that our experiences are literally like frames in a movie—just like a single frame, an experience contains colors and shapes, but it does not contain change, motion or succession. Barry Dainton (2010a) calls such a position *Phenomeno-temporal Antirealism*.

Despite this, the snapshot view does not entail the rejection of temporal experiences. Indeed, as will be discussed in the last part of the chapter, it is also possible to subscribe to *Phenomeno-temporal Realism* within the framework of the snapshot view. Nevertheless, because such possibility is very rarely mentioned (see Dainton 2010a), and the only existing well-developed version of the snapshot view denies the reality of temporal experiences, the two are not usually separated. To separate the snapshot view from this more restrictive form, which denies the reality of temporal experiences, I will follow Dainton and refer to the latter as *the cinematic model*.

Contemporary philosophers, however, almost take for granted the phenomenology related to temporally extended events. Thus, they maintain that we can experience change, motion and other dynamic events with the same immediacy that we experience colors and shapes. As a result, the cinematic model is outright rejected. Given that the model is not usually separated from the snapshot view, the latter is also rejected. Consequently, most philosophers argue that the contents of our experiences are temporally extended, i.e., not confined in practically durationless moments, as the simple view holds. This idea is known as *the doctrine of the specious present*.

The idea that an experience covers a temporal interval allows experiential contents that appear (for a subject) to occur at different times to be parts of a single experience. In this framework, the experience of one flash succeeding another can be explained as follows: At the time we experience the latter flash, the first flash lingers in our consciousness as past or preceding content. Because we are conscious of both flashes during the same specious present, we also experience the succession. Correspondingly, James (1890, 574, his italics) argued that "It is only as parts of this *duration-block* [i.e., specious present] that the relation of *succession* of one to the other is perceived." Similarly, listening to a melody does not reduce to hearing one note at a time in isolation. Rather, the previous notes still linger in our consciousness in some way when we hear that which is currently being played.

If our experiences indeed cover an extended interval, it follows that the contents of experiences appear to one as temporally (or dynamically) structured. Otherwise, all things within one specious present would be experienced as simultaneous. This does not mean that the experience itself would be temporally structured in a sense that it has temporal parts—only that, to a subject, the experiential contents within one specious present appear as if embedded in a temporal or dynamic structure. Nevertheless, this apparent temporal structure is usually considered separate from the contents embedded within it. In James' (1890, 630) words, the contents of a specious present are "in a constant flux… Meanwhile, the specious present, the intuited duration, stands permanent, like the rainbow on the waterfall, with its own quality unchanged by the events that stream through it."

Saying that the contents of an experience are temporally extended is more a description of temporal experiences than an explanation of them (see, e.g., Gallagher 2009; Mölder 2014b). In addition, an explanation of how the specious present itself is implemented is required. This is usually understood as the task of explaining the relationship between the objective temporal properties of a specious present and its apparent temporal structure. (A related task, namely how the contents of one specious present appear as temporally structured, is less discussed and will be elaborated upon in the third section.)

The provided explanations come in two main models, which form the remaining two groups of the philosophical theories of time consciousness. The first is the *retentionalist model* (or *intentionalist model*), according to which experiences take place, objectively speaking, in snapshots. However, true to the doctrine of the specious present, the contents of experiences are temporally extended. In more concrete terms, our experience of succession is thought to come about by having two experiential contents appear to be in succession on a single near-momentary experience. This is achieved when the first experiential content is presented as something that just occurred (retained content) while the other is presented as current content (primal image). The competing view, the *extensionalist model*, maintains that both the experiences and their contents are temporally extended. Thus, our experience of succession comes about when two experiential contents which really take place in succession are perceived as the contents of a single experience. So, what separates the two models is their stance on the relationship between the properties of an experience and its contents. Whereas the retentionalist model maintains that our experiences have longer subjective duration than they in fact have—experiences are (near-)momentary while their contents are temporally extended—the extensionalist model maintains that experiences and their contents share an identical temporal structure.

9.2.2   Temporal Isomorphism and Minimal Delay

The two other theses of the simple view are those of temporal isomorphism and minimal delay. The thesis of temporal isomorphism claims that the contents of our experiences and the neural states that underlie them share the same temporal properties. Thus, if the thesis is correct, the order in which stimuli is experienced to occur is the same in which the processing related to contents is completed.[3] The thesis of minimal delay concerns the temporal relation between external events and our experiences of them. It takes as its starting point the necessary delays in perceiving the events and assumes that such a delay is not compensated for in any way. Accordingly, although we appear to experience events immediately—as they happen—the contents of our experiences are always slightly delayed. Together, these two theses imply that the timing of our experiences is simply a matter of neural latencies. Each stimulus (or their features) are processed in parallel, and once the processing is completed—i.e., once the perceptual end point is reached—the stimuli (or their features) are experienced and they are experienced to occur in that very moment (e.g., not half a second ago).[4]

This position, i.e., the combination of the thesis of temporal isomorphism and the thesis of minimal delay, is challenged by the postdiction effects. These are effects in which a stimulus presented at a certain moment (e.g., objectively speaking at t_2) influences how we experience what occurred before the presentation of the stimulus (t_1). The postdiction effect that has drawn the most attention from philosophers is that of apparent motion (e.g., Arstila 2015b, Grush 2005, 2008; Dainton 2008; Hoerl 2015), whereas scientists have focused more on the metacontrast masking and the flash-lag effects.

Visual apparent motion experiment consists of two brief, spatially separate flashes (flash F_1 in location L_1 and flash F_2 in location L_2) presented in succession and with an empty screen between the two flashes. In such experiments, subjects often report seeing one stimulus (rather than two) moving from L_1 to L_2. This is thus an illusion of movement caused by two stationary stimuli, not an illusion of perceived temporal properties per se. Yet, the mysterious part of the phenomenon is temporal: subjects report perceiving motion before the second stimulus. This is

---

[3] The thesis of temporal isomorphism is not the same as the inheritance principle, which states that an experience possesses the same temporal properties as those which are apparently presented in the experience (Phillips 2014a, b). For example, while the thesis of temporal isomorphism concerns the temporal relation between an experiential content and its neural basis, the inheritance principle concerns the relationship between an experience and what is being experienced. That is, the inheritance principle does not take a stance on the neural processing. Then again, since the thesis of temporal isomorphism concerns the experiential contents, not experiences per se, it does not take a stance on the temporal structure of experiences. Moreover, whereas the snapshot view makes the inheritance principle trivially true, it does not make the thesis of temporal isomorphism true. Finally, while the inheritance principle has been used in arguments for the extensionalist model and against the retentionalist model, the thesis of temporal isomorphism is neutral between the models. (One could argue, for example, that the thesis holds for the primal images but not for retained contents because the latter are not really experiential contents.) See also Soteriou (2010), Hoerl (2013), Lee (2014). The claims discussed in these papers are not exactly the same as the inheritance principle, however. For example, Lee (2014) discusses *the mirroring view* and explicitly associates it with the idea that temporal experiences unfold over time. Hence, unlike the inheritance principle, the mirroring view would be incorrect as regards the snapshot view.

[4] This position is (often implicitly) held by scientists working on, for example, the perceptual simultaneity (Kopinska and Harris 2004), duration estimation and reproduction (Reutimann et al. 2004; Wittmann et al. 2010), and the flash-lag effect (Whitney and Murakami 1998; Whitney et al. 2000). For other examples, see Pfeuty et al. (2005); Arnold and Wilcock (2007); Arstila (2015a).

puzzling because the movement from L_1 towards L_2 cannot begin before L_2 is somehow determined. This means that the second stimulus must have been processed to some extent before the motion processing can begin. Accordingly, it is reasonable to assume that the processing of the second stimulus also ends before the apparent motion related processing has been completed. Thus, the simple view conflicts with these reports, as it predicts that the second flash should be experienced before the motion itself is experienced.

One response to this problem is to reject the thesis of temporal isomorphism. Just as our experiences can represent the color blue without the experience itself being blue, it is possible that the temporal properties as experienced differ from the temporal properties of the experiences. Thus, it could be that the experience of succession, for instance, does not require one to have two experiences in succession. Of course, this then means that the temporal properties of events need to be indicated or (re)presented in some fashion (e.g., by means of separate content). Accordingly, this position has been called *the temporal indicator view* (Mölder 2014a). Although this view is less widely held among scientists than the braintime view, it is not without supporters (e.g., Eagleman and Sejnowski 2007).

The first proposal along these lines, by Dennett and Kinsbourne (1992), is compatible with the thesis of instantaneous contents. It simply states that while, objectively, the experience of the second flash can occur before the experience of the movement, subjectively, the order can be reversed. This is because the latter is determined by the time markers (temporal indicators) that accompany these contents (flash and movement). Laurie A. Paul (2010) also appears to argue for this position, as her description of apparent motion does not incorporate the doctrine of the specious present either.

However, the best-developed position that rejects the thesis of temporal isomorphism also rejects the thesis of instantaneous contents. This is Rick Grush's trajectory estimation model. In short, he argues that our experience at t_2 can include, say, contents that represent interval $t_{0-2,}$ and that our experience at t_3 has contents that represent interval t_{1-3}. While both experiences include contents covering t_1 and t_2, those moments are represented in two separate experiences. This means that the interval t_{1-2} "can be re-interpreted" (Grush 2007) and hence, the way in which t_{1-2} is constructed in experiences which are taking place at t_2 and t_3 can be different. In particular, our experience at t_2 may represent that an empty screen was presented at t_1, while at t_3 we experience that there was movement at t_1. Thus, in this explanation, the empty screen is initially experienced in the apparent motion experiments. However, once the second flash is registered and sufficiently processed, the experience of an empty screen is rewritten to represent (apparent) movement.[5]

While the rejection of the thesis of temporal isomorphism is compatible with both the snapshot view and the retentionalists model, it is not compatible with the extensionalist model. The reason for this is that, in this alternative position, the

---

[5] Grush's view resembles Dennett and Kinsbourne's idea of Orwellian revision. However, Grush maintains that the experience of the empty screen is revised, whereas in Orwellian revision it is the memory of the empty screen that is revised.

experience of the empty screen is really already in the past at the time when the need for the revision arises. Thus, the extensionalist model needs to account for the apparent motion by other means. This can be done by rejecting the thesis of minimal delay and maintaining that our experiences are delayed more than the neural processing necessitates. One reason to hold this view is based on the idea that our experiences are underdetermined by sensory signals and that in order to make experiences more accurate, our sensory system makes use of the sensory signals of events occurring before and after each signal. Thus, our experiences of events that happen in t_1 would be influenced by the events that occur at t_0 and t_2, which in turn requires that the neural correlates of experiences of events at t_1 are delayed until the events at t_2 are taken into account (Eagleman and Sejnowski 2007; Eagleman 2010).

In this framework, the experiences of apparent motion are explained by postulating that the information related to F_2 is taken into account before we experience the empty screen. Although at one point we have registered F_1 and the empty screen, we never become conscious of them as such. Instead, "as soon as the second flash [F_2] registers, our visual system reaches the conclusion that the likely source is a moving light, and this is what we experience" (Dainton 2010b). In other words, the order of the processing related to conscious experiences corresponds with the experienced order of events—the thesis of temporal isomorphism is held—whereas the pre-experiential processing corresponds with the temporal structure of stimuli. Unlike the previous alternative, this one is compatible with all three positions on temporal experiences because the differences between them concern the nature of experiences and their contents, whereas in this alternative the revisions occur pre-experientially.

In short, the problem that the postdiction effects pose for the simple view is due to the combination of the thesis of temporal isomorphism and the thesis of minimal delay—it appears that one can subscribe to one but not both. Neither of the presented solutions is ideal, however. On the one hand, the notion of temporal indicators remains underdeveloped (see the next section). On the other hand, the idea of added delay in perception is implausible based on what we know of latencies in neural processing (Arstila 2015b; Dennett and Kinsbourne 1992) and it has been argued that delays in perception could be costly (Grush 2007).[6]

Consequently, one might be tempted to reject the correctness of the subjects' reports in the post-diction experiments. Hence Dennett (1992, 44), for example, rejects the idea of "filling in," namely that subjects' experiences include a "continuous (or even roughly continuous) representation of the motion" between locations L_1 and L_2. Christoph Hoerl (2015) agrees with Dennett's view in this respect, but also proposes that subjects still have an experience of pure motion (i.e., a feeling of

---

[6] Grush argues that our behavior in the world would be more effective if the processing delays were compensated for. Nijhawan (1994) likewise proposes that our sensory system extrapolates the trajectory of a moving stimulus, and this extrapolated position is what we experience. For a similar suggestion, see also (Changizi et al. 2008). It is worth noticing, however, that the compensated processing delays themselves cannot account for the postdiction effects when the effects occur in situations that cannot be predicted beforehand.

movement without an experience of something changing its location continuously as a function of time). However, the rejection of the idea of filling-in appears questionable in the light of empirical evidence. For example, when subjects experience motion in apparent motion experiments, there is a continuous representation of motion in the primary visual cortex (Larsen et al. 2006; Sterzer et al. 2006). Likewise, the trajectory of apparent motion causes the same kind of motion masking as the real motion (Yantis and Nakama 1998; Schwiedrzik et al. 2007). In regards to the other postdiction effects, it is worth noting that Dennett and Kinsbourne (1992) are also incorrect in their claim that psychophysics does not tell us whether the target stimulus is initially perceived in the metacontrast masking experiments (Breitmeyer et al. 2004; Todd 2009). Consequently, it is doubtful that the simple view can be saved by denying these reports concerning the postdiction effects.[7]

9.3   How Is Subjective Time Constructed?

The thesis of temporal isomorphism and the thesis of minimal delay touch upon the issue of the timing of experiences in relation to the objective measures of timing: How does the experienced time of events relate to (i) the real time of events and (ii) the time of neural processes underlying the experiences? If the thesis of instantaneous contents is rejected, one can also ask another question as regards the timing of experiences: how is the time when the experiential content is experienced determined when considered purely from the subjective point of view?

In effect, this is a question about the nature of the apparent temporal structure of experiences because the experiential contents in question are embedded in such a structure, and this structure enables us to have an experience whose contents appear as being in some temporal relation to each other (e.g., one content preceding another). Once we have an explanation of how the temporal properties of experiential contents within specious present are expressed in our phenomenology, we also have an explanation of how we can have an experience with two contents in a way in which one of them appears to us as preceding the other.

Concerning the question of how the contents within specious present are experienced as temporally ordered, the first response appeals to the idea of a necessary dynamic character of experiences, i.e., the experienced flow or passage of time. This flowing character is assumed to be common to all experiences, and thus Dainton (2000, 114) suggests "perhaps this is why a strictly durationless sensory experience, existing all by itself, seems impossible to conceive." Pelczar agrees, and argues that each experience possesses some kind of dynamic content (e.g., change, succession, or something as enduring)—all conscious experiences include "earlier and later parts or phases" (Pelczar 2010, 52).

---

[7]For a more detailed discussion on the apparent motion and different explanations for it, see (Arstila 2015b).

Furthermore, Dainton argues that this intrinsic dynamic character explains our temporal phenomenology. The flowing character makes the apparent temporal structure of experiences directionally asymmetrical. This means that the dynamic character is understood as a sort of mental momentum that automatically orders items within the specious presents. Hence, the structure is explained by appealing to the idea of the dynamic and directed characteristic of experiences.

This response is rather disappointing, however. While it is true that it explains how the contents within specious present are experienced as temporally ordered, the explanation relies upon the dynamic temporal nature of specious presents which is a "fundamental and inexplicable feature of conscious experience" (Pelczar 2010, 58). This makes the response more like a description of the apparent temporal structure of experiences rather than an explanation of it. Whereas previously our temporal experiences remained unexplained, now the central feature of the structure that explains our experiences is unexplained. This "solution" simply pushes the problem to another level of explanation.[8]

Moreover, the similar "explanation" would hardly be satisfactory in regards to the apparent spatial structure of experiential contents. Patients with visual orientation disorder, for example, see objects and recognize their shapes. However, they are often unable to estimate how far away objects are, and they cannot estimate the relative size and position of objects placed before them. It is therefore unsurprising that they repeatedly run into objects—even walls!—although they are able to describe and recognize the objects by sight (Holmes 1918). This disorder brings forth two issues. First, some mechanism is responsible for the apparent spatial structure of our (visual) experiences, and thus, there is a story to be told concerning how this structure is achieved. One aspect of the story is whether experiential contents are embedded in some pre-existing subjective coordinate system or whether the spatial structure of a visual field is subordinate to the experienced spatial relations between the contents. Second, there is an open question about whether all experienced spatial properties should be treated similarly, given that this disorder does not prevent patients from seeing some spatial properties.

If such questions regarding the apparent spatial structure of our experiences can be asked (and in fact answered), then similar questions can justifiably be asked concerning the apparent temporal structure as well. For example, it is reasonable to expect an answer to the question of whether the apparent temporal structure (or the intrinsic dynamism) is primary to the experiential contents—as James, Gallagher and Dainton appear to argue in most places[9]—or whether it is subordinate to the experiential contents—as Dainton and Gallagher's discussion on the flexibility of

---

[8] In accordance with this, Gallagher (2009, 200) points out that "[if] we say that the phenomenal contents have an intrinsic flow structure, is that anything more than saying that consciousness itself just has an intrinsic flow structure?"

[9] E.g., James (1890, 630) argues that the "intuited duration, stands permanent" although its contents are in constant flux. Likewise, Gallagher (2009, 200) argues that according to the retentionalist model, which he holds, retention is "not a particular thing in consciousness" but a structural aspect of consciousness that together with other aspects "is taken to be one of the things that require explanation." See also Dainton (2000, 2008).

the duration of specious present suggests. Likewise, it can be asked whether all temporal experiences should be treated similarly or whether some temporal phenomena can be experienced without specious present. Moreover, it is also interesting to ask whether two contents could be part of the same specious present without us being able to discern their temporal order. In some theories (as in Dainton's and Pelczar's view) this is not possible, whereas in others (possibly in Grush's view) some kind of comparison might be needed.[10]

The second way to account for the apparent temporal structure of experiences allows us to make progress on these issues. This response appeals to explicit temporal properties of subjective time which are similar to those Dennett and Kinsbourne suggested as a solution to the postdiction effects. According to this response, the contents of experience are accompanied by a time marker that represents when they were perceived. Consequently, the time markers form and order the apparent temporal structure of experiences.

Before this view can be properly evaluated, however, the notion of time markers needs to be developed more. For example, the idea can be interpreted in two ways. On the one hand, the time markers could be fixed by temporal coordinates, which would mean that there is a pre-existing temporal coordinate system on which perceived objects are located. This is one interpretation of retentionalists' notions of "now," "just-past," and "past." On the other hand, the time markers could also be relative, meaning that the temporal location of an experiential content is determined in relation to other experiential contents—the experiential contents always appear to be simultaneous or succeed each other. The totality of such relations then organizes the apparent temporal structure of specious present. Pelczar appears to hold a view similar to this, and it is presumably compatible with the extensionalist model as well. (That is, adopting a notion of time markers does not necessarily mean that the thesis of temporal isomorphism is rejected.)

Although both positions need to be developed more,[11] it seems safe to say that the lack of certain deficits argues against both and the existence of mechanisms responsible for time markers in general. Insofar as time marking is caused by some mechanism, it is susceptible to breaking down at some point, with corresponding loss. For example, due to deficits in naming color, recognizing faces, and being able to understand spoken language, scientists have postulated the existence of specific

---

[10] Thus the theories of the first class need to explain empirical results which suggest that we can tell that two auditory stimuli are asynchronous, but cannot tell their temporal order. This can be done, for example, by arguing that performance in these experiments is not based on the experienced temporal properties per se. Instead, they could be based on (i) a difference in the perceived spatial locations of synchronous and synchronous stimuli, or (ii) the two stimuli appearing different in some other respect.

[11] The first one, for example, must answer the question of whether or not there needs to be a mechanism responsible for interpreting the time markers and determining the order or time of events. The second one, for example, must tell how a single experiential content can be experienced to have duration, given that in this case there are no relative temporal markers that bring about the temporal structure. Clearly, these issues do not refute the view but only illustrate that the notion of time markers remains underspecified by philosophers (and scientists).

neural mechanisms related to these abilities. In a similar way, if time markers (the time as represented) determine when something is subjectively experienced to happen, one would assume that corresponding deficits exist. As regards the mechanism for allocating absolute or literal time markers for each experiential content, this would mean that there should be cases in which time markers will always be somehow mixed. For example, one would always experience—separated from cognition—things in the past. Or all temporal markers could be incoherent, and consequently, all the experiences could be in temporal disarray. If the time markers are understood in relative terms, then there should be cases in which subjects will never experience simultaneity or temporal order. Concerning both possible forms of the time marker, it could also be that no experiential content is time marked at all. Given that there are no known empirical cases that resemble these hypothetical cases, the picture provided by the operation of time markers remains unsubstantiated.

The third explanation for the nature of specious present is consistent with those previously mentioned in that all contents of a specious present are experiential (or as it is sometimes expressed, sensory). Differing from the previous explanations, however, it holds that the contents are presented under different temporal modes. (Dainton calls this *modal conception of specious present.*) In other words, the claim that we experience one content "as present" and another as "just past" should not be understood as having two contents with the same phenomenal presence but accompanied by different, explicit time markers. Instead, the apparent temporal structure of specious presence is brought about by temporal modes; we experience A as preceding B because A is presented as "having occurred already" and B is presented as "currently occurring."

Although this view has been held, or at least entertained, by some of the best philosophers working on time consciousness, it has proven most difficult to state clearly what the temporal modes of presentation are. C.D. Broad (1938), for example, characterizes the temporal modes in terms of different degrees of presentedness, but he never really defines what presentedness means. Husserl (1991), on the other hand, is explicit in how the temporal modes of presentation (especially retention) differ from memory or imagination. He also explains that the temporal modes of presentation are not a matter of the vivacity experiences, because we can experience a weak stimulus and a strong stimulus as simultaneous—the experienced difference in their vividness does not make one of them more "current" and the other more "past." Nevertheless, such claims do not amount to a positive characterization of temporal modes of presentation, and it is indeed something that Husserl did not provide either. The difficulty of explaining the temporal modes of presentation is possibly due to the fact that, for Broad and Husserl, the modes cannot be reduced to or explained by means of other modes of consciousness. However, if this is the case, then this explanation of apparent temporal structure would be as explanatory disappointing as the explanation provided by Dainton.

In the previous three explanations there has been an assumption that the contents within a specious present are experiential. Thus, the task has been to explain how experiential contents that belong to the same specious present can appear to a subject as being in some temporal order. The fourth and final view on the nature of apparent temporal structure holds that the contents in question are intentional, not experiential. They are something that can be thought of as providing context for current experiential contents. Gallagher (2009, 201) expresses this idea, which originates from Husserl's later view, as follows:

> [R]etention does not keep a set of fading images in consciousness. Rather, at any moment what we perceive is embedded in a temporal horizon. What I see is part of or a continuation of, or a contrasting change from what went before, and what went before is still intentionally retained so that the current moment is seen as a part of the whole movement. Consciousness retains the just past with the meaning or significance of having just happened.

Most philosophers, if not all, would agree that we usually have a sense of what just happened and that past perceptions can influence our current perception. The problem with this response is that it can also be accepted by philosophers who subscribe to the thesis of instantaneous contents. For example, Mellor (1998, 144), who is one of these philosophers, argues that "[f]or me to see e precede e^*, my seeing e^* must include something like a memory-trace of my seeing e. It need not be explicit or a conscious memory, but some trace of the earlier perception must somehow be incorporated in the later one." Hence he argues, just like Gallagher, that prior experiences can influence our current experiences. The difference between him and Gallagher is that for Mellor, past perceptions do not need to be conscious. Mellor's position is also empirically sound in the light of priming studies, in which an unconscious perception of stimulus can influence current experiences.

In other words, we do not need to perceive past events consciously in order for them to influence how we perceive current events. Moreover, even if we did perceive them consciously, there is no reason why the effect could not be due to memory effects. Accordingly, one can agree with Gallagher's assertion that the significance of what just happened influences what is currently experienced without accepting the doctrine of the specious present. This means that the assertion is neutral as regards the doctrine of the specious present. As a result, it does not address how the apparent temporal structure of consciousness would be constituted.

To summarize, temporal experiences have been explained by means of specious present. In order for an explanation to be truly explanatory, it must also explain how contents within specious present are experienced as temporally ordered. Four different proposals have been put forward, but all of them provide inadequate explanations concerning the temporal organization of the contents within a specious present. It is worth emphasizing that this shortcoming cannot be used as an argument against the extensionalist model or the retentionalist model. This is because, although there is no good explanation for the apparent temporal structure, it does not mean that such an explanation could not emerge in the future—and if such an explanation emerges, it could be compatible with both models, as most of the current proposals are.

9.4 Simple View Vindicated

As mentioned before, for reasons related to temporal phenomenology, philosophers have been rather univocal in their rejection of the thesis of instantaneous contents. Moreover, because of the postdiction effects, they reject either the thesis of temporal isomorphism or the thesis of minimal delay. Thus, philosophers reject at least two of the three theses that comprise the simple view.

I think the simple view can be defended, however. Such defense comes in the form of two other views, which in my opinion are theoretically sound and at least as empirically well-grounded as their alternatives. The first one, *the dynamic snapshot view*, explains the temporal phenomenology in the framework of the snapshot view—in the framework that the thesis of instantaneous contents affords us. The second one, *the non-linear latency difference view*, explains the postdiction effects in a way that is compatible with the thesis of temporal isomorphism and the thesis of minimal delay.

9.4.1 Dynamic Snapshot View

The dynamic snapshot view, as its name implies, subscribes to the thesis of instantaneous contents. This thesis has been rebuffed by most philosophers because, it has been claimed, it leads to Phenomeno-temporal Antirealism. Hence, a philosophical model endorsing the thesis needs to either provide a convincing argument of why there is no temporal phenomenology, or demonstrate how the thesis can be compatible with the realism about temporal phenomenology. While the cinematic model takes the first route, and has had little success in doing so, the dynamic snapshot view attempts to provide the demonstration of compatibility referred to above. That is, according to the dynamic snapshot view, we have immediate experiences of change, motion and other temporal phenomena, just like most philosophers claim. This means that a snapshot can (but does not have to) include contents that a frame in a movie does not allow (namely, temporal phenomenology).

The main problem here is, of course, that the temporal phenomenology cannot be explained in the same way as in the extensionalist and retentionalist models. Because the dynamic snapshot view maintains that the contents of our experiences are not temporally extended, it cannot appeal to the idea that a single experience includes contents that subjectively appear to occur at different times. Instead, the dynamic snapshot view holds that such contents are not required for temporal phenomenology to occur.

The dynamic snapshot view explains the temporal phenomenology by means of "pure" phenomenology. The meaning of this is best explained with examples, one of which has already been mentioned in relation to Hoerl's view on the apparent motion experiments. To remind, Hoerl holds the view that subjects of apparent motion experiments have a feeling of movement without an experience of something changing its location continuously as a function of time. Such experience of motion is called pure motion. Whether such experiences occur in the apparent motion experiments is open to debate, but they are reported to occur in similar kind of experiments in which the interstimuli interval is zero milliseconds (these are sometimes called the pure motion experiments). Other motion illusions corroborate with the separation of motion "qualia" and the perceived change of the location of an object. In waterfall illusions, for example, an object appears to move and not move at the same time, whereas in the rotating snake illusion a stationary stimulus brings about an experience of movement.

The next step is to explain an ordinary experience of motion by means of the phenomenology of pure motion. Here we can follow Robin Le Poidevin (2007). By drawing from psychology, Le Poidevin argues that two independent neural mechanisms are involved in the waterfall illusion. The first mechanism detects motion while the second detects changes in the object's position.[12] Because the mechanisms are independent, the two can give incompatible impressions. In the waterfall illusion, for example, the first mechanism gives us the impression of movement while the second gives the impression that the object's position remains the same. (As we have the experience of motion without seeing anything move, this is a case of pure motion.) Notably, these two mechanisms also figure in Le Poidevin's explanation of the experience of ordinary motion, but in this case the second mechanism gives the impression that the positions of objects change.

The importance of accounting for the motion phenomenology by means of pure motion, which in turn depends on an independent mechanism, is this: the experience of motion is explained in a framework where the experiential contents can be, subjectively speaking, confined to an instant. This is due to the fact that, as the waterfall illusion exemplifies, we can have an experience of motion without an object appearing as being in different places at different times.

The dynamic snapshot view holds that all temporal phenomenology can be explained in a similar fashion, namely by appealing to the existence of mechanisms specific to different types of temporal phenomenology. Thus, our experiences of causality, change, motion, succession and so forth would be due to mechanisms separate from each other, and subsequently also separate from more general mechanisms such as working memory. This is where the dynamic snapshot view differs from Le Poidevin's position, as he accounts for temporal experiences other than motion by appealing to the memory.

Both claims—that other temporal phenomenology could also be "pure" and that such phenomenology is due to separate mechanisms—corroborate with empirical results. For example, it has been recently argued that our awareness of change consists of two separate things: One is the gut feeling that something has changed. Scientists call this "sensed change." Then, there is the awareness of what it is that has changed. This is called "seen change." Because the two are separate, people can

---

[12] The dissociation of experienced motion and position of an object is well supported by the studies showing that it is possible to experience stimulus as moving in one direction while its position is experienced as shifting in the opposite direction (Bulakowski et al. 2007).

have the experience of sensed change even though they have no visual experience of what it is that has changed (Rensink 2004; Busch et al. 2010). That is, analogously to the case of pure motion, where we have experiences of motion in the absence of any seen change in an object's position in space, in this case we have experiences of change in the absence of perceived change in an object's properties. Thus, the sensed change is best understood as an experience of pure change.

Psychologists also separate the experiences of causality (they call it perceptual causality) from attributed causality (causality that we judge to have occurred between two events). The brain imaging data suggests that the mechanism that accounts for our experiences of perceptual causality is analogous to the mechanism that provides the sensed change—both result from purely visual processes. Seen change and attributed causality, on the other hand, depend on a more central and general mechanism of working memory. Thus, although there have been no unequivocal investigations into whether or not pure causality exists,[13] the mechanism behind perceptual causality supports the possibility of its existence. Moreover, just as motion perception is generally regarded to be largely modular (independent of other visual processes), Fonlupt (2003) argues that the mechanism of perceptual causality is modular as well.

It is worth emphasizing that the dynamic snapshot view does not forbid the existence of ongoing memories, that is, experiential contents related to the past. Instead, the claim simply asserts that such mental states play no role in establishing our temporal phenomenology. In other words, it is possible (and even probable) that in the usual situations in which we have an ongoing experience we can also have some kind of memory of what just occurred. However, according to the dynamic snapshot view, it is a mistake to conclude from this that the two types of mental states are intrinsically related in a way that memory is required for the temporal phenomenology to occur.

The noteworthy issue as regards the pure temporal phenomenology is that we can have it without having an experience whose contents appear to us as temporally spread. Pure motion, for example, can be experienced even though an object does not appear to us as being in different places at different times. Because pure change does not depend on having an experience of what the change was, an experience of pure change does not require the past experiential contents to be part of the current experience. Obviously, in the usual cases, we have had a succession of experiential contents, which are in turn the reason why we have the temporal experiences in the first place, but their influence on the current perception can be unconscious

---

[13] This issue depends on what is meant by pure causality. If it means the impression of causality in the absence of the perception of any "causing" stimulus, then it has not been investigated. However, if pure causality refers to the impression of causality in situations where we would not normally claim that cause-effect relationship holds, then it is shown to exist. Consider, for example, perceptual causality experiments in which subjects are shown one moving stimulus (A) and two stationary stimuli (B and C). If A collides with B and B begins to move, it is easy to see why people say that A caused B to move. However, if it is C (not B) that begins to move when A collides with B, then the cause-effect relationship is more susceptible. Yet, people report having an experience of causality in these latter cases as well.

(reminding us of the earlier remarks on Mellor and the priming studies). Having pure temporal phenomenology therefore does not require the doctrine of the specious present to be correct.

Furthermore, provided that the experiences of motion can be explained by appealing to pure motion, it is also justified to explain our experience of change by appealing to pure change and experience of causality by appealing to pure causality. Therefore, temporal phenomenology can be accounted for even if one holds the thesis of instantaneous contents to be true. Contrary to the claims made by those who endorse the doctrine of the specious present, accepting the snapshot model does not necessarily mean rejecting the reality of temporal experiences.

9.4.2 Non-linear Latency Difference View

As we saw above, the combination of the theses of minimal delay and temporal isomorphism has been difficult to reconcile with the postdiction effects. If the simple view is correct, then it appears unexplainable how the latter of two stimuli with the same latency can influence the perception of the first stimulus. As a result, one or both theses have been rejected.

Both theses can be subscribed to, however, once it is recognized that the arguments against the simple view are based on a (too simple-minded) view that the processing in sensory systems always proceeds linearly, in a feed-forward manner. The simple view does not necessitate this view on neural processing, however. Instead, the processing could also incorporate the possibility of non-linear influences. By doing so, the postdiction effects can be explained in the framework of the simple view, or so I shall argue next. The central assumption of such a position, which I call *the non-linear latency difference view*, is that the perceptual end point is defined by the means of reentrant activation of the primary visual cortex (V1).

In general, there appears to be three possibilities what the perceptual end point could be. The first one is grounded on the fact that different features of the stimuli are processed in different areas of the visual cortex. According to this alternative, we become conscious of a feature once the processing related to it is completed in the area that is specialized in processing it. Thus Semir Zeki (Zeki and Bartels 1999; Zeki 2003, 2007) argues that we become conscious of colors once the processing in visual cortical area V4 is completed, and conscious of motion once the processing in visual cortical area V5 is completed. While this alternative relies on the feedforward processing from V1 to later cortical areas, the other alternatives define the perceptual end point in terms of reentrant processing. In these later cases, the perceptual end point would be reached at the moment of activation of the primary visual cortex due to reentrant processing that originates from the later cortical areas. Here we need to separate two alternatives. The perceptual end point could be defined in terms of local reentrant loops, which originate within the visual cortex. Or, it could be defined in terms of global reentrant loops, which originate from later cortical areas, namely from the frontal lobe.

For our purposes, it is enough to assume that it is the local reentrant loops that determine the perceptual end point. This assumption receives support from the empirical results showing that such reentrant processing is required for the processing of even such elemental features as figure-ground perception (Lamme et al. 2002), surface segmentation (Scholte et al. 2008) and responses related to gratings (Shapley 2004). Likewise, motion perception depends on local reentrant processing—if the reentrant activation from V5 to V1 is disrupted, we do not have a perception of motion regardless of V5 activation (Pascual-Leone and Walsh 2001; Silvanto et al. 2005). Indeed, most neurophysiological theories of consciousness postulate that the reentrant processing is necessary for perception to occur (e.g., Dehaene et al. 2006; Kouider 2009; Lamme 2006).[14]

Even if the local reentrant processing determines the perceptual end point, it does not mean that the processing is non-linear. The reentrant processing enables the violation of linearity however. This happens when the area of V1 that is activated by the reentrant processing is also activated at roughly the same time by the feedforward processing originating from retina. In this case, our perception of the first presented stimulus (the cause of the reentrant processing) would be influenced by the latter presented stimulus (the cause of the feedforward processing) and not merely by the shorter latency of the latter.

Such influence can come in two forms. First, the feedforward processing can be fused together with the reentrant processing—the outcome being a combination of both. By using TMS, for example, it has been found that although the V5 modulated activation of V1 is necessary for motion perception, the experience resulting from such activation is also influenced by the properties of V1 neurons (Silvanto et al. 2005). Second, if the activation of V1 by the feedforward processing is much stronger than that of the reentrant processing, and the properties of the used stimuli are suitable, the feedforward processing can inhibit or even disrupt at least some of the processes that depend on the local reentrant loop. One example of such is the figure-ground separation (Lamme et al. 2002).

The idea that the feedforward sweep and the local reentrant processing together determine our experiential content has the interesting consequence that the reentrant processing, which is necessary for perception to occur, does not need to be specific about its cause. To put this somewhat differently, if the idea is correct, the reentrant processing can bring about the perception of things other than those which caused the reentrant processing in the first place. This allows us to explain the postdiction

---

[14]The views about the function of the two types of reentrant loops differ. In Victor Lamme's theory (2004, 2006) local reentrant processing brings about phenomenally conscious states, which is what Ned Block calls phenomenality (2007, 2011). In the global workspace theory, it amounts to unconscious perception (Dehaene et al. 2006; Dehaene and Changeux 2011). In both theories, global reentrant loops bring about cognitive access to the sensory qualities that have been processed within localized reentrant loops. Given the difference regarding the nature of the local reentrant processing, the explanation of the postdiction effects based on this difference may already concern the level of unconscious perception.

effects. Because philosophers have focused on the apparent motion, I will only elaborate on this phenomenon.[15]

To remind, the puzzling part of the apparent motion is how we can experience motion before the second stimulus, given that the motion processing requires information about the location of the second stimulus. What the (mainly conceptual and philosophical) debate over the phenomenon has not acknowledged is, however, that for the purpose of motion processing it is enough if the retinotopic location of the second stimulus is determined.[16] In practice, this means that the (apparent) motion processing can begin at the same time that the processing of the second stimulus begins—namely in the retina.

Usual latency differences do not explain the apparent motion however, because the measured latency difference between moving and stationary stimuli in V1 is only around 20 ms. The non-linear latency difference view allows, however, for another possibility concerning the latency differences: the activation of V5 could be due to processing that bypass V1. After all, the view does not take a stance on the cause of the activation of the later cortical areas, and thus these areas could be stimulated by the sensory signals that bypass V1. Moreover, this possibility is not merely hypothetical because, although most of the information from the retina reaches the visual cortex via V1, V5 also receives visual inputs that do not come through V1 (Sincich et al. 2004). Since such information bypasses V1, a moving stimulus can activate V5 at roughly the same time as V1, or even sooner (Ffytche et al. 1995). The would mean that V5 is activated much earlier than any other area of the visual cortex specialized in the processing of particular features—when V5 is activated, other areas still need to receive an input from V1. Consequently, the processing of visual motion can take place faster than the processing of motionless stimuli.

Such direct activation of V5 due to processing bypassing V1 has been shown to occur when one uses stimuli similar to those used in the apparent motion experiments (Blythe et al. 1986; Azzopardi and Hock 2011). As a result, motion processing in the cortex can begin even before the sensory signals resulting in the perception of the second stimulus reach the cortex in the apparent motion experiments. Given that the apparent motion stimuli can induce V5 activation, which in turn activates

---

[15] It should be mentioned though that the idea that the recurrent processing plays a role in apparent motion also receives support from the fact that such processing has been postulated to play a role in other postdiction effects as well. For instance, many theories of the metacontrast masking incorporate it, see Bridgeman (1980), Enns and Lollo (1997, 2000), Di Lollo et al. (2000), Visser and Enns (2001), Lamme et al. (2002), Fahrenfort et al. (2007), and Ro et al. (2003). See Arstila (forthcoming) for the more thorough explication of the non-linear latency difference view and how it accounts for the flash-lag effect and the metacontrast masking.

[16] In the retinotopic coordinate system, the location of a stimulus represents the location of corresponding cells in the retina. This system needs to be separated from the egocentric coordinate system that corresponds to the experienced location of things. When you are reading this text and your eyes move, for example, the retinotopic positions of the words and the page change. Nevertheless, you do not experience them as moving because, in an egocentric coordinate system, they continue to have the same positions in relation to yourself. The retina and early visual areas (including V1 and V5) are retinotopic.

V1 by means of reentrant processing in a mere 20 ms after the activation of V5 (Muckli et al. 2005; Larsen et al. 2006; Wibral et al. 2009), there is ample time for us to perceive (apparent) motion before the processing related to the perception of the second stimulus is completed.

It is worth emphasizing that this explanation rests upon the idea that we perceive motion and the second stimulus once the reentrant processing related to them terminate in V1. Hence, this explanation does not depend upon the separation of the moment when something is experienced to occur and the moment the neural processes realize the experiences. This means that the thesis of temporal isomorphism can be accepted. The explanation does not depend upon any added delays in neural processing either (quite the contrary), which means that the thesis of minimal delay can be accepted as well. Therefore, this explanation is compatible with the simple view, as well as being based on empirical findings that are independent of the interests that motivated the formulation of the non-linear latency difference view in the first place.

9.5 Conclusions

Philosophers have approached the issue of the timing of experiences mainly through the question of how the experiences of temporal phenomena can be explained. The widely accepted view among philosophers is that this can only be done by means of the doctrine of the specious present. Accordingly, the philosophical issues regarding the timing of experiences relate to the postdiction effects and the apparent temporal structure of experiences.

Thus, the timing of experiences means two different things in the context of philosophical debates. For one, there are the questions concerning the relationship between the experience and (i) its external cause or (ii) the neural processes underlying the experience. Unlike scientists who address such questions in relation to, say, simultaneity or temporal order judgments (as illustrated by Yarrow and Arnold's Chap. 10 of this volume; Yarrow et al. 2011) or EEG markers for consciousness (Sergent et al. 2005; Del Cul et al. 2007; Koivisto and Revonsuo 2010; Railo et al. 2011), philosophers have focused on particular postdiction effects (mainly visual and tactile apparent motion, i.e., cutaneous rabbit). Secondly, the timing of experiences can also be understood as a question of how the apparent temporal structure of specious present is constituted. This question, which concerns the phenomenology of experiences, is more specific to philosophers' debates and is largely ignored by scientists. The obvious reason for this is the fact that the question only becomes relevant if one endorses the doctrine of the specious present, which is something that scientists rarely do.

Overall, it can be concluded that because of their commitment to Phenomenotemporal Realism and because of the results related to the postdiction effects, philosophers' views concerning the timing of experiences contradict the views held by most scientists. However, as illustrated by the dynamic snapshot view and the non-linear latency difference view, this rift between disciplines is not necessary. On the contrary, if these two views are correct, the issues identified by philosophers as important to the matters at hand can be addressed in a framework that is also compatible with current scientific positions.

Acknowledgments I am grateful for Christoph Hoerl, Dan Lloyd, Julian Kiverstein, Kielan Yarrow, and Marc Wittmann for stimulating discussions on these and related issues over the years. I want to thank also two anonymous referees for their thorough and helpful comments.

References

Arnold, Derek H., and Paul Wilcock. 2007. Cortical processing and perceived timing. *Proceedings of the Royal Society B: Biological Sciences* 274: 2331–2336.

Arstila, Valtteri. 2011. Further steps in the science of temporal consciousness? In *Multidisciplinary aspects of time and time perception*, ed. Argiro Vatakis, Anna Esposito, Maria Giagkou, Fred Cummins, and Georgios Papadelis, 1–10. Dordrecht: Springer.

Arstila, Valtteri. 2015a. Defense of the brain time view. *Frontiers of Psychology* 6(1350). doi:10.3389/fpsyg.2015.01350.

Arstila, Valtteri. 2015b. Theories of apparent motion. *Phenomenology and the Cognitive Sciences*. doi:10.1007/s11097-015-9418-y.

Arstila, Valtteri. Forthcoming. Keeping postdiction simple. *Consciousness and Cognition*.

Azzopardi, Paul, and Howard S. Hock. 2011. Illusory motion perception in blindsight. *Proceedings of the National Academy of Sciences of the United States of America* 108: 876–881.

Block, Ned. 2007. Consciousness, accessibility, and the mesh between psychology and neuroscience. *Behavioral and Brain Sciences* 30: 481–499; discussion 499–548.

Block, Ned. 2011. Perceptual consciousness overflows cognitive access. *Trends in Cognitive Sciences* 15: 567–575.

Blythe, Isobel M., Jane M. Bromley, Christopher Kennard, and K.H. Ruddock. 1986. Visual discrimination of target displacement remains after damage to the striate cortex in humans. *Nature* 320: 619–621.

Breitmeyer, Bruno G., Tony Ro, and Neel S. Singhal. 2004. Unconscious color priming occurs at stimulus- not percept-dependent levels of processing. *Psychological Science* 15: 198–202.

Bridgeman, Bruce. 1980. Temporal response characteristics of cells in monkey striate cortex measured with metacontrast masking and brightness discrimination. *Brain Research* 196: 347–364.

Broad, Charles Dunbar. 1938. *An examination of McTaggart's philosophy*. Cambridge: Cambridge University Press.

Bulakowski, Paul F., Kami Koldewyn, and David Whitney. 2007. Independent coding of object motion and position revealed by distinct contingent aftereffects. *Vision Research* 47: 810–817.

Busch, Niko A., Ingo Fründ, and Christoph S. Herrmann. 2010. Electrophysiological evidence for different types of change detection and change blindness. *Journal of Cognitive Neuroscience* 22: 1852–1869.

Changizi, Mark, Andrew Hsieh, Romi Nijhawan, Ryota Kanai, and Shinsuke Shimojo. 2008. Perceiving the present and a systematization of illusions. *Cognitive Science* 32: 459–503.

Dainton, Barry. 2000. *Stream of consciousness: Unity and continuity in conscious experience*. London: Routledge.

Dainton, Barry. 2008. The experience of time and change. *Philosophy Compass* 3(4): 619–638.

Dainton, Barry. 2010a. Temporal consciousness. In *Stanford encyclopedia of philosophy*. http://plato.stanford.edu/entries/consciousness-temporal/. Accessed 14 June 2014.

Dainton, Barry. 2010b. Temporal consciousness; Supplement: Interpreting temporal illusions. In *Stanford encyclopedia of philosophy*. http://plato.stanford.edu/entries/consciousness-temporal/temporal-illusions.html. Accessed 14 June 2014.

Dehaene, Stanislas, and Jean-Pierre Changeux. 2011. Experimental and theoretical approaches to conscious processing. *Neuron* 70: 200–227.

Dehaene, Stanislas, Jean-Pierre Changeux, Lionel Naccache, Jérôme Sackur, and Claire Sergent. 2006. Conscious, preconscious, and subliminal processing: A testable taxonomy. *Trends in Cognitive Sciences* 10: 204–211.

Del Cul, Antoine, Sylvain Baillet, and Stanislas Dehaene. 2007. Brain dynamics underlying the nonlinear threshold for access to consciousness. *PLoS Biology* 5: e260.

Dennett, Daniel. 1992. "Filling in" versus finding out: A ubiquitous confusion in cognitive science. In *Cognition: Conceptual and methodological issues*, ed. L. Herbert Jr., Paulus Willem van den Broek, and David C. Knill, 33–49. Washington, DC: American Psychological Association.

Dennett, Daniel, and Marcel Kinsbourne. 1992. Time and the observer. *Behavioral and Brain Sciences* 15: 183–247.

Di Lollo, Vincent, James T. Enns, and Ronald A. Rensink. 2000. Competition for consciousness among visual events: The psychophysics of reentrant visual processes. *Journal of Experimental Psychology: General* 129: 481–507.

Eagleman, David M. 2010. How does the timing of neural signals map onto the timing of perception? In *Space and time in perception and action*, ed. Romi Nijhawan and Beena Khurana, 216–231. Cambridge: Cambridge University Press.

Eagleman, David M., and Terrence J. Sejnowski. 2000. Motion integration and postdiction in visual awareness. *Science* 287: 2036–2038.

Eagleman, David M., and Terrence J. Sejnowski. 2007. Motion signals bias localization judgments: A unified explanation for the flash-lag, flash-drag, flash-jump, and Frohlich illusions. *Journal of Vision* 7: 1–12.

Enns, James T., and Vincent Di Lollo. 1997. Object substitution: A new form of masking in unattended visual locations. *Psychological Science* 8: 135–140.

Enns, James T., and Vincent Di Lollo. 2000. What's new in visual masking? *Trends in Cognitive Sciences* 4: 345–352.

Fahrenfort, Johannes J., H. Steven Scholte, and Victor A.F. Lamme. 2007. Masking disrupts reentrant processing in human visual cortex. *Journal of Cognitive Neuroscience* 19: 1488–1497.

Ffytche, Dominic H., C.N. Guy, and Semir Zeki. 1995. The parallel visual motion inputs into areas V1 and V5 of human cerebral cortex. *Brain* 118: 1375–1394.

Fonlupt, Pierre. 2003. Perception and judgement of physical causality involve different brain structures. *Brain Research* 17: 248–254.

Gallagher, Shaun. 2009. Consciousness of time and the time of consciousness. In *Elsevier encyclopedia of consciousness*, ed. William Banks, 193–204. London: Elsevier.

Grush, Rick. 2005. Internal models and the construction of time: Generalizing from state estimation to trajectory estimation to address temporal features of perception, including temporal illusions. *Journal of Neural Engineering* 2: S209–S218.

Grush, Rick. 2007. Time and experience. In *Philosophie der Zeit*, ed. Thomas Müller, 27–44. Frankfurt am Main: Klostermann.

Grush, Rick. 2008. Temporal representation and dynamics. *New Ideas in Psychology* 26: 146–157.

Hoerl, Christoph. 2013. A succession of feelings, in and of itself, is not a feeling of succession. *Mind* 122: 373–417.

Hoerl, Christoph. 2015. Seeing motion and apparent motion. *European Journal of Philosophy* 23(3): 676–702. doi:10.1111/j.1468-0378.2012.00565.x.

Holmes, Gordon. 1918. Disturbances of visual orientation. *The British Journal of Ophthalmology* 2(9): 449–468.

Husserl, Edmund. 1991. *On the phenomenology of the consciousness of internal time (1893–1917)*. Dordrecht: Kluwer Academic Publishers.

James, William. 1890. *The principles of psychology*. New York: Dover.

Johnston, Alan, and Shinya Nishida. 2001. Time perception: Brain time or event time? *Current Biology* 11: R427–R430.

Kiverstein, Julian, and Valtteri Arstila. 2013. Time in mind. In *Blackwell companion to the philosophy of time*, ed. Adrian Bardon and Heather Dyke, 444–469. Oxford: Wiley-Blackwell.

Koivisto, Mika, and Antti Revonsuo. 2010. Event-related brain potential correlates of visual awareness. *Neuroscience and Biobehavioral Reviews* 34: 922–934.

Kopinska, Agnieszka, and Laurence R. Harris. 2004. Simultaneity constancy. *Perception* 33: 1049–1060.

Kouider, Sid. 2009. Neurobiological theories of consciousness. In *Encyclopedia of consciousness*, vol. 2, ed. William P. Banks, 87–100. Oxford: Academic.

Lamme, Victor A.F. 2004. Separate neural definitions of visual consciousness and visual attention; a case for phenomenal awareness. *Neural Networks* 17: 861–872.

Lamme, Victor A.F. 2006. Towards a true neural stance on consciousness. *Trends in Cognitive Sciences* 10: 494–501.

Lamme, Victor A.F., Karl Zipser, and Henk Spekreijse. 2002. Masking interrupts figure—Ground signals in V1. *Journal of Cognitive Neuroscience* 14: 1044–1053.

Larsen, Axel, Kristoffer H. Madsen, Torben E. Lund, and Claus Bundesen. 2006. Images of illusory motion in primary visual cortex. *Journal of Cognitive Neuroscience* 18: 1174–1180.

Le Poidevin, Robin. 2007. *The images of time: An essay on temporal representation*. Oxford: Oxford University Press.

Lee, Geoffrey. 2014. Extensionalism, atomism, and continuity. In *Debates in the metaphysics of time*, ed. L. Nathan Oaklander, 149–173. London: Bloomsbury.

Mellor, David Hugh. 1998. *Real time II*. London: Routledge.

Mölder, Bruno. 2014a. Constructing time: Dennett and Grush on temporal representation. In *Subjective time: The philosophy, psychology, and neuroscience of temporality*, ed. Valtteri Arstila and Dan Lloyd, 217–238. Cambridge: MIT Press.

Mölder, Bruno. 2014b. How philosophical models explain time consciousness. *Procedia-Social and Behavioral Sciences* 126: 48–57.

Muckli, Lars, Axel Kohler, Nikolaus Kriegeskorte, and Wolf Singer. 2005. Primary visual cortex activity along the apparent-motion trace reflects illusory perception. *PLoS Biology* 3: e265.

Nijhawan, Romi. 1994. Motion extrapolation in catching. *Nature* 370: 256–257.

Pascual-Leone, Alvaro, and Vincent Walsh. 2001. Fast backprojections from the motion to the primary visual area necessary for visual awareness. *Science* 292: 510–512.

Paul, Laurie A. 2010. Temporal experience. *Journal of Philosophy* 107: 333–359.

Pelczar, Michael. 2010. Must an appearance of succession involve a succession of appearances? *Philosophy and Phenomenological Research* 81: 49–63.

Pfeuty, Micha, Richard Ragot, and Viviane Pouthas. 2005. Relationship between CNV and timing of an upcoming event. *Neuroscience Letters* 382: 106–111.

Phillips, Ian. 2014a. The temporal structure of experience. In *Subjective time: The philosophy, psychology, and neuroscience of temporality*, ed. Valtteri Arstila and Dan Lloyd, 139–158. Cambridge: MIT Press.

Phillips, Ian. 2014b. Experience of and in time. *Philosophy Compass* 9: 131–144.

Railo, Henry, Mika Koivisto, and Antti Revonsuo. 2011. Tracking the processes behind conscious perception: A review of event-related potential correlates of visual consciousness. *Consciousness and Cognition* 20: 972–983.

Rensink, Ronald A. 2004. Visual sensing without seeing. *Psychological Science* 15: 27–32.

Reutimann, Jan, Volodya Yakovlev, Stefano Fusi, and Walter Senn. 2004. Climbing neuronal activity as an event-based cortical representation of time. *The Journal of Neuroscience* 24: 3295–3303.

Ro, Tony, Bruno G. Breitmeyer, Philip Burton, Neel S. Singhal, and David Lane. 2003. Feedback contributions to visual awareness in human occipital cortex. *Current Biology* 11: 1038–1041.

Scholte, H. Steven, Jacob Jolij, Johannes J. Fahrenfort, and Victor A.F. Lamme. 2008. Feedforward and recurrent processing in scene segmentation: Electroencephalography and functional magnetic resonance imaging. *Journal of Cognitive Neuroscience* 20: 2097–2109.

Schwiedrzik, Caspar M., Arjen Alink, Axel Kohler, Wolf Singer, and Lars Muckli. 2007. A spatio-temporal interaction on the apparent motion trace. *Vision Research* 47: 3424–3433.

Sergent, Claire, Sylvain Baillet, and Stanislas Dehaene. 2005. Timing of the brain events underlying access to consciousness during the attentional blink. *Nature Neuroscience* 8: 1391–1400.

Shapley, Robert. 2004. A new view of the primary visual cortex. *Neural Networks* 17: 615–623.

Silvanto, Juha, Alan Cowey, Nilli Lavie, and Vincent Walsh. 2005. Striate cortex (V1) activity gates awareness of motion. *Nature Neuroscience* 8: 143–144.

Sincich, Lawrence C., Ken F. Park, Melville J. Wohlgemuth, and Jonathan C. Horton. 2004. Bypassing V1: A direct geniculate input to area MT. *Nature Neuroscience* 7: 1123–1128.

Soteriou, Matthew. 2010. Perceiving events. *Philosophical Explorations* 13: 233–241.

Sterzer, Philipp, John-Dylan Haynes, and Geraint Rees. 2006. Primary visual cortex activation on the path of apparent motion is mediated by feedback from hMT+/V5. *NeuroImage* 32: 1308–1316.

Todd, Steven J. 2009. A difference that makes a difference: Passing through Dennett's Stalinesque/ Orwellian impasse. *The British Journal for the Philosophy of Science* 60: 497–520.

Visser, Troy A.W., and James T. Enns. 2001. The role of attention in temporal integration. *Perception* 30: 135–145.

Whitney, David, and Ikuya Murakami. 1998. Latency difference, not spatial extrapolation. *Nature Neuroscience* 1: 656–657.

Whitney, David, Ikuya Murakami, and Patrick Cavanagh. 2000. Illusory spatial offset of a flash relative to a moving stimulus is caused by differential latencies for moving and flashed stimuli. *Vision Research* 40: 137–149.

Wibral, Michael, Christoph Bledowski, Axel Kohler, Wolf Singer, and Lars Muckli. 2009. The timing of feedback to early visual cortex in the perception of long-range apparent motion. *Cerebral Cortex* 19: 1567–1582.

Wittmann, Marc, Alan N. Simmons, Jennifer L. Aron, and Martin P. Paulus. 2010. Accumulation of neural activity in the posterior insula encodes the passage of time. *Neuropsychologia* 48: 3110–3120.

Yantis, Steven, and Takehiko Nakama. 1998. Visual interactions in the path of apparent motion. *Nature Neuroscience* 1: 508–512.

Yarrow, Kielan, Nina Jahn, Szonya Durant, and Derek H. Arnold. 2011. Shifts of criteria or neural timing? The assumptions underlying timing perception studies. *Consciousness and Cognition* 20(4): 1518–1531. doi:10.1016/j.concog.2011.07.003.

Zeki, Semir. 2003. The disunity of consciousness. *Trends in Cognitive Sciences* 7: 214–218.

Zeki, Semir. 2007. A theory of micro-consciousness. In *The Blackwell companion to consciousness*, ed. Max Velmans and Susan Schneider, 580–588. Oxford: Blackwell.

Zeki, Semir, and Andreas Bartels. 1999. Toward a theory of visual consciousness. *Consciousness and Cognition* 8: 225–259.

Chapter 10
The Timing of Experiences: How Far Can We Get with Simple Brain Time Models?

Kielan Yarrow and Derek H. Arnold

Abstract When questioned, we are generally able to provide a coherent narrative regarding the order in which recent events happened. In considering this ability, many theorists have appealed to the idea that our perception of physical event timing might be related to the corresponding timing of neural events (i.e. brain time). However, a number of findings indicate that our perception does not slavishly follow from brain time, which might lead us to disregard the whole notion that the time of neural events is important. In this chapter we will suggest that this is premature. We will outline some simple models in which brain time matters, and discuss ways in which they would need to be developed to deal with the realities of our perceptual experiences. Our main point is not that these models are necessarily correct, but rather that theorists need to make alternative accounts similarly concrete and implementable before they will provide a compelling alternative.

10.1 Introduction

Temporal sequencing is at the heart of many of our experiences. The world around us is in flux, and we have a constant sense of the order in which key events are occurring. How this impression arises is a big question, so we will start by limiting our scope. We are going to talk mainly about the more constrained problem of how observers manage to determine the order (or simultaneity/successiveness) of pairs of events arising in different sensory modalities. In fact, we will mainly consider quite bland experimental stimuli: Punctate sights and sounds (e.g. flashes and beeps). In common with other reductionist endeavours, we are hoping that determining how humans solve this constrained problem will tell us something useful

K. Yarrow (✉)
Department of Psychology, City University London,
Northampton Square, London EC1V 0HB, UK
e-mail: kielan.yarrow.1@city.ac.uk

D.H. Arnold
School of Psychology, The University of Queensland, Brisbane, QLD, Australia

© Springer International Publishing Switzerland 2016
B. Mölder et al. (eds.), *Philosophy and Psychology of Time*, Studies in Brain and Mind 9, DOI 10.1007/978-3-319-22195-3_10

about the more general question of temporal consciousness. Needless to say, even the simpler problem is unresolved.

The question of how people establish the order of events has a venerable heritage, which we are not going to review in any great detail here. Instead, our approach will be as follows. First we will make some brief comments about a well-known paper on the perception of relative time authored by Dennett and Kinsbourne (1992). This paper laid bare some important considerations for those interested in perceived time, and is a terrific read for those new to the area. Importantly, this paper put out a strong challenge to the idea that our perception of the timing of events in the world should follow from the timing of corresponding events in the brain.

However, at the outset we want to suggest that "brain time" accounts of timing perception, which presume a link between the timing of events in the brain and timing perception, have value. In particular, we will spend the majority of this chapter describing various models which suggest that the latencies of neural propagation towards a decision centre inform our perceptual decisions about the causative events. We have selected these models because we are not aware of any similarly well-specified and quantitatively precise models in which brain time is held to be irrelevant. Note, however, that we believe that brain time is supplemented, and perhaps even ignored, in particular circumstances. What we argue is that brain time is a good place to start.

We will finish by considering a few ways in which simple models like these would need to be developed to deal with some of what we know about temporal perception. We use "brain time," "neural time" and "latency" fairly interchangeably when suggesting that the timing of events in the brain informs many perceptual decisions. Our key point is that when you have a working computational model, rather than a vague theory, it makes it easier to think about how it is lacking and/or how you might further develop it.

10.2   Time and the Observer

Many theorists have suggested that the time it takes for signals to propagate through the central nervous system and reach some critical part of the brain (e.g. sensory cortex) is an important determinant of the time at which observers perceive the causative events to occur (e.g. Frohlich 1923; Paillard 1949; Sternberg and Knoll 1973; Whitney and Murakami 1998; Zeki and Bartels 1998). We will refer to this general idea as the brain time or latency account. Dennett and Kinsbourne (1992) provided a cogent critique of this notion. They argued that when asked what just happened, the brain has at its metaphorical fingertips all the information accrued up to the moment of the question. Thus it is likely to use all of this information to make the judgement, without necessary reference to what it might have decided at some earlier point. Hence we are not in thrall to a succession of real-time snapshots, each made using the information arriving at that particular moment. This seems pretty sensible as a general framework for an adaptive brain.

On the way to making this point, Dennett and Kinsbourne also argue that there is little reason to ever represent the time at which an event happened by the time at which the corresponding neural activity occurred: Just as we date letters by the date that is written on them, rather than by when they arrive, the brain could somehow assign temporal labels to incoming events. This is a good conceptual point, albeit one that does not sit particularly comfortably with much preceding empirical evidence. However, the existence of possible alternative representational schemes does not mean that brain time doesn't actually matter for perceived time (see Phillips 2014, for a robust defence of this "naive" position). Nor does it go very far towards elucidating exactly what those alternative schemes would look like (although in fairness Dennett and Kinsbourne do suggest that something akin to the low-level registration of images from the two eyes that supports stereo vision might be suitable).

To our minds, a key consideration is whether theorists can specify, in detail, a neurocognitive model that can turn sensory inputs into decisions about the relative timing of events. This can certainly be done by allowing that propagation times towards a decision centre inform the first stages of temporal order perception (as we outline in the following sections). We are not completely clear what the equivalent first stages of a model eschewing brain time entirely might look like (although we would certainly be interested to find out). In essence, we favour a hybrid position, where brain time matters, but is not all that matters. We see a fairly clear path from brain time models towards this hybrid, but we don't currently see any such path leading in the opposite direction.

10.3   Latency Models of Temporal Judgements

From the 1960s onwards, ideas derived from signal detection theory (Green and Swets 1966) began to be applied to the problem of perceived order. Briefly, the key idea in signal detection theory is that the brain is noisy. Hence, the problem of detecting a signal, such as a weak sound, becomes a problem of discriminating signal plus noise from noise alone. In this example, the sensory dimension is stimulus intensity; the more intense the signal, the easier the discrimination. Signal detection theory also makes clear the distinction between perception (based on the internally represented magnitude of a sensory dimension, e.g. perceived intensity) and interpretation (e.g. making a judgement by determining whether perceived intensity exceeded a decision threshold). This division between sensory and decision processes is doubtless too simple to reflect accurately all that goes on during perceptual judgements. However, by recognising an interpretative stage intervening between raw percept and subjective report or judgement, signal detection theory has profoundly affected subsequent thinking in the field of psychophysics.

In the case of perceived timing, the logical step in applying signal detection theory was to make the sensory dimension the time between two signals that are being judged, often referred to as the stimulus onset asynchrony (SOA). The basic idea is as follows. Signal A (e.g. a light) takes time to propagate through the brain to a decision centre, and likewise for signal B (e.g. a sound). The average propagation times might vary, giving one signal a systematic head start (e.g. a shorter neural pathway for sound than light). Importantly, propagation latencies are noisy: repeated instances of a physically identical input will give rise to latencies that vary from trial to trial. For simplicity, this noise is often assumed to follow a Gaussian (i.e. bell-shaped) distribution.

On any given trial, then, what arrives at the decision centre is a subjective SOA, i.e. the delay between signals, which is composed of an objective asynchrony based on arrival times at sensory receptors, plus any difference in their mean propagation times through the brain, and some noise. At the decision centre, a decision rule is applied. For example, a simple rule would be "say light came first if the subjective SOA is greater than zero, else say sound came first." For this kind of decision rule, the dividing point between categories (zero in this example) is known as a decision criterion. The noise in the signals means that if the same objective SOA is repeated on several trials (as is typical in psychophysical experiments) the decision will vary from trial to trial. This fact leads to imperfect performance (i.e. observers sometimes get it wrong, particularly when the SOA is near zero).

These ideas were discussed in detail in an excellent paper by Sternberg and Knoll (1973), who considered the complete class of models, including the simple decision rule described above and several more complicated alternatives, and called this class the "general independent channels model." The key ideas are illustrated in Fig. 10.1, which also illustrates the predictions of this model (with the simple decision rule outlined above). It is a "psychometric function" with a characteristic "s" shape. These kinds of psychometric function are commonly observed in experiments, providing some support for the model. The model also makes predictions about performance when observers are asked slightly different questions (for example whether sound and light were simultaneous or not; Fig. 10.1c, Schneider and Bavelier 2003).

Where these sorts of models do not quite capture the data, they can be adjusted in various ways to provide a better fit. For example, Ulrich (1987) considered a "general threshold model," an extension of Sternberg and Knoll's general independent channels model that allowed the latency noise to come from distributions other than a Gaussian, and also allowed for the possibility that the decision criterion might vary from trial to trial (see Yarrow et al. 2011, 2013 for an application of this model to simultaneity judgements). In a different kind of variant of the general independent channels model, Garcia-Perez and Alcala-Quintana (2012) incorporated various kinds of keying error (i.e. hitting the wrong button by mistake) to provide an improved fit to data.

The key point for our purposes here is that these models all assume that brain time (i.e. the latency for neural activity triggered by sensory events to reach a decision centre) is a critical determinant of perceptual decisions. Furthermore, these models are sufficiently well specified to predict patterns of response in psychophysical experiments. Note that while the decision centre posited in these models looks like a kind of homunculus, it is a homunculus that we can at least imagine simplifying as a set of dumb neural processes which deal with one specific mental

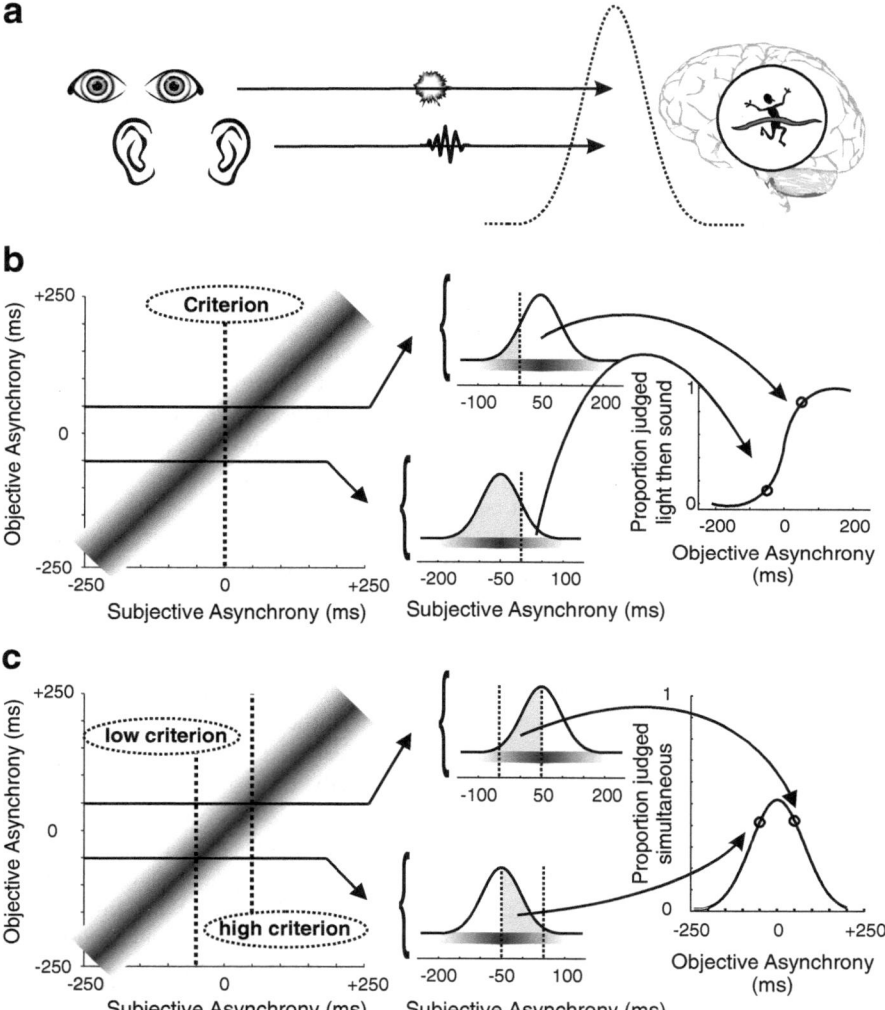

Fig. 10.1 Illustration of how latency models predict order judgements. (**a**) Schematic of a latency model. Neural activations representing two sensory signals travel along independent pathways to a comparator, where a decision about their relative timing is made. Signals each have a mean latency to reach the decision centre, but latencies vary from trial to trial according to a probability distribution (Gaussian in this example). (**b**) Schematic of a temporal order judgement under this kind of model. The process outlined in part *a* generates an orderly mapping between objective asynchronies (i.e. those at the sense organs) and subjective asynchronies (i.e. those at the decision centre) across repeated trials of an experiment (*left*). In this example, the two neural pathways are assumed to be of the same length (so the function relating them passes through zero) but the presence of latency noise (shown by the *shading* around the function, with *darker* regions having higher probability densities) means that for a given objective asynchrony, a range of subjective asynchronies will be generated across multiple trials (*centre*). These facts predict the proportion of judgements favouring one particular order that an observer will make at each objective asynchrony, generating their predicted psychometric function (*right*). (**c**) Schematic of a simultaneity judgement. The process is the same as outlined in part *b*, but two criteria are required to reach a decision (because simultaneity has both low and high limits) yielding a psychometric function that is bump shaped (formally, it is the difference of two cumulative Gaussians)

operation (as outlined next). Hence this homunculus is not the occupant of the Cartesian theatre so derided by Dennett and Kinsbourne (1992), with the implication of an infinite regress without explanatory power.

10.4   Possible Neural Implementations for Latency Models

The models outlined in the previous section are psychological process models (i.e. they represent the steps in cognition leading to behaviour) but they are pitched at a rather high level, so that we don't get much sense of the nitty gritty. However, we can map these ideas (or at least some quite closely related ideas) onto the brain in various ways to generate proposals about plausible mechanisms. Although models mutate somewhat as we do this, they retain the notion that transmission times affect perception, and the mutation might be no bad thing.

To start with, what do we mean by a decision centre? In brain terms, we would have to think about a population of neurones. Confining ourselves to the audiovisual case, this population would need to receive inputs from both auditory and visual systems. Indeed, given the generality with which we can reach temporal decisions about stimuli, these neurones would need to have access to all manner of visual and auditory inputs. Broadening our scope beyond the audio-visual case, if there is but one such centre, it would need to receive inputs from the myriad of sensory modalities from which we can judge timing, including olfaction, taste, proprioception etc., and it would even have to receive information about the timing of sub-vocal thought, as subjectively it seems we can time the impression of reaching "6" while mentally counting from one to ten, and judge the timing of this sensation relative to other physical events. One can see why, faced with this explicit suggestion, one might be tempted to invoke the concept of a Cartesian theatre. However, what limited evidence there is on this point tends to point towards several decision centres. For example, points of subjective simultaneity (the objective SOA at which two signals seem most synchronous) in auditory-visual, visuo-tactile, and auditory-tactile temporal order judgements don't add up to zero (when their magnitudes are appropriately signed). This suggests a kind of neural latency intransitivity that one would not predict if all modalities converged at a common location to determine relative time (Sternberg and Knoll 1973).

What an individual neurone sees, or hears, is known as its receptive field.[1] Looking at the visual system, receptive fields tend to get larger, in functional terms, as we move along the visual pathways from the eyes toward higher cortical regions of the brain. This is consistent with a pooling of information from lower-level neurones, which have more constrained receptive fields, onto higher-level target neurones (Hubel 1988). Hence we can conceive of high-level neurones that receive visual input from the entire visual scene (and, somewhat equivalently, receive audi-

---

[1] We recognise that our choice of phrasing represents a mereological fallacy, but we find that the perceptual metaphor is instructive (because it is easy to grasp).

tory input capturing all frequencies of sound that can be perceived). An obvious problem would be that such wholesale pooling would mean our decision population could be overpowered in a situation with several visual or auditory events occurring close to one another in space and/or time. The obvious solution would be a mechanism of selective attention, to filter out everything unrelated to current interests. Indeed, there is evidence to suggest that temporal judgements become impossible when multiple competing visual events are present, unless selective attention can be used to home in on the particular visual event that is to be timed (Fujisaki et al. 2006).

The next component of latency models that needs fleshing out is the way in which a subjective SOA is actually determined at the decision centre. Here, we probably require some kind of transformation from a temporal code to a different sort of representation, with the most obvious candidate being spike-rate coding within a population of neurones that each prefer a different SOA (i.e. a conversion from a temporal to an essentially spatial coding; spatial because SOAs are now distributed over space in the brain). There are several existing models explaining analogous processes that could be adapted.

Perhaps closest to the spirit of a latency model is the delay line model used as an explanation for how we localise sounds based on inter-aural time differences (ITD; Jeffress 1948). Sounds from the left will hit our left ear slightly before they hit our right ear, providing a strong cue to localisation. In Jeffress' model, a population of neurones each receives an input from both the left and right ears, but the neural axons from each ear can be of different lengths. Because action potentials travel along axons at finite speeds, the length of each axon imposes a delay, and if the axons from each ear are of different lengths, the delays might be different. Critically, neurones at the decision centre (in this case the superior olivary complex) integrate and fire according to a logical "And" rule, i.e. they only fire if both of their inputs arrive at the same time. By varying the length of the two axons connecting the ears to the integrating neurone, a preference for a particular ITD (which corresponds to a constrained region of auditory space) can be created. Across a full population, all conceivable ITDs, arising from localisation-related sound delays, could be represented.

Differences in arrival times for sound striking the two ears are in the microsecond, rather than the millisecond, range. This makes delays imposed by changes in the length of a single axon appropriate. However, the principle could easily be extended to longer timescales by using longer (multi-synapse) pathways. Indeed, a closely related account has been used to explain the spatiotemporal correlation that underlies motion detection in the visual system, the so called Reichardt detector (e.g. Van Santen and Sperling 1985).

Where this "delay and compare" principle has been explicitly invoked to account for subjective timing judgements, modelling has tended to follow from literature on visual motion perception. For example, in motion perception literature, signals propagating through the visual system are often modelled by a form of low-pass temporal filtering, which means that an input is delayed and smeared in time as it progresses through the visual brain. It is widely believed that there are two (or more) "channels" in vision with different temporal filtering properties, such that the same input is simultaneously represented by two (or more) patterns of activity, and each is delayed to a different extent (e.g. Fredericksen and Hess 1998). Conceptually, this provides a means by which a stimulus at one location can be delayed and compared to itself at a second location (in order to imply motion).

Returning to perceived time, Burr et al. (2009) proposed that similar low-pass filtering operations could be applied to visual and auditory inputs before integration at a decision centre. They were concerned only with perceived simultaneity vs. successiveness, so only needed to propose one kind of filter, applied to both modalities. In their model, the outputs from each modality's filter were combined according to a non-linear rule (squaring in this case). The size of the combined response deviated from the response that a synchronous pair would evoke with a magnitude that was proportional to the detection thresholds observed experimentally. A rather similar idea was proposed earlier by Stelmach and Herdman (1991), but using a different kind of non-linear rule. To extend a model like this to recover order as well, a modeller would need to add at least one filter, with a different time constant and thus a different degree of delay. This would allow them to begin to build an elementary (opponent) population code, one that combined inputs from each possible pairing of filters (e.g. four eventual outputs with two auditory and two visual filters) and compared the outputs in a second stage. The modeller would also need to think carefully about how these combined responses should be "normalised" to deal with issues such as changes in signal intensity.

Leaving aside the issue of exactly how a delay is imposed, the idea of a population code for perceived crossmodal time has recently been explored by Roach et al. (2011). Their modelling began at the level of the population code itself, which, in order to fit their data, contained 27 neural units with SOA preferences spanning the range from −650 to +650 ms. Drawing on findings from areas such as visual orientation perception, neurones were assumed to have Gaussian tuning curves, meaning that while each unit spiked most vigorously for its preferred SOA, spike rates only fell off gradually as SOAs diverged from this preferred value. The whole population was interpreted collectively (or "decoded") in order to recover an estimate (equivalent to what we have been calling the subjective SOA).

In Roach et al.'s paper, a Bayesian decoding scheme was used (Jazayeri and Movshon 2006) which recovers the most likely SOA given the noisy spiking activity of the population. The result: An output (i.e. a subjective SOA) that can be simulated for any input value, and then interpreted with a decision rule as required, with this process repeated across multiple trials to match experimental data. Hence time has now been represented in a way that is divorced from the times at which neural activity occurs, somewhat as Dennett and Kinsbourne suggest. However, it is not at all obvious how we might reach this stage of representation without brain time ever having mattered, and this model has at least an implicit commitment to brain time. This is because the most obvious process by which the component neural units could become selective for different SOAs is if they respond to signals that arrive with a characteristic offset (similar to the Reichardt detectors outlined previously).

10.5   Experimental Evidence That Brain Time Matters

Differential neural latency has been postulated to account for various perceptual effects in the experimental literature. Examples include the flash-lag effect (Whitney and Murakami 1998) and the perceived asynchrony of simultaneous changes in colour and motion (Moutoussis and Zeki 1997). However, alternative interpretations have also been postulated (e.g. Nijhawan 1994; Eagleman and Sejnowski 2000; Nishida and Johnston 2002). Given the voluminous literature on these effects, we do not intend to assess them in further detail here. Instead, we will briefly mention one experimental effect which appears to demonstrate that neural latencies matter, and for which we are not aware of an equally compelling alternative explanation (although other accounts certainly exist; Aschersleben 2002). The effect arises during synchronisation tapping (e.g. Dunlap 1910), where a metronome provides a timing signal and the participant must tap along.

In this scenario, with an auditory metronome and no auditory or visual feedback about finger contacts, participants tend to tap slightly early, i.e. they introduce a negative asynchrony between their tap and the auditory click. This is consistent with an attempt to align the tactile sensation from the tap with the auditory sensation from the click, assuming that the time for the tactile stimulus to reach (and affect) a decision centre is longer than for a click. More importantly, the asynchrony is of greater magnitude if tapping with the foot than with the hand (consistent with a longer afferent pathway from the foot; Fraisse 1980) and there is a negative asynchrony between foot and hand taps when produced together without any metronome (Paillard 1949). Because manipulations of sensation intensity (e.g. more forceful tapping) also matter, it is likely that accumulation to a threshold at the decision centre must be considered alongside simple nerve latencies (Aschersleben 2002). Indeed, the latency models discussed earlier have often been adapted in a similar manner, by blending them with sensory accumulation models popular for explaining choice reaction times (e.g. Miller and Schwarz 2006; Cardoso-Leite et al. 2007). This does not, however, remove the fundamental dependency of time perception from the timing of neural events in these models.

10.6   Expanding Latency Accounts

Lest we get carried away, it is worth reiterating at this point that latency models certainly do not offer a complete account of perceived event timing. However, it's equally unclear that this means they need to be abandoned entirely. In fact, there are quite reasonable ways in which they might be supplemented and expanded to provide a more comprehensive account of judgements about order and successiveness. This is really what we are promoting here: A view in which brain time is at the heart of many everyday perceptual experiences, but should be considered a start point, rather than an end point, from which more complete models of particular perceptual phenomena can be constructed. In this section we will outline four literatures which suggest possible modifications to latency accounts. This selection is not systematic or exhaustive, and mainly reflects some areas where we happen to have previously done work.

Firstly, consider a report by Yarrow et al. (2006). This paper built on previous work investigating the stopped-clock illusion: The momentary sensation, when glancing at a clock with a silently advancing second hand, that the clock has stopped. The illusion appears to be linked to a slightly extended perceived duration for stimuli that start during a rapid eye movement (a saccade): It is as though the stimulus onset is being judged to occur when the saccade begins, rather than when the new stimulus is actually fixated (Yarrow et al. 2001; see Yarrow et al. 2010 for review). Hence, when the second hand happens to have ticked just as we look towards a clock, it subsequently seems to take longer than a second to move again (because of the added time).

To test this account directly, Yarrow et al. (2006) used a temporal order judgement between a beep and a peri-saccadic visual event. Specifically, a cross (presented at the saccadic target location) changed shape to a square, and this visual event happened while the eyes were in motion. Participants judged whether the visual change had come before or after a beep, which was presented at various times. This same temporal order judgement was also carried out in a control condition, without a saccade. As predicted, the point of subjective simultaneity (between beep and visual event) differed between control and saccadic conditions (and did so even more with a longer saccade) consistent with the mid-saccadic event having been perceived to occur at the start of the saccade.

Clearly, this is not consistent with the output of a latency model as described so far, because the visual stimulus has unexpectedly received a sizeable head start in the saccadic situation. However, the saccadic situation is a special case, and one in which a number of mechanisms appear to be compensating for the loss of visual information engendered by the rapid translation of the eye (e.g. Ross et al. 2001). Rather than abandoning the latency framework, we might reasonably speculate about whether a different brain signal, perhaps one relating to the efferent command to move the eyes, is being swapped with the usual visual signal in this unusual situation (Yarrow et al. 2004). In essence, a special situation might demand a special inferential strategy.

Moving on, a second literature that might seem problematic for a latency model is that on Bayesian multisensory integration (e.g. Ernst and Banks 2002). To take a pertinent example, Ley et al. (2009) used a task in which participants had to make temporal judgements about the order of two stimuli presented to the right and left in rapid succession. Stimuli could be tactile, auditory, or both. As predicted based on a wider literature concerning multisensory integration, performance in the bimodal condition was better than either unimodal condition. In fact, data were consistent with stimuli having been optimally combined, such that the perceived time of each bimodal stimulus was represented by a weighted average of timing estimates from single-modality components. This would seem to require complex operations in which time has been converted to a spatial code prior to the moment at which left and right hand stimuli are compared (c.f. Knill and Pouget 2004), a violation of latency models.

However, there is an equally plausible interpretation of this result, in which two unimodal left/right order judgements are made (for example by separate within-modality latency models for sound and touch) and the two subjective SOAs are only then combined into a final estimate concerning the bimodal stimuli. What we are postulating here is a set of additional operations after latency models have yielded a preliminary output, in order to take account of all available information. This information might be additional and complementary sensory data from another modality, as described above. Another possible source would be an informed expectation about timing based on recent sensory history, a so called Bayesian "prior" (e.g. Miyazaki et al. 2006). Indeed, it seems very likely that a large amount of context is considered and integrated whenever we form opinions about relative time (see also Scheier et al. 1999; Roseboom et al. 2011), but this can generally be considered as an additional stage following on from latency models.

A third literature of interest concerns an effect known as temporal recalibration (Fujisaki et al. 2004; Vroomen et al. 2004). This effect arises when participants adapt to a train of stimuli which collectively imply a particular temporal relationship between two modalities (e.g. repeated audiovisual pairs where beeps lag flashes by 200 ms). Such exposure results in observers modifying their temporal judgements, such that the relationship to which the observer has been repeatedly exposed now seems more synchronous that it did prior to adaptation. Doesn't this shift in perceived relative time imply that, under a latency account, the latency with which one or both signals propagate through the brain has changed? And doesn't that seem pretty unlikely, particularly in the space of a couple of minutes of adaptation, or possibly after even a single audio-visual presentation (see Van der Burg et al. 2013)?

Well, yes, this does seem fairly unlikely (at least to us), but no, latency models don't need to assume that this is what happens. The independent channel and general threshold models derived from signal detection theory provide an alternative account—these observations might be explicable in terms of changes in the decision criteria used to interpret sensory evidence (see Yarrow et al. 2011, 2013 for discussion). This would provide flexibility, even in the case of a hard-wired latency difference. Meanwhile, the population code model of Roach et al. (2011) was actually posited in order to account for exactly this kind of data. It does so on the basis that the population of neurones can be locally adapted (i.e. the responses of neurones with preferences near to a repeatedly experienced adaptor are transiently suppressed). This biases the decoding process, which estimates SOAs on each trial, yielding changed estimates. Hence recalibration studies don't really suggest that neural latencies are irrelevant, only that they are not the only determinants of judgements we make about relative timing.

Before concluding this chapter, it seems worthwhile to very briefly consider one final set of findings which overlap with the studies of explicit timing that we have been considering so far, but which might be considered conceptually distinct. There is now a very large literature investigating how information is integrated from separate sources, and it is clear that the relative timing of two events provides a strong cue about whether they belong together and should be combined. For example, in the McGurk illusion, viewing a speaking face can profoundly bias the perceived sound of certain phonemes, but the biasing effect falls off quite rapidly when visual and auditory information sources are desynchronised (Van Wassenhove et al. 2007). Hence time matters for integration. How is the brain representing time in order to determine if integration occurs?

We would suggest that the underlying process is very similar to the one we have outlined for explicit temporal judgements: Integration probably occurs when the two signals arrive at a critical region within a critical window, i.e. brain time matters. Of course the neural location of integration may be very different from the location at which explicit judgements about time are formulated, but the principle is probably similar. Indeed, we are aware of only one formal model for this class of experimental effects that deals with the temporal relationship between signals (Colonius and Diederich 2004). It does so by using a latency model, with an integration window taking the place of decision criteria, and the simple assumption that if the signals arrive at the relevant brain region within the integration window, integration occurs.

10.7 Conclusions

In this chapter, we have laid out what we believe to be the case about the perception of relative time: That most if not all current models of temporal judgements (at least those which actually warrant the label "model," in the sense of their being able to make quantitative predictions about time perception) assume that brain time matters. Of course, brain time is not all that matters, which means that these models are incomplete. There is clearly much work to be done in evaluating and developing alternative models of time perception, and some if not all brain time based accounts will turn out to be wrong. However, an incomplete model can be built upon. On the other hand, taking the position that brain time plays no role in time perception appears somewhat futile, in light of the wealth of contradictory evidence. A primary focus on non brain time linked influences seems rather like building a house from the roof down; until somebody has firmly established the nature of the mechanisms that provide the foundations for such time perception, ruminations concerning additional influences promise to be inconclusive.

What we are saying is perhaps not terribly controversial. We are not claiming that brain time determines perceived time. However, we are claiming that brain time informs perceived time, and that its influence is likely to arise prior to other modulatory processes that apply in specific cases. In our view, that makes brain time a pretty good place to start when considering models of temporal perception. Actually, it's the only one we can make sense of right now. However, if you are reading this and can conceive (concretely) of a way in which temporal relationships could be encoded in a manner that is completely divorced from the timing of brain activations, we would certainly welcome your suggestions!

References

Aschersleben, Gisa. 2002. Temporal control of movements in sensorimotor synchronization. *Brain and Cognition* 48: 66–79.

Burr, David, Ottavia Silva, Guido M. Cicchini, Martin S. Banks, and Maria C. Morrone. 2009. Temporal mechanisms of multimodal binding. *Proceedings of the Royal Society of London. Series B: Biological Sciences* 276: 1761–1769.

Cardoso-Leite, Pedro, Andrei Gorea, and Pascal Mamassian. 2007. Temporal order judgment and simple reaction times: Evidence for a common processing system. *Journal of Vision* 7: 11.

Colonius, Hans, and Adele Diederich. 2004. Multisensory interaction in saccadic reaction time: A time-window-of-integration model. *Journal of Cognitive Neuroscience* 16: 1000–1009.

Dennett, Daniel C., and Marcel Kinsbourne. 1992. Time and the observer: The where and when of consciousness in the brain. *Behavioral and Brain Sciences* 15: 183–247.

Dunlap, Knight. 1910. Reactions on rhythmic stimuli, with attempt to synchronize. *Psychological Review* 17: 399–416.

Eagleman, David M., and Terrence J. Sejnowski. 2000. Motion integration and postdiction in visual awareness. *Science* 287: 2036–2038.

Ernst, Marc O., and Martin S. Banks. 2002. Humans integrate visual and haptic information in a statistically optimal fashion. *Nature* 415: 429–433.

Fraisse, Paul. 1980. Les synchronisations sensori-motrices aux rythmes [The sensorimotor synchronization of rhythms]. In *Anticipation et comportement*, ed. Jean Requin, 233–257. Paris: Centre National.

Fredericksen, R. Eric, and Robert F. Hess. 1998. Estimating multiple temporal mechanisms in human vision. *Vision Research* 38: 1023–1040.

Fröhlich, Friedrich. 1923. Über die Messung der Empfindungszeit [On the measurement of sensation time]. *Zeitschrift für Sinnesphysiologie* 54: 58–78.

Fujisaki, Waka, Shin Shimojo, Makio Kashino, and Shin'ya Nishida. 2004. Recalibration of audio-visual simultaneity. *Nature Neuroscience* 7: 773–778.

Fujisaki, Waka, Anskar Koene, Derek Arnold, Alan Johnston, and Shin'ya Nishida. 2006. Visual search for a target changing in synchrony with an auditory signal. *Proceedings of the Royal Society of London. Series B: Biological Sciences* 273: 865–874.

Garcia-Perez, Miguel A., and Rocio Alcala-Quintana. 2012. Response errors explain the failure of independent-channels models of perception of temporal order. *Frontiers in Psychology* 3: 94.

Green, David M., and John A. Swets. 1966. *Signal detection theory and psychophysics*. New York: Wiley.

Hubel, David H. 1988. *Eye, brain, and vision*. San Francisco: W.H. Freeman.

Jazayeri, Mehrdad, and J. Anthony Movshon. 2006. Optimal representation of sensory information by neural populations. *Nature Neuroscience* 9: 690–696.

Jeffress, Lloyd A. 1948. A place theory of sound localization. *Journal of Comparative and Physiological Psychology* 41: 35.

Knill, David C., and Alexander Pouget. 2004. The bayesian brain: The role of uncertainty in neural coding and computation. *Trends in Neurosciences* 27: 712–719.

Ley, Ian, Patrick Haggard, and Kielan Yarrow. 2009. Optimal integration of auditory and vibrotactile information for judgments of temporal order. *Journal of Experimental Psychology: Human Perception and Performance* 35: 1005–1019.

Miller, Jeff, and Wolfgang Schwarz. 2006. Dissociations between reaction times and temporal order judgments: A diffusion model approach. *Journal of Experimental Psychology: Human Perception and Performance* 32: 394–412.

Miyazaki, Makoto, Shinya Yamamoto, Sunao Uchida, and Shigeru Kitazawa. 2006. Bayesian calibration of simultaneity in tactile temporal order judgment. *Nature Neuroscience* 9: 875–877.

Moutoussis, Konstantinos, and Semir Zeki. 1997. A direct demonstration of perceptual asynchrony in vision. *Proceedings of the Royal Society of London. Series B: Biological Sciences* 264: 393–399.

Nijhawan, Romi. 1994. Motion extrapolation in catching. *Nature* 370: 256–257.

Nishida, Shin'ya, and Alan Johnston. 2002. Marker correspondence, not processing latency, determines temporal binding of visual attributes. *Current Biology* 12: 359–368.

Paillard, Jacques. 1949. Quelques données psychophysiologiques relatives au déclenchement de la commande motrice [Some psychophysiological data relating to the triggering of motor commands]. *L'Année Psychologique* 48: 28–47.

Phillips, Ian. 2014. The temporal structure of experience. In *Subjective time: The philosophy, psychology, and neuroscience of temporality*, ed. Valtteri Arstila and Dan Lloyd, 139–158. Cambridge: The MIT Press.

Roach, Neil W., James Heron, David Whitaker, and Paul V. McGraw. 2011. Asynchrony adaptation reveals neural population code for audio-visual timing. *Proceedings of the Royal Society of London. Series B: Biological Sciences* 278: 1314–1322.

Roseboom, Warrick, Shin'ya Nishida, Waka Fujisaki, and Derek H. Arnold. 2011. Audio-visual speech timing sensitivity is enhanced in cluttered conditions. *PLoS ONE* 6: e18309.

Ross, John, Maria C. Morrone, Michael E. Goldberg, and David C. Burr. 2001. Changes in visual perception at the time of saccades. *Trends in Neurosciences* 24: 113–121.

Scheier, C.R., Romi Nijhawan, and Shin Shimojo. 1999. Sound alters visual temporal resolution. *Investigative Ophthalmology and Visual Science* 40: S792.

Schneider, Keith A., and Daphne Bavelier. 2003. Components of visual prior entry. *Cognitive Psychology* 47: 333–366.

Stelmach, Lew B., and Christopher M. Herdman. 1991. Directed attention and perception of temporal order. *Journal of Experimental Psychology: Human Perception and Performance* 17: 539–550.

Sternberg, Saul, and Ronald L. Knoll. 1973. The perception of temporal order: Fundamental issues and a general model. In *Attention and performance IV*, ed. Sylvan Kornblum, 629–686. London: Academic.

Ulrich, Rolf. 1987. Threshold models of temporal-order judgments evaluated by a ternary response task. *Perception and Psychophysics* 42: 224–239.

Van der Burg, Erik, David Alais, and John Cass. 2013. Rapid recalibration to audiovisual asynchrony. *The Journal of Neuroscience* 33: 14633–14637.

Van Santen, Jan P.H., and George Sperling. 1985. Elaborated Reichardt detectors. *Journal of the Optical Society of America A* 2: 300–320.

Van Wassenhove, Virginie, Ken W. Grant, and David Poeppel. 2007. Temporal window of integration in auditory-visual speech perception. *Neuropsychologia* 45: 598–607.

Vroomen, Jean, Mirjam Keetels, Beatrice de Gelder, and Paul Bertelson. 2004. Recalibration of temporal order perception by exposure to audio-visual asynchrony. *Brain Research. Cognitive Brain Research* 22: 32–35.

Whitney, David, and Ikuya Murakami. 1998. Latency difference, not spatial extrapolation. *Nature Neuroscience* 1: 656–657.

Yarrow, Kielan, Patrick Haggard, Ron Heal, Peter Brown, and John C. Rothwell. 2001. Illusory perceptions of space and time preserve cross-saccadic perceptual continuity. *Nature* 414: 302–305.

Yarrow, Kielan, Helen Johnson, Patrick Haggard, and John C. Rothwell. 2004. Consistent chronostasis effects across saccade categories imply a subcortical efferent trigger. *Journal of Cognitive Neuroscience* 16: 839–847.

Yarrow, Kielan, Louise Whiteley, Patrick Haggard, and John C. Rothwell. 2006. Biases in the perceived timing of perisaccadic perceptual and motor events. *Perception & Psychophysics* 68: 1217–1226.

Yarrow, Kielan, Patrick Haggard, and John C. Rothwell. 2010. Saccadic chronostasis and the continuity of subjective visual experience across eye movements. In *Space and time in perception and action*, ed. Romi Nijhawan and Beena Khurana, 149–163. Cambridge: Cambridge University Press.

Yarrow, Kielan, Nina Jahn, Szonya Durant, and Derek H. Arnold. 2011. Shifts of criteria or neural timing? The assumptions underlying timing perception studies. *Consciousness and Cognition* 20: 1518–1531.

Yarrow, Kielan, Ingvild Sverdrup-Stueland, Warrick Roseboom, and Derek H. Arnold. 2013. Sensorimotor temporal recalibration within and across limbs. *Journal of Experimental Psychology: Human Perception and Performance* 39: 1678–1689.

Zeki, Semir, and Andreas Bartels. 1998. The asynchrony of consciousness. *Proceedings of the Royal Society of London. Series B: Biological Sciences* 265: 1583–1585.

Part V
Time and Intersubjectivity

Chapter 11
Time in Intersubjectivity: Some Tools for Analysis

Bruno Mölder

Abstract Proper timing and other temporal factors are often viewed as important for rhythmic and synchronised social interaction. The chapter attempts to clarify what roles temporal properties play in interpersonal coordination. The tools for investigating the role of time in intersubjectivity are taken from Craver's account of explanation, which I extend to intersubjective processes. A distinction is made between causal relevance, constitutive relevance, temporal constraints and background conditions. With the help of these tools, various cases of intersubjective coordination are scrutinised: the interaction between mother and infant and Thomas Fuchs' phenomenological accounts of schizophrenia and depression, where the disturbance is supposed to involve both intersubjectivity and temporality. Thanks to these fine-grained distinctions we can give specific and different verdicts about the role of time in each particular case.

As the words imply, social interaction is action which is social. It takes place between people. Any action unfolds in time but the temporal coordination of mutual action is especially significant in the social case. To respond properly to other's actions one cannot act too early nor too late. A successful social interaction requires proper timing. Curiously, the role of time in relation to other people has not received much philosophical discussion, and where it has been discussed, the role has not been delineated precisely enough.

The present chapter attempts to take some steps towards reaching clearer verdicts concerning these issues. The plan for the chapter is the following. After a short introduction to the topic, I sharpen the tools, so to speak, and adapt Carl Craver's notions to the present area of study. Then I apply these tools to basic cases of social cognition where time has been considered important: the online interaction between the mother and the infant and psychopathology. With respect to the latter, I concentrate mainly on Thomas Fuchs' recent phenomenological account of schizophrenia and depression, which in his view involve disturbances in both intersubjectivity and temporality. My main aim is not to defend these accounts but to show that temporal

B. Mölder (✉)
Institute of Philosophy and Semiotics, University of Tartu, Ülikooli 18, 50090 Tartu, Estonia
e-mail: bruno.moelder@ut.ee

© Springer International Publishing Switzerland 2016
B. Mölder et al. (eds.), *Philosophy and Psychology of Time*, Studies
in Brain and Mind 9, DOI 10.1007/978-3-319-22195-3_11

factors can play more than one role when we relate to other people and that it is possible to reach very specific verdicts concerning the role of time in social cognition and to illustrate this with some case studies. Since this chapter brings together different traditions, in the course of discussion I also reflect on the terms used, bringing out some differences and commonalities.

11.1   Introduction: Terminology and Basic Issues

A wide variety of terms is used about the psychological relations we bear to other people. Before we can embark on the study of the relationships between time and other people, a small introduction to this terminology is needed. "Social cognition" is an overarching term that covers all those ways one could relate to other people. Under this term falls understanding others' mental states, their actions as well as what they mean; *social understanding* in short, or "mindreading" or "mentalizing" as it is sometimes called. In addition, there is *social interaction* with others, which may, but does not have to involve the attribution or the grasp of mental states. Social interaction can also be based on a more fundamental relation to others than mindreading. I take it that conceiving of "social cognition" in such a broad way so that it includes both intellectual abilities for social understanding and enactive abilities for social interaction is pretty standard (see, e.g. Spaulding 2012, 431), although there may be people who use those terms in a different way.

A term widely used in these and related contexts is "intersubjectivity." It has seldom been given an explicit definition but my overall impression is that it is often employed in the general sense of involving a relation to the other. In other words, it means one subject relating to another subject. However, it has also been used in more specific senses, some of which will be outlined in the following. The term itself originates from the phenomenological tradition, having been used notably by Edmund Husserl but also by many other phenomenologists. In Husserl's phenomenology, intersubjectivity appears in various contexts (Zahavi 2003, 2005). I am trying to summarise these ideas plainly without employing technical terms. In one sense, intersubjectivity involves the experience of other subjects from the experiencing subject's first-person perspective. In other words, intersubjectivity can be understood as empathy, the experience of other's feelings in their embodied expressions and the realisation that the other is a subject just like oneself. However, Husserl theorised about intersubjectivity also in a more basic sense. Namely, phenomenology is an attempt to explicate and describe the structure of consciousness where these structural conditions are meant to make possible the kinds of experiences that we have. In addition, phenomenology also studies the constitution of objects through consciousness; that is, what makes it the case that the objects appear to us in the way they do. In this regard, Husserl took intersubjectivity to be a condition of the possibility of objectivity and object-hood. He called this "transcendental intersubjectivity." The idea is that it belongs to the very essence of an object that it can also be perceived by others. An object has parts that are hidden to me but I presume that another observer could in principle see those hidden parts. Thus, the possibility of another is already involved in the essence of an object (Zahavi 2003, 119). Transcendental intersubjectivity also secures the objectivity of the world: as objects can be perceived by others, they are not created by my subjective consciousness alone (Zahavi 2003, 115). Each subject lives in a world shared with other subjects. Transcendental intersubjectivity is thus the very basic condition of our experience of the world and as such it also underlies one's empathic relation to others.

The notion of intersubjectivity also features prominently in the work of Colwyn Trevarthen who has been interested in infant's pre-linguistic communication and sensitivities to other people. He distinguishes between *primary* and *secondary* intersubjectivity (Trevarthen and Hubley 1978; Trevarthen 1979). Primary intersubjectivity includes the earliest and most basic kinds of interaction between infants and other people which is in place already at birth, whereas secondary intersubjectivity arises at around the age of 9 months and is exhibited in new repertoires such as sharing attention to objects and grasping others' intentions. More recently, these notions have been taken up by Gallagher and Hutto (2008), who view primary intersubjectivity as involving capacities for social interaction without mindreading and secondary intersubjectivity as the emergence of a practical grasp of common contexts.

Trevarthen (1979, 322) explicates intersubjectivity as the ability to adjust one's own subjectivity to that of others. Elsewhere he (Trevarthen 1999a, 415) provides a somewhat more general definition of intersubjectivity as "the process in which mental activity—including conscious awareness, motives and intentions, cognitions, and emotions—is transferred between minds." He traces his use of the term back to Jürgen Habermas via a chapter by Joanna Ryan (1974), who wrote about pre-linguistic communication (Trevarthen 2008, ix). Habermas was concerned with explaining verbal communication, and in his view this also requires, beside linguistic competence, certain dialogue universals that fix the common objects and pragmatic features of the conversation (Ryan 1974, 187). Trevarthen (1979, 347) preferred "intersubjectivity" to "interpersonal" as the former seems to be more concerned with psychological processes, for "it does specify the linking of subjects who are active in transmitting their understanding to each other." It is somewhat ironic that Habermas' notion of intersubjectivity is taken as a point of departure here for that is strongly linked with language, whereas Trevarthen applied the term to capacities that precede linguistic abilities. Habermas' notion has even been called "linguistic intersubjectivity," stressing the idea that for him, the intersubjectivity between people is based on the language they have in common (Zahavi 2005, 147).[1] On the other hand, there is a commonality between Habermas' and the phenomenological notion of transcendental intersubjectivity too: both take it as something fundamental that underlies other capacities. However, that aspect is not in the foreground in Trevarthen's case, unless one considers more sophisticated cognitive abilities that presume primary and secondary intersubjectivity. I will return to this notion later but at present some clarifications concerning the notion of time are required.

---

[1] Of course, when pre-linguistic processes are analysed in terms of an analogy with language, as a kind of "protolanguage," the irony is not big (see Halliday 1979).

In the case of time, the basic distinction lies between *objective time* and *subjective time*. Objective time is time as it flows (if it does!) independently from us and it is studied by physics and metaphysics. Subjective time concerns all those ways we experience and represent time or temporal properties of events and things around us. Thus understood, it includes both time as it phenomenally appears to us, whether we notice this or not, and temporal relations as they are represented by our perceptual and cognitive systems. The latter need not be phenomenally present but they are nevertheless a matter of subjective depiction, not a direct uptake of objective temporal relations. Is there *intersubjective time* beside objective and subjective time? Gratier and Apter-Danon (2009, 309) think so. They suggest that acknowledging intersubjective time helps to make sense of mother-child interaction and group improvisation when playing music.[2] They argue that it is unique in that it arises only in live interaction between people. They also claim that intersubjective time can be studied in relation to events the occurrence of which can be measured by clocks, but it also has a strong experiential component. In this respect, it has common elements both with objective time and subjective time. However, I tend to think that the objective-subjective distinction is a more basic one as it draws a line between what is there independently from us and what is mind-dependent. As intersubjective time depends on subjects' minds, it falls into the side of subjective time. True, it is a very special kind of subjective time, as it involves more than one subject, but they are subjects nevertheless. Also subjective time, insofar the respective experiences manifest themselves in behaviour, can be measured with respect to events that can be timed by clocks. So in this respect, the difference is not one of a kind.

Another term that one may encounter in such contexts is "temporality." Dan Lloyd (2012) makes an explicit distinction between temporality and timing. The latter is time *qua* that which is the focus of attention in various commonly performed tasks (something which has been studied in cognitive science). Temporality, on the other hand, is more elusive. It makes up a recurrent theme in phenomenological research, especially in Husserl's work where it is conceived of as a necessary condition of all awareness. Temporality is part of the very structure of consciousness. According to Husserl (1991), acts of consciousness have a complex structure. They contain functional parts that bring together the past, the present and the future. Every act involves the *primal impression* of what one is conscious of, the *retention* of that of which one was conscious just a moment ago and the *protention* or the anticipation of what one will be conscious just in a moment. The presence of these parts is intentional (see also Gallagher 1998). They furnish the meaning or sense of conscious contents but they do not compose conscious contents as some have thought sense-data or sensory material compose them. Husserl's account of time consciousness is widely acknowledged; for instance, it also forms the backbone of

---

[2] Creating live music is often given as an example of a synchronised interaction. Not all members of the band are doing the same thing but they adjust their own playing to that of the others. This kind of mutual responsiveness is what is meant by synchrony here. Trevarthen (1999b, 157) takes this analogy a step further and writes about musicality as the innate "psycho-biological need in all humans." See also Malloch and Trevarthen (2009).

Thomas Fuchs' approach to psychopathology, which is analysed in Sects. 11.3.2 and 11.3.3.

However, Lloyd (2012, 695) unpacks temporality as "the continual background awareness of passing time." It seems to me that this understanding of temporality is too narrow. First, although Husserl's model of the constant structure of consciousness is an account of temporal consciousness, it is not directly an account of the "awareness of passing time." It is one thing to say that consciousness flows. It is another thing to say that the passage of time is consciously given, even in the background. Second, temporality is not implied for us only in the background. When it belongs to the structure of consciousness, it should also be part of explicit conscious acts as well. For example, Fuchs (2013) who proceeds from Husserl's notion distinguishes between explicit and implicit temporality, allowing that temporality can also be experienced explicitly. I also think that Lloyd (2012, 697) allows this possibility, for he claims that "we encounter temporality when background expectations of duration are violated." Presumably it is still the same temporality once it comes to be the focus of attention. Thus, I do not think that remaining in the background and one's being aware of the passage of time need be part of the notion of temporality. "Temporality" can also be used in a more general sense, perhaps even as another term for the phenomenal side of subjective time. Barry Dainton (2006, 114) uses the term in that way when he talks about *phenomenal temporality* or "of how time manifests itself within consciousness."

Having outlined some key concepts in the area, I turn now to the basic issues that can be raised with respect to subjective time and intersubjectivity. Taking the terms in a broad sense, subjective time or temporality and other people can be approached from two different directions. We can ask for

(1) the role of subjective time in intersubjectivity

and

(2) the role of intersubjectivity in subjective time.

Arguably, this distinction is contentious. If phenomenologists are right, then temporality and intersubjectivity are closely intertwined and any attempt to separate them only disrupts our experience, which is lived through as both temporal and intersubjective. However, I regard this distinction as useful for achieving some clarity among the mass of different works on these issues. It also allows us to bring out the direction of influence in the relationship between these two broad fields. After all, the question "What role does subjective time play in understanding other people and interaction with them?" differs considerably from the question "What role do other people play in our sense of time?"

In the present chapter I mostly deal with (1); that is, I look at some possible roles subjective time can play in intersubjectivity. In particular, I examine the role of subjective time in the mother-child social interaction and in phenomenological accounts of schizophrenia and depression, where temporality is given a prominent position. The other direction will remain beyond the focus of my discussion. I only point to some ways in which such a direction has been conceived in the literature.

Some phenomenologists claim that the very structure of temporal consciousness is intersubjective (Gallagher 1998; Rodemeyer 2006). However, it seems that such intriguing claims are made possible by construing the notion of intersubjectivity very broadly so that almost anything can become an "other," even oneself or one's consciousness. For example, protention, which is one of the tripartite functions of consciousness in the Husserlian picture can be seen as intersubjective at its core because it reaches out beyond present consciousness, involving, in effect, "the openness to the other" (Rodemeyer 2006, 189). One's own self can be the other too. Gallagher (1998, 113), in his discussion of George Herbert Mead, points out that understanding the future involves having a sense that one's self will become the other with respect to itself. Once "intersubjectivity" is understood in a more narrow way that involves relating to other people, such broad claims may become more difficult to uphold.

In psychology, Sylvie Droit-Volet and her co-workers have investigated various ways in which subjective time can be influenced by emotions. More specifically, they have found that the duration estimations of faces with various emotional expressions depend on the exhibited emotion. For example, duration is overestimated for faces exhibiting anger and fear, but not for faces displaying disgust. This is explained by the fact that anger indicates possible aggression and the expression of fear shows that there is some potential danger. Both lead the organism to prepare for a possible attack but seeing that the other person is disgusted does not prepare one for action (Droit-Volet and Gil 2009). As concerns social interaction, Droit-Volet and Gil (2009) point out that we tend to mirror the rhythms of other people's action. Such mirroring involves changes in bodily processes, which in turn influence one's duration estimations. For example, mirroring older people makes one slower overall and it makes one to take durations to be shorter than they actually are (see also Droit-Volet et al. 2013). However, such effects depend on various factors and the issues here are no doubt complicated. This is just one example of how intersubjectivity could colour our subjective time, not the whole picture. In what follows, I will examine influence from the other direction; that is, what is the role that subjective time can play in intersubjective processes. But before I can move on to this, more precise tools are needed to distinguish between various roles. The next section is devoted to the task of honing such tools.

11.2 Tools for Analysis

There is no established inventory of roles that various temporal factors can play in intersubjectivity, so we need to borrow it from another field. There is a very detailed account of explanatory and other relations in neuroscience by Carl Craver (2007), which, I think can be appropriated for present purposes. In what follows, I will first outline Craver's account and then discuss how it could be employed to clarify the relation between time and intersubjectivity.

Craver (2007) aims to develop an overarching account of explanation in the tradition of mechanistic explanation. The general idea of that tradition is that an explanation is given by providing a description of the mechanism. Craver (2007, 139) conceives of the phenomenon that is to be explained as quite broad: it is "some behavior of a mechanism as a whole." What explains this is then a comprehensive description of the mechanism. The mechanism is described in action; that is, it is unveiled how the components of a mechanism interact in order to produce the phenomenon to be explained. Accordingly, the components of a mechanism, their activities as well as the organisation of the former two elements are all crucial to the mechanistic explanation. Central to this account is the elaborate notion of constitutive relevance, which is needed in order to distinguish explanatorily relevant components from other parts of the mechanism. Craver distinguishes between constitutive and causal relevance. They are related but different relations—in order to understand the former we need to know what the latter is.

Craver's account of causal relevance is inspired by James Woodward's (2003) theory (see Craver 2007, 93–104). The latter is a manipulationist approach to causation, which explicates it through possible manipulations. The relata of causal relations are variables, things that can have different values.[3] On this view, one variable is causally relevant to another variable if a manipulation of its value leads to a change in the value of that other variable. A successful manipulation, which alters a variable's value, is called an "intervention." Also non-human interventions are allowed in this picture; thus the account can accommodate purely natural causal relations. If "X" and "Y" stand for variables, the account can, in a nutshell, be presented as follows:

X is *causally relevant* to Y if the direct intervention on X alters Y.

Such an intervention is subject to several conditions that are intended to ensure that nothing else changes Y: (1) Y is not directly intervened upon; (2) no causal intermediaries between X and Y are manipulated directly; (3) intervention is not correlated with some other cause of Y; (4) intervention switches off the influence of other potential causes of X (Craver 2007, 96, based on Woodward and Hitchcock 2003).

Craver also employs the manipulationist approach to account for the constitutive relevance. By "constitutive," he means something related to constituents or components; the constitutive relation is thus the relation between the components and the whole system. His point is that constitutive relevance requires the part-whole relation and the mutual manipulability of the components and the whole. Using M for the whole mechanism and ψ for its activity and C for a component with an activity φ, the idea of constitutive relevance can be expressed in the following way (Craver 2007, 154):

---

[3] This is not such a big break from more traditional approaches to causation as it may seem. Craver (2007, 95) notes that more usual causal relata such as events or processes can be easily replaced by variables as those can be understood in terms of their potential values (for example, an event can occur or fail to occur).

C's φ-ing is *constitutively relevant* for M's ψ-ing in case:
(1) C is part of M
(2) C's φ-ing and M's ψ-ing are mutually manipulable (an intervention on C's φ-ing changes M's ψ-ing and an intervention on M's ψ-ing changes C's φ-ing).

The requirements on intervention are similar to those outlined in the case of causal relevance. Note the importance of mutual manipulability. If the activities of C and M are not mutually manipulable, then the relation cannot be constitutive relevance. This reflects the point that constitutive relevance defines what it is to be a component of a mechanism. True components stand in a bidirectional relationship with the whole system. Craver (2007, 153) stresses that this distinguishes constitutive relevance from causal relevance as the latter goes in one direction only. He also brings out additional differences between causal and constitutive relevance. Firstly, the relata of causation do not stand in a part-whole relationship like the relata of the constitutive relationship. Secondly, causes occur before their effects, whereas constitution always takes place at the same time.

Another notion that is important in the present context is that of *temporal constraint*. As already noted, Craver stresses that mechanisms are not simply aggregates of their parts, but are organised along three dimensions: in terms of their activities, spatial properties and time. The temporal dimension pertains to the rate, duration, order, frequency and other time-related features of a mechanism's activity (Craver 2007, 138). Those features, when they are prescribed by the real world, constrain the range of potential mechanisms, so that it becomes possible to find out which of them can be actual. Temporal constraints are not separate parts of the mechanism. They belong to the full description of the organisation of a mechanism's components.

Finally, there are also *background conditions* for the working of the mechanism. To use a simple example from Craver (2007, 143): a normally functioning heart is a background condition for reading. Although damage to the heart influences one's reading activity, it cannot be said that the heart or the heart's functioning is a component of the mechanism that is responsible for the reading ability. It is thus important to distinguish constitutive components from mere background conditions. Craver (2007, 146–152) has described several experimental strategies used in neuroscience to ascertain what are the proper components; these include experiments employing interference, stimulation and activation. He claims that these can also be used to distinguish components from background conditions. Basically, interference experiments involve interfering with components to see if there are any effects on the whole system. That is, interference with C's φ-ing is expected to change M's ψ-ing. Stimulation experiments involve stimulating some component to make a change in the whole mechanism. That is, an intervention that stimulated C to φ is expected to lead to M's ψ-ing or stop M's ψ-ing if C is an inhibiting component of M. Activation experiments involve activating the mechanism in order to see if there will be any changes in its components. In other words, an intervention on M's ψ-ing is expected to change C's φ-ing. To find out if we are dealing with a background condition, Craver (2007, 157) suggests a strategy that *combines interference and stimulation* experiments: although blocking the background condition might inter-fere with M's ψ-ing, but stimulating the background condition has no effect on M's ψ-ing. This can be seen with the example of reading: blocking the heart functioning inhibits reading, but heart stimulation does not elicit reading. Thus, if a condition behaves in such a way, one can assume that it is a background condition, not a con-stitutive component. Another way to distinguish background conditions is by using an *activation* experiment: a condition is a background condition if activating M's ψ-ing has no effect on it. In addition, background effects are usually *nonspecific* and *nonsubtle* (Craver 2007, 158). That is, changing the background condition has non-specific (or wide) effects throughout the whole system. However, subtle changes to the background condition tend to have effects that are far from subtle (such as the effect of the tampering of the heart on reading.)

De Jaegher et al. (2010) have distinguished the roles that *social interaction* could play in social cognition. Since time could play similar roles, it would be useful to compare them with those I have extracted from Craver's account. They make a distinction between a *contextual factor*, an *enabling condition* and a *constitutive element*. Some of them more or less coincide with Craver's notions, but they are much less specific than Craver's categories. For instance, a constitutive element is an element that "is part of the processes that produce" the whole phenomenon. The basic idea is conveyed by Craver's constitutively relevant relations, but the latter account includes precise proposals about how to detect it. An enabling condition is a condition the absence of which keeps the phenomenon from happening, so it is essentially a necessary condition. As such, it can be viewed as one kind of back-ground condition. For example, the absence of a heartbeat prevents reading. But presumably not all background conditions need be necessary ones. Finally, their definition of a contextual factor is somewhat difficult to understand. As stated, it leaves the impression that it is a sufficient condition: "F is a contextual factor if variations in F produce variations in F." However, it cannot be solely sufficient for the phenomenon and informally they claim that it is "simply something that has an effect on X" (all quotations from De Jaegher et al. 2010, 443). As such it is a bit crude, for all above-discussed relations satisfy this condition. Perhaps what is really meant is something that *only* has an effect and nothing more. Then the factor can be viewed as a broadly causal one, but as one that has not been specified further or distinguished from more direct causal conditions.

In sum, based on Craver, we can distinguish between the following relations:

(1) *causal relevance*
(2) *constitutive components*
(3) *temporal constraints*
(4) *background conditions*, which may include enabling conditions

But do they also apply to time and intersubjectivity? Craver's account is tailored for multilevel explanations in neuroscience, where the material part-whole relationship is important. In addition, the mechanisms under discussion are robust biological ones that differ from more abstract descriptions of mechanisms that could be found among intersubjective processes. Indeed, the main example he is using is that of the action potential mechanism. However, I presume that this model could have a wider application. For when employing this definition in the case of time and intersubjectivity, there are various options. (A) One could allow for talk about mechanisms in an extended sense so that intersubjective occurrences could also be seen as instantiating an abstract mechanism at work. In addition, one could also stipulate part-whole relationships in such more abstract cases. After all, we do speak of the part-whole relation in the case of abstract entities. For example, we can say that a protagonist's actions are part of the narrative or that Friday and Thursday are part of the week. (B) We can deny that we can talk properly about mechanisms in the case of intersubjectivity, for they are not robust enough, and their descriptions do not usually reach the biological or neural level. However, we could still view accounts of intersubjective processes that involve temporal factors as mechanism sketches with several gaps that still need to be filled out. Such sketches are couched in functional terms, which leave their concrete realizations open (cf. Craver 2007, 113; Piccinini and Craver 2011). In the case of sketches of mechanisms, we can still distinguish between causal and constitutive components and background conditions, although the results of the discussion would depend upon further research, which turns sketches into full descriptions of actual mechanisms. (C) We can deny that we are dealing with proper mechanisms as they are found in neuroscience, but hold that the basic tenets of the manipulationist approach are still applicable. All we need for this to be the case is events or activities that can be seen as having different values, and that can be responsive to manipulation. Given some analogue notion of parthood, the outlined tools can still be employed.

As all those options allow employing Craver's tools for the topic of time in intersubjectivity, we do not really need to choose between them at present. In what follows, I assume that the above-mentioned relations can also be applied outside neuroscientific explanations and I will analyse some cases in terms of these relations. I do not claim that this inventory exhausts all possible ways in which time can be related to intersubjectivity, but this looks like a good list to start with. The following case studies can, in turn, also be viewed as an examination of these tools. We will see in the course of these studies how well, if at all, the tools honed in this section apply outside neural and biological explanations.

11.3 Case Studies

In this section, I turn to three cases of intersubjectivity, the explanations of which pay particular attention to time. The time under discussion is mostly subjective time as well as temporality, the phenomenal side of subjective time. I look at the mother-infant interaction and a phenomenological account of psychopathologies such as schizophrenia and depression. I examine these fields only from the viewpoint of the relation of time and intersubjectivity, asking what kind of role does time play in these intersubjective processes or the preferred explanation of those processes.

11.3.1 Mother-Infant Interaction

The interaction between mother and infant has been intensively studied. By now, it has been established that already from birth an infant relates to people differently compared to inanimate objects. An infant has some intersubjective capacities already from birth and these are the basis on which more elaborate abilities will be built upon during the development (for more details, see Trevarthen 1984, 1986; Chap. 12 of this volume). In the 1970s, when research on communication between an infant and its mother was gaining ground, it soon became apparent that timing is an important factor in such communication. Successful communication requires a mutual coordination of action such as the proper timing of gestures and other expressions, and in general, responding to other's behaviour.

One of the early studies of such interactions is Brazelton et al. (1974). They observed patterns in looking and attending at each other and looking away and not attending between mothers and infants over a period of 4 months when infants were 2–20 weeks old. They found that the rhythm of mutual looking and not looking was crucial. Attention to each other takes place in cycles in which both are responding to each other's cues. Attention alternates rhythmically with nonattention. Thus it is not the case that the mother and the infant are always doing the same thing. Rather what is important is that they are responsive to one another. Brazelton et al. (1974) conclude that when the mother's rhythm is adjusted to the infant's rhythm, the interaction was deemed as positive but negative where there was a mismatch of rhythms.[4]

A well-known example of what happens when the rhythm gets broken is an experiment by Murray and Trevarthen (1985). They created the double video live-replay paradigm, which advances upon the previous still-face paradigm in which mothers break the interaction by keeping a still face for some period. In their experiment, the mother and an infant (between 6 and 12 weeks) interact through a television system. They were filmed and the result was shown in real time through the TV monitor. Murray and Trevarthen showed the infant a video of the mother's previously recorded action, which disrupted the coordination between them. The delayed presentation of the mother's action leads to the infant's distress: "the feeling conveyed by the infant's behavior … is one of detached, unhappy puzzlement or confusion" (Murray and Trevarthen 1985, 191).

What is the role of time in this example? Although the delay under discussion can be measured by a clock and thus is in some way related to objective time, the time playing a role in interaction can still be regarded as subjective time. What matters is how the infant and the mother represent each other's responses. So even if interaction is a process that unfolds in objective time, what affects the participants of the interaction is the subjective side of time and temporal features. In other

---

[4] There is a question of whether reciprocity is real or illusory. Some have suggested that the illusion of reciprocal interaction comes from the fact that the mother adjusts her behaviour to infant's spontaneous movements leaving the impression that the infant responds to mother (Papoušek and Papoušek 1995, 127). However, I examine the case under the assumption that it is as it seems. For a corroboration of Murray and Trevarthen's results, see Nadel et al. (1999).

words, what one can manipulate in such experiments is the clock time, which is objective in the sense that it can be objectively measured, but which is culturally tainted, since clocks are not natural kinds. Moreover, the effects of manipulations of clock time are made possible by the fact that these periods of time are "taken in" or "lived through" by an organism whose reactions to temporal features depend on the organism's own structure and organisation. Accordingly, the temporal features become part of subjective time as they modify the actions of the organism.

Now we can apply the tools outlined in the previous section to this example, asking for the role of time in rhythmically synchronised interaction. When we view the mother-infant interaction as a system of its own, then it is subject to temporal constraints. There must be an optimal range of responses for the interaction partners which keeps up the synchrony. A longer delay from one or the other may lead to de-synchronisation and thus to a decline in the quality of interaction.[5] Such constraints need to be taken into account, when manipulating the temporal properties of interaction. Thus, a temporal constraint (as measured in clock time) is the most likely candidate indeed, but let us discuss briefly also other options.

The representation of time is not manipulated directly in this experiment, but temporal properties are intervened upon by manipulating other features such as the presentation of the video. Thus time cannot be easily viewed as a causally relevant factor. In any case, time cannot be the sole cause since successful interaction presumes many other factors; the interacting partners, for example. Constitutive relevance requires a part-whole relation, which does not really apply to time in the mother-infant interaction case even when parthood is understood very liberally. In addition, constitutive relevance involves mutual manipulability. What we learn from this experiment is that manipulating temporal properties of interaction changes the quality of the interaction. It is not clear if manipulating the quality of interaction would change its temporal properties as this was not done in the experiment. This would be the case, for example, when the interaction is modified so that if it is felt as negative then this influences the rhythm of interaction. *Prima facie*, this is likely, but the current case does not allow us to decide on this. As for the background condition, then the interference and stimulation strategy does not apply here: interference with temporal features of the interaction influences the quality of interaction or may stop it altogether, but it is difficult to imagine what would constitute stimulating temporal properties of the interaction. Activation criterion fails, for activating interaction likely has effect on its temporal features, if for no other than for the reason that before the interaction there was no temporal synchrony between the mother and the infant. Depending on the nature of the intervention, the tampering of temporal features may have quite subtle and specific effects on the interaction. A small delay in responding does not terminate the whole process of interaction, but it may influence its perceived quality. Therefore, the background condition also cannot be regarded as a suitable role.

---

[5] Fuchs (2013, 81) suggests that this range is 200–800 ms. He refers to Papoušek and Papoušek (1995), although I did not manage to find such data in that particular chapter. Notwithstanding this, a range in such a vicinity seems plausible.

Since other candidate relations do not really apply here as we saw, the verdict for this case that was concerned with the role of time in mother-infant interaction is that it plays the role of a *temporal constraint*, which determines whether the interaction is synchronous and hence has the felt positive quality. Temporal constraints have to be mentioned to give a constitutive account of the whole mechanism, but they are not distinct parts themselves.

11.3.2 The Phenomenological Account of Schizophrenia

The next two cases are taken from a phenomenological approach to psychopathology, which attempts to clarify and explain psychopathologies using the vocabulary derived from the phenomenological tradition. Although several phenomenologists have analysed psychopathologies and have often related those to disturbances in time consciousness, I chose Thomas Fuchs's accounts for my case study because he lays emphasis on the intersubjective nature of such disorders with an impaired sense of time. He even uses the phrase "intersubjective temporality" to stress the point that understanding subjective time requires reference to intersubjectivity (Fuchs 2013, 76). As such, his accounts of schizophrenia and depression are good paradigms for analysing the role of subjective time in intersubjectivity.[6]

Before we can move on to Fuchs' account of schizophrenia, his more general phenomenological background assumptions need a brief introduction. He draws a distinction between two basic ways in which subjective time comes to play a role in our lives: "implicit" and "explicit" temporality (Fuchs 2005; 2013, 77–81). *Implicit temporality* is a kind of awareness of time in which time is lived through pre-reflectively when one is immersed in some action. In this mode, one is not aware of any temporal features on its own, but it is a mode that makes more explicit forms of temporal experience possible. According to Fuchs, implicit temporality has two preconditions in structural features of consciousness. First, it requires the intentional structure and operation of internal time consciousness. This is understood in the already mentioned Husserlian terms as involving integration or—as it is called—the "transcendental synthesis" of protention, primal impression and retention in an act of consciousness. The integration takes place automatically, and it secures the diachronic continuity and unity of consciousness that phenomenologists, including Fuchs, regard as closely connected to pre-reflective self-consciousness, which is the most fundamental and primitive sense of self. The idea is that whenever one is aware of some new event, this is placed into the context of the things one was aware of just previously, which includes the awareness of one's previous awareness. Second, it requires conation, by which Fuchs means the affective and motivational

---

[6] Since the clinical aspects are not my main focus, I just proceed from Fuchs' notions of schizophrenia and depression, without attempting to compare them to standard classifications such as DSM or criticizing his approach. For a recent critique of Fuchs, see Ratcliffe (2012) who argues that depression is more varied than Fuchs conceives it.

force—an implicit readiness to act. This is the source of the sense of being alive and the sense of agency. By contrast, *explicit temporality* is becoming explicitly aware of time. Of course, we do not become aware of objective time as such. Instead, we become aware of "changes in the temporalization of our existence which results from its relation to the rhythms and processes in which our life is embedded from the very beginning" (Fuchs 2013, 83). Fuchs claims that this happens mostly when the implicit flow of time gets disrupted. It is usually accompanied by unpleasant feelings or even shock in case of the present time. Those feelings could also be directed towards the future (anticipation, impatience) or the past (regret, longing).

As concerns the intersubjective side of subjective time, Fuchs (2005, 196) relates these two modes of time to synchrony. Synchrony and the lack of it can occur in both individual and social life. An individual can fall out of synchrony with its environment (when lacking food, for instance) or with other people (by being too quick or slow). In line with his idea that time becomes explicit in the case of some disturbance, Fuchs connects explicit time with the lack of synchrony and implicit temporality with synchrony that has not been broken: "While implicit temporality is characterized by synchronization with others, explicit temporality arises in states of desynchronization (acceleration or retardation)" (Fuchs 2005, 196).

Fuchs (2001, 2007, 2013) argues that schizophrenia and depression are both due to disturbances in implicit temporality. Put briefly, in the case of schizophrenia the disturbance lies in time consciousness, whereas for depression the conative aspect is central. I will discuss depression in the next section. Here let us have a closer look at Fuchs' view on schizophrenia. He conceives of schizophrenia as a failure in the synthesis of time consciousness. In particular, it involves a malfunction of protention: one fails to anticipate events properly, and those seem to occur suddenly and unexpectedly. This in turn leads to troubles with retention and self-consciousness. Experiences are retained with small delays; incompatible thoughts and associations are not suppressed. As the intentional integration of time consciousness fails, also self-consciousness becomes fragmented and the sense of agency becomes corrupt (Fuchs 2013, 85–88). On the intersubjective plane, persons with schizophrenia lack "intercorporal affective resonance" with other people: they fall out of synchrony with others as their expressions are puzzling for the patients and, in turn, the patients expressive behaviour becomes strongly reduced or eccentric (Fuchs 2013, 92). The disturbance in intersubjectivity is compensated with explicit rituals and rules or as in the case of a schizophrenic delusion, the behaviour of others is reinterpreted in an inflexible delusional framework (Fuchs 2013, 93). Time is central for this account: since the implicit temporality is impaired, the normally implicit synchrony with others is impeded, and the patients try to compensate this with explicit attention to their movements as well as to the timing of events around them.

Fuchs summarises his phenomenological account of schizophrenia in the following way:

> the fundamental disorder … of schizophrenia consists in a *weakening and temporal fragmentation of basic self-experience*. It appears in pre-morbid or chronic phases as a lacking sense of self-coherence which undermines the habitual conduct of life and needs to be compensated for through rational reconstruction at the explicit time level. In acute phases,

it manifests itself in an increasing fragmentation of the intentional arc, and of the self-coherence linked with this on the micro-level of time consciousness, resulting in the appearance of major self-disturbances (such as thought withdrawal or insertion, hallucinations and delusions of influence). In all phases, this disturbance of self-constitution is accompanied by profound desynchronisations of intersubjective temporality which culminate in delusion as a "frozen reality," detached from the ongoing intersubjective constitution of a shared world. (Fuchs 2013, 94, italics in original)

As we can see, the basic impairment is located in implicit temporality, which ruins the coherence of basic self-consciousness. This is "accompanied by" the loss of intersubjective synchrony leading to the impairment of interpersonal interaction in general. I'll base my analysis on this description. As the topic of the present study is the role of time in social cognition, I am looking at Fuchs account of schizophrenia insofar as it is conceived as a condition in which temporality and intersubjectivity come together in intricate ways. In particular, the question I am interested in is what role does temporality play in intersubjective coordination, which is impaired in schizophrenia. I am not analysing the role of temporality in the schizophrenia as such in Fuchs's account. The mechanism under consideration is one that is responsible for intersubjective synchrony, provided that we can view this as a mechanism.

It may be said that the relation is not one of constitutive relevance. Implicit temporality is indeed described as part of the mechanism of intersubjective coordination. However, it cannot be a constitutive relation because the mutual manipulability criterion is not satisfied. Affecting the synthesis of time consciousness would affect intersubjective coordination, but there is no evidence (at least in the theoretical account given) that manipulating intersubjective coordination would affect the synthesis of time consciousness. Once the synchrony is lost, interaction stumbles and explicit temporality may come to the fore. But I do not see how it would disrupt the basic protention, primal impression and retention structure and functioning of consciousness. If the direction of influence from intersubjectivity to temporality were possible, it could be used in therapy but that does not seem to be the case.

Another candidate is causal relevance. Fuchs talks somewhat vaguely about "accompanying," but it should be safe to assume that an intervention on temporality would also intervene on intersubjective coordination. Nevertheless, the relation is causal only if the intervention satisfies certain above-mentioned requirements. For instance, it depends on whether one can manipulate the synthesis of time consciousness directly without intervening on other factors that may impinge on intersubjective synchrony, e.g. self-coherence (conditions 2 and 4). It is far from obvious that this is possible.

What about temporality as a background condition? This looks like the most plausible candidate. First of all, manipulating implicit temporality would have nonspecific and nonsubtle effects. It would impact much more than just intersubjective synchrony. The activation criterion applies too. Recall that in order to qualify as a background condition, activating the mechanism would have to have no effect on it. Indeed, in this case activating intersubjective processes, that is, creating a synchrony between people does not have an effect on the synthesis of time consciousness. The interference plus stimulation strategy might be applicable, although some doubts can be harboured about this. As required, interfering with temporality interferes with interpersonal synchrony among other things, and stimulating temporality, if that can be done, does not elicit intersubjective coordination. However, it is not altogether clear what would stimulating the synthesis of inner time consciousness involve, for presumably this kind of passive synthesis takes place all the time. But given that at least two criteria are in place, and they were non-exhaustive suggestions anyway, we can come to the verdict that in Fuchs conception of schizophrenia, temporality plays the role of a *background condition* in intersubjective coordination.

11.3.3 The Phenomenological Account of Depression

Finally, let us have a look at Fuchs' account of depression. According to Fuchs (2013, 95, italics omitted), depression is "the result of an intersubjective desynchronisation, and … a disturbance of conation." Desynchronisation initiates the condition, whereas the failing conation is its basic element. Fuchs describes the triggering factor of depression as one's failing to be up-to-date. One gets stuck with the events that happened in the past; that is, one falls out of synchrony with one's current environment, including social environment. The next step is desynchronisation on the biological level, characterised by "loss of conation" in the organism. Fuchs (2013, 98) points out that the synthesis of time consciousness is not impaired in depression, what is lost is the drive or "the energetic or dynamic quality of the flow which allows us to "hold pace" with the sequence of events, or causes us to lag behind." On the one hand, the past becomes overwhelming for the patient, on the other hand, the future does not appear as a plane of possibilities, as something that could bring about changes to the present situation. This in turn aggravates the intersubjective impairment (falling even more out of synch with other people). In Fuchs' terms,

> melancholic depression is mostly triggered by a desynchronisation of the individual from his environment, which then develops into a physiological desynchronisation. As an inhibition of vitality, it then proceeds to include the conative basis of experience and thus also the basic self-affection. The resulting retardation of lived time reinforces decoupling from the social environment. (Fuchs 2013, 100)

This yields the following sequence: losing synchrony with one's environment leads to the disturbance of conation, which results in additional intersubjective desynchronisation. What could be analysed here regarding the relationship between temporality and intersubjectivity? Fuchs characterises depression as a condition where the intersubjective synchrony is disturbed from the start. When the condition progresses, the impairment gets only worse. So there is no clear case where time would play a separable role in the progression of the loss of intersubjective resonance. As already mentioned, Fuchs indeed takes depression to be "the result of an intersubjective desynchronisation." In this notion as in the notion of intersubjective temporality, both aspects—intersubjectivity and temporality—are principal components. This would be an example, where temporality plays the role of a constitutive component. More specifically, temporal properties as implicated in synchronisation and desynchronisation are *constitutively relevant* for the process of intersubjective synchrony. They are part of it, and there is mutual manipulability. A change in temporal properties of synchrony between people obviously changes intersubjective interaction, regardless of whether that interaction involves synchrony or not. Appropriate change in the interaction or communication between people changes the properties of synchrony. Synchrony may be lost and may require resynchronisation or when there is no interaction, one cannot talk about synchrony at all.

Another relation that can be scrutinised here is the relation of conation to the mechanism that upholds proper intersubjective functioning. Recall that for Fuchs (2013, 78), conation is a "prerequisite for implicit temporality" besides the basic structure of time consciousness. He describes it also as "the conative-affective dynamics of implicit temporality" (Fuchs 2013, 96, italics omitted). When this is impaired, the organism slows down and loses its motivating energy, so to say. Fuchs (2013, 96) characterises the situation in terms of "psychomotor inhibition, thought inhibition, and … a slow-down or standstill of lived time", but also as "an increasing rigidity of the lived body." He uses etiological terminology; words like "increase" and "reinforce":

> With the fundamental loss of conation, the depressive psychopathology further increases the social desynchronisation. Vain attempts to keep up with events and obligations reinforce the feeling of remanence. To this is added the loss of intercorporal resonance: Whereas conversations are normally accompanied by the synchronisation of bodily gestures and gazes, the depressive's expression remains frozen and his emotional attunement with others fails. (Fuchs 2013, 96–97, footnote omitted)

This makes it reasonable to look for a *causal* relation. A properly functioning conative drive is causally relevant for intersubjectivity provided that a direct intervention on conation affects intersubjective processes. This can indeed be observed in this model. When conation is lost, intersubjective desynchronisation follows. Admittedly, the description provided is not sharp enough to allow a judgement to be made as to whether all conditions for direct intervention are met. In a sense, this is understandable, for psychiatric conditions are usually quite messy and involve the interplay of various factors. Thus, the relation between conation and social cognition need not come out as clearly as one could expect in fields that deal with more concrete entities, which can be controlled more easily. However, it is not nonspecific and general enough to pass for a background condition. Even if it is not the only cause, it may well be one of the main causal factors. Unlike constitutive relevance, the relation of causal relevance is unidirectional. Thus, an intervention on intersubjective processes (such as restoring the synchrony, for instance) should not bring along a change in the conative drive. I presume that this is the case here, at least when this intervention is singular, not consistently recurring. The therapeutic treatments of depression that Fuchs (2001, 184–185) proposes strive for re-establishing the environmental and intersubjective synchrony, but it does not seem that the causal direction goes from intersubjectivity to conation. On the contrary, intersubjective functioning is repaired through an attempt to restore the conative aspects.

11.4   Conclusion

This chapter constitutes an attempt to bring some more clarity to discussions of time in intersubjectivity. To this end, I borrowed a toolbox from neuroscientific explanation and I applied it to accounts of intersubjective processes. Drawing from Carl Craver's conception, I proceeded from the distinction between causal relevance, constitutive relevance, temporal constraints and background conditions. I looked closely at some cases in which time has been deemed important, mainly with respect to the social synchrony between people and I tried to reach specific verdicts with the specified tools. Besides helping to bring some clarity to the topic of time in intersubjectivity, I hope that this chapter also demonstrates the broader viability of constructs originating from the examination of practices in neuroscience. Of course, cases in which I experienced difficulties when applying certain tools were duly pointed out.

I came to the following verdicts. For the quality of the interaction as characterised by the synchronised rhythm between mother and infant, the role of time was that of the temporal constraint. As conceived in Fuchs' account of schizophrenia, the synthesis of time consciousness stands to intersubjective synchrony as a background condition. In Fuchs' understanding of intersubjective synchrony, temporality is a constitutively relevant component of it. Finally, in Fuchs's account of melancholic depression, conation as one side of implicit temporality is causally relevant for intersubjective synchrony. We have thus an example for every outlined role. This illustrates an obvious, but perhaps not always noticed, point that time fulfils different roles with respect to various manifestations of intersubjectivity.

Acknowledgments Work on this chapter was supported by the COST Action TD0904 (Time in Mental Activity), institutional research funding IUT20-5 of the Estonian Ministry of Education and Research and a grant ETF9117 from the Estonian Science Foundation. I am grateful to Colwyn Trevarthen for reading suggestions and to Alexander Davies for checking my English.

References

Brazelton, T. Berry, Barbara Koslowski, and Mary Main. 1974. The origins of reciprocity: The early mother-infant interaction. In *The effect of the infant on its caregiver*, ed. M. Lewis and L.A. Rosenblum, 49–76. London: Wiley.

Craver, Carl. 2007. *Explaining the brain: Mechanisms and the mosaic unity of neuroscience.* Oxford: Clarendon.

Dainton, Barry. 2006. *Stream of consciousness: Unity and continuity in conscious experience*, 2nd ed. London: Routledge.

De Jaegher, Hanne, Ezequiel Di Paolo, and Shaun Gallagher. 2010. Can social interaction constitute social cognition? *Trends in Cognitive Sciences* 14(10): 441–447.

Droit-Volet, Sylvie, and Sandrine Gil. 2009. The time-emotion paradox. *Philosophical Transactions of the Royal Society* 364: 1943–1953.

Droit-Volet, Sylvie, Sophie Fayolle, Mathilde Lamotte, and Sandrine Gil. 2013. Time, emotion and the embodiment of timing. *Timing & Time Perception* 1: 99–126.

Fuchs, Thomas. 2001. Melancholia as a desynchronization: Towards a psychopathology of interpersonal time. *Psychopathology* 34: 179–186.

Fuchs, Thomas. 2005. Implicit and explicit temporality. *Philosophy, Psychiatry & Psychology* 8: 323–326.

Fuchs, Thomas. 2007. The temporal structure of intentionality and its disturbance in schizophrenia. *Psychopathology* 40: 229–235.

Fuchs, Thomas. 2013. Temporality and psychopathology. *Phenomenology and Cognitive Sciences* 12: 75–104.

Gallagher, Shaun. 1998. *The inordinance of time*. Evanston: Northwestern University Press.

Gallagher, Shaun, and Daniel D. Hutto. 2008. Understanding others through primary interaction and narrative practice. In *The shared mind: Perspectives on intersubjectivity*, ed. J. Zlatev, T. Racine, C. Sinha, and E. Itkonen, 17–38. Amsterdam: John Benjamins Publishing Company.

Gratier, Maya, and Gisèle Apter-Danon. 2009. The improvised musicality of belonging: Repetition and variation in mother-infant vocal interaction. In *Communicative musicality: Exploring the basis of human companionship*, ed. Stephen Malloch and Colwyn Trevarthen, 301–327. Oxford: Oxford University Press.

Halliday, Michael A.K. 1979. One child's protolanguage. In *Before speech: The beginning of human communication*, ed. Margaret Bullowa, 171–190. Cambridge: Cambridge University Press.

Husserl, Edmund. 1991. *On the phenomenology of the consciousness of internal time (1893–1917)*. Trans. J. B. Brough. Dordrecht: Kluwer Academic Publishers.

Lloyd, Dan. 2012. Neural correlates of temporality: Default mode variability and temporal awareness. *Consciousness and Cognition* 21(2): 695–703.

Malloch, Stephen, and Colwyn Trevarthen (eds.). 2009. *Communicative musicality: Exploring the basis of human companionship*. Oxford: Oxford University Press.

Murray, Lynne, and Colwyn Trevarthen. 1985. Emotional regulation of interactions between two-month-olds and their mothers. In *Social perception in infants*, ed. T.M. Field and N.A. Fox, 177–197. Norwood: Ablex.

Nadel, Jacqueline, Isabelle Carchon, Claude Kervella, Daniel Marcelli, and Denis Réserbat-Plantey. 1999. Expectancies for social contingency in 2-month-olds. *Developmental Science* 2(2): 164–173.

Papoušek, Hanuš, and Mechthild Papoušek. 1995. Intuitive parenting. In *Handbook of parenting: Biology and ecology of parenting*, vol. 2, ed. Marc H. Bornstein, 117–136. Mahwah: Lawrence Erlbaum.

Piccinini, Gualtiero, and Carl Craver. 2011. Integrating psychology and neuroscience: Functional analyses as mechanism sketches. *Synthese* 183: 283–311.

Ratcliffe, Matthew. 2012. Varieties of temporal experience in depression. *Journal of Medicine and Philosophy* 37: 114–138.

Rodemeyer, Lanei M. 2006. *Intersubjective temporality: It's about time*. Dordrecht: Springer.

Ryan, Joanna. 1974. Early language development: Towards a communicational analysis. In *The integration of a child into a social world*, ed. M.P.M. Richards, 185–213. London: Cambridge University Press.

Spaulding, Shannon. 2012. Introduction to debates on embodied social cognition. *Phenomenology and Cognitive Sciences* 11: 431–448.

Trevarthen, Colwyn. 1979. Communication and cooperation in early infancy: A description of primary intersubjectivity. In *Before speech: The beginning of human communication*, ed. Margaret Bullowa, 321–347. Cambridge: Cambridge University Press.

Trevarthen, Colwyn. 1984. How control of movements develops. In *Human motor actions: Bernstein reassessed*, ed. H.T.A. Whiting, 223–261. Amsterdam: Elsevier.

Trevarthen, Colwyn. 1986. Development of intersubjective motor control in infants. In *Motor development in children: Aspects of coordination and control*, ed. M.G. Wade and H.T.A. Whiting, 209–261. Dordrecht: Martinus Nijhof.

Trevarthen, Colwyn. 1999a. Intersubjectivity. In *The MIT encyclopedia of cognitive sciences*, ed. R.A. Wilson and F.C. Keil, 415–419. Cambridge: The MIT Press.

Trevarthen, Colwyn. 1999b. Musicality and the intrinsic motive pulse: Evidence from human psychobiology and infant communication. *Musicae Scientiae* 3(1): 157–213.

Trevarthen, Colwyn. 2008. Shared minds and the science of fiction: Why theories will differ. In *The shared mind: Perspectives on intersubjectivity*, ed. J. Zlatev, T. Racine, C. Sinha, and E. Itkonen, vii–xii. Amsterdam: John Benjamins Publishing Company.

Trevarthen, Colwyn, and Penelope Hubley. 1978. Secondary intersubjectivity: Confidence, confiding and acts of meaning in the first year. In *Action, gesture and symbol: The emergence of language*, ed. Andrew Lock, 183–229. London: Academic.

Woodward, James. 2003. *Making things happen*. New York: Oxford University Press.

Woodward, James, and Christopher Hitchcock. 2003. Explanatory generalizations, part I: A counterfactual account. *Noûs* 37(1): 1–24.

Zahavi, Dan. 2003. *Husserl's phenomenology*. Stanford: Stanford University Press.

Zahavi, Dan. 2005. *Subjectivity and selfhood: Investigating the first-person perspective*. Cambridge: The MIT Press.

Chapter 12
From the Intrinsic Motive Pulse of Infant Actions to the Life Time of Cultural Meanings

Colwyn Trevarthen

Abstract Research on the timing of spontaneous actions made from birth—for self-sensing of the infant's own body posture and movements, for perceptual apprehension of objects or events in the outside world, and for intimate communication with a parent—proves that innate measures of time regulate prospective control of body actions of a coherent human Self. Infants are born capable of synchronising with, imitating and complementing expressions of a parent's feelings and interests in proto-conversational exchanges. Affective appraisals of imagined consequences for the well-being of body and mind are signalled by expressive movements of emotion that are sympathetically shared. Within a few weeks infants join with rhythms of play in narratives of expressive display, learning how to participate in conventional rituals that tell stories of a "proto-habitus." Before the end of the first year the child shares conventional use of objects and develops proto-linguistic symbolic coding of actions that identify objects to be shared in cooperation with familiar companions.

Infants' movements have universal temporal parameters of "musicality," as cultivated in the "imitative arts" of theatre, dance, song and music. The same measures of "pulse," "quality" and "narrative" are also evident in serial ordering of steps in completion of practical projects and in propositional representations. This primary sensori-motor intelligence and its intersubjective "vitality dynamics" must be considered as fundamental in any philosophy of the nature of our experience of time and its cultivated uses and measures. Understanding of inner "life time" benefits from a neuro-psychology of movements and their prospective perceptual control, a science developed over the past century.

C. Trevarthen (✉)
School of Philosophy, Psychology and Language Sciences, The University of Edinburgh, Dugald Stewart Building, 3 Charles Street, Edinburgh EH32 9AD, Scotland, UK
e-mail: C.Trevarthen@ed.ac.uk

© Springer International Publishing Switzerland 2016
B. Mölder et al. (eds.), *Philosophy and Psychology of Time*, Studies in Brain and Mind 9, DOI 10.1007/978-3-319-22195-3_12

12.1 Introduction: Life Time Awareness and the Invention of Human Common Sense

On the one side there is a given environment with organisms adapting themselves to it. The other side of the evolutionary machinery, the neglected side, is expressed by the word creativeness. The organisms can create their own environment. For this purpose the single organism is almost helpless. The adequate forces require societies of cooperating organisms. (Whitehead 1926, 138)

It is by natural signs chiefly that we give force and energy to language; and the less language has of them, it is the less expressive and persuasive. ... Artificial signs signify, but they do not express; they speak to the understanding, as algebraical characters may do, but the passions, the affections, and the will, hear them not: these continue dormant and inactive, till we speak to them in the language of nature, to which they are all attention and obedience. (Reid 1764/1997, IV, II Of Natural Language)

This paper is concerned with the descriptive natural science of spontaneous, self-generated, human movements, their "subjective" timing and their "inter-subjective" coordination. It also addresses how the associated feelings of joy in harmony of companionship, or distress with conflict or isolation are expressed, and imitated, in the dynamics of movement. Attention is focused on the organization and regulation of movements in infancy, where innate principles of self-regulation for action, and how they are animated to learn, are clearly evident.

I do not assume that perceptions of measures of time originate in experience of external "objective" events, or that stimuli from the environment cause movements. I will, therefore, not review experimental research on the abilities of infants to discriminate the time course of artificially presented visible or audible stimuli, computational theories of sensory-motor control, or models of learned cognitive processes and mental representations hypothesized to direct skilled actions of manipulation, speech, writing or mathematics. These I conceive to be assimilated as cultural "tools," by adaptation of the temporal patterns of primary motor-affective intelligence to meet the demands of life with other human beings in a cooperative, technical and political world with its changing history of explanations.

Research with infants indicates that an Intrinsic Motive Pulse of "life time" is the essential property of human action-with-awareness; a foundation for aesthetic appreciation of the intentions and feelings of effort as our bodies try to move well, and also for appreciation of the sympathetic expressions of moral feeling in communication with others (Papoušek 1996a, b; Stern 1974, 2000, 2004, 2010; Trehub 1990; Trevarthen 1984, 1999, 2001, 2005b, 2008, 2009a, 2012a, b, 2015). Transmission of meaning for our cultural intelligence, for what Adam Smith called the "imitative arts" (Smith 1777), and for education of our children, depends on "mimesis" in company with the rhythms and emotions of human expression, sharing the significance or value we give to the histories discovered by our will to live (Donald 2001; Frank and Trevarthen 2012; Thelin 2014; Trevarthen 2005a, b, 2009a, 2012b, c, 2015).

We learn to use and celebrate the "inner time" of imaginative movement within a human body, which has many parts and which is informed by many senses, to know the world and to share its pleasures. We also invent tools to anticipate and record environmental events, using observation and manipulation to represent structure and change in a story of physics, and to create and perpetuate this meaning with science. Geniuses like Plato, Galileo, Newton, Bacon, Lavoisier, Descartes and Kant, Becquerel, Pierre and Marie Curie, built rational conceptions of an "outer time" measured with conventional units of inertia or energy—to describe inorganic objects falling, burning, rolling on earth, or rotating in the cosmos; the inertia of masses and of chemical atoms interacting and radiating energy. Their science conceives potent forms persisting and changing, or machines working—out of body, out of mind, out of the life world, and out of the living personal inner time with which each thinker imagines (Abram 1996; Thelin 2014). Einstein was aware of the problem. In a letter to the mathematician Jacques Hadamard about his process of mathematical invention, he declared that experiences of bodily movement came first, admitting that, "the words of language, as they are written or spoken, do not seem to play any role in my mechanism of thought" (Hadamard 1945, Appendix). Hadamard related Einstein's insight to the psychology of William James (1890).

The scientist, occupied with a particular problem, inclines to detach their focused and systematically applied consciousness of an external reality, and the communication of what is found, from the sympathetic conviviality of their social life, with its innate aesthetic and moral impulses and feelings that anticipate moving cooperatively with grace and responsibility in the natural world (Trevarthen 2005b, 2009a). In the end, however, mathematics, physics, and chemistry must distinguish properties and changes of objects by intuitive comparison to properties of human action.

Contrived cultural knowledge begins with the eager curiosity children show for taking experience in company. The references of language to which a child shows interest are made and remembered with sense of time and energy in body movement that is active at birth (Lenneberg 1967; Trevarthen and Delafield-Butt 2013a). They depend on the "autopoiesis," or self-making, of individual human lives, and "consensuality," the mutual influence of life processes between them in different degrees of intimacy (Maturana and Varela 1980; Maturana et al. 1995).

12.1.1 New Methods and New Knowledge

Microanalysis of purposeful actions of the human body as it masters new ways of moving in complex sequences to gain awareness of objects and for communication reveals an Intrinsic Motive Pulse, generated in the brain before birth and active through infancy (Beebe et al. 1985; Stern 1974; Stern et al. 1982; Trevarthen 1974, 1999, 2001, 2015). This time sense shows up in imitative "mirroring" of the forms of expressive movements in "felt immediacy" (Bråten 1988, 1998), exhibiting "attunement" to the affective qualities in expressive movements (Stern et al. 1985). A "negotiated" or "improvised" common time results from deliberate intersynchrony between matching subjective time sense of separate individuals in expressive movements that become complementary and mutually supportive. Failure to participate in shared time results in breakdown of mutual confidence and loss of meaningful or goal-directed social action and awareness. Its maintenance and protection is essential for the development of language and all other symbolic communication.

Brain science identifies deep and ancient systems in animals for controlling inner life processes and development of the brain, extending sensory-motor and cognitive consciousness of body form and its field of agency with acquired experiences of outer "objective" events which are picked up in measured intervals of internally generated time. The same human time sense motivates cultural practices, from the celebratory rituals and cosmic beliefs of hunter-gatherer societies to the lexico-grammatical and semantic functions of language, and the technical and scientific systems that support productivity and social systems and structures in large modern industrial cultures. All human action-with-awareness and all products of intersubjective story-telling and cooperation, including both educational and psychothera-peutic practices, depend on the broad range of inner life times that regulate actions, awareness, imagination and memory, and what we can learn to do throughout development. The rhythmic foundations of this motor intelligence are now shown to be established in fetal stages, with the first movements.

12.1.2 Taking Account of Motor Time for Intelligence and Sociability

Research on life times needs methods that enable accurate, unselective recording to describe natural or "free" initiatives of organisms. Experimental selection to obtain responses in particular "controlled" physical conditions of imposed stimulation, as is appropriate for testing functions of machines or the properties of non-living substances, is not the way to gain description of the pace and patterning of what organisms are adapted to do by their natural action in their habitual environment. Fields of biology that take life time into account include evolutionary biology, ethology, especially of the social behaviours of different species, field ecology and adaptation of growth and behaviour to diurnal and seasonal events, anthropology of human arts and rituals, functional social linguistics, and study of the aesthetics of music, dance and drama. Play among young animals has marked rhythmicity, which regulates both social affections and learned collaborations. Methods of teaching and education require sensitivity to the learner's pace of practice and response (Erickson 2009).

I will focus on evidence from the natural history of an Intrinsic Motive Pulse in human communicative behaviour in infancy, relate these to the brain science of motor activities and their emotional regulation, and review the preparatory developments in the body and brain of embryo and fetus. These motivate the transition to communication that enables the child's mastery of cultural tools and rituals, including language (Trevarthen 2001, 2015).

I show evidence that the infant is both a seductive companion and a prescient talker, with ambition to share songs, conversations, games and stories, moving with what Dan Stern (2010) has called "vitality dynamics," and creating serially ordered expressions of interest and emotion which lay the foundation for development of both linguistic syntax and rational intelligence (Lashley 1951). The baby advances in the first year in powers of "proto-language," expressing both affection for companions and interest in experiences of the world, including how objects might be recognized for use, richly supported by the intuitive parenting of a shared awareness for mental states transmitted in body movements (Halliday 1979; Papoušek 1996a, b; Trevarthen 1979, 1986, 1990, 2004a, 2009a, 2012b, c, Trevarthen and Delafield-Butt 2013a). All this is achieved before vocabulary and grammar are consolidated in formal symbolic forms and conventional syntax.

These extraordinary ways that the human animal develops "languaging" as a "consensual" way of life (Maturana et al. 1995) can be related to the intricate preparations of body and brain for such ingenious and sociable intelligence that are formed "autopoietically" before birth (Maturana and Varela 1980), as well as to the particular communicative behaviours that other intelligent social species employ to animate cooperate in adaptation to life's opportunities (von Uexküll 1957; Sebeok 1994; Wallin et al. 2000).

Life forms "use" the environment *creatively*, as Whitehead says, and *sociably* as Darwin observed in the famous last paragraph of *On the Origin of Species*, in which he celebrates the shared vitality of a "tangled bank" of plants with birds, insects and worms finding their place for life (Darwin 1859). Animals move to conserve their vital energy, guiding body parts to meet prescribed affordances of the environment by prospective control of the consequences of moving (Sherrington 1906; Gibson 1979; Lee 2009). Every action is a purposeful and integrated transformation of the whole body—a Self-related activity, ready to make instantaneous use of the senses, prospectively.

Humans have hyper-motility of exceedingly complex bodies, which requires more elaborate innate principles of sensori-motor control with feeling (Donald 2001). We are highly sociable, and become part of a consensual world of life in movement. From infancy, we have unique powers of moving to express our imaginative thoughts and to share experience with emotion, and for cooperative education of our collective intelligence (Trevarthen et al. 2014).

12.1.3 *"Communicative Musicality": Tempo and Phrasing in Human Vitality and Its Sharing*

Studies attending to how the infants respond to the musical sounds in their mother's voice revealed the importance of an innate sense of rhythm and melody in human body movement (Trehub 1990; Papoušek 1996a, b). Scientific theories of mental growth, especially of the ability of the human brain to learn and use language, must attend to this special way to signal thinking and feeling (Trevarthen 1999, 2015). Stephen Malloch applied his expert knowledge of musical physics to my recordings of mother-infant proto-conversations to define the primary parameters of intuitive time-keeping and expressive modulations of "musicality" for "communicating the vitality and interests of life" (Malloch and Trevarthen 2009a). He identified *pulse, quality* and *narrative*, defined as follows, all specifying regulation of actions and awareness in subjective and intersubjective life time:

1. *Pulse* is the regular succession of discrete behavioural steps through time, representing the "future-creating" process by which a subject may anticipate what might happen and when.
2. *Quality* consists of the contours of expressive vocal and body gesture, shaping time in movement. These contours can consist of psychoacoustic attributes of vocalizations—timbre, pitch, volume—or attributes of direction and intensity of the moving perceived in any modality.
3. *Narratives* of individual experience and of companionship are built from the sequence of units of pulse and quality found in the jointly created gestures—how they are strung together in affecting chains of expression. These "musical" narratives allow adult and infant, and adult and adult, to share a sense of sympathy and situated meaning in a shared sense of passing time (Malloch and Trevarthen 2009b, 4; Malloch 1999).

In *Rhythms of the Brain*, Buzsaki (2006) notes the strong responses to music that are recorded in many areas of the brain and he makes this evaluation of what it is the mind is seeking to experience.

> Perhaps what makes music fundamentally different from (white) noise for the observer is that music has temporal patterns that are tuned to the brain's ability to detect them because it is another brain that generates these patterns. (Buzsaki 2006, 123)

Two brain scientists end a review on events that may be recorded throughout the brain when a person is appreciating music as follows:

> tomographic analysis of MEG responses to real music demonstrates that very large areas of the brain are activated when we listen to music. These activations differ in the left and right hemispheres; the left hemisphere is more engaged when regular rhythms are encountered. The activity in different brain areas reflects musical structure over different timescales; auditory and motor areas closely follow the low-level, high-frequency musical structure. In contrast, frontal areas contain a slower response, presumably playing a more integrative role. All of these results show that listening to music simultaneously engages distant brain areas in a cooperative way across time. This might be one reason why music has such a profound impact on humans. (Turner and Ioannides 2009, 171)

12.2 Development of Communication in Infants

12.2.1 An Example: Rhythms of Human Biology Born for Conversation

To illustrate innate human life time and its sharing, I present an analysis the musician and acoustics expert Stephen Malloch (1999) has made of a dialogue of simple sounds made by a 2-month premature baby and her father who was holding her against his body, "kangarooing" in an intensive postnatal care unit in Amsterdam

(Fig. 12.1). Malloch's measurements reveal how they each command the muscles of their chest, vocal organs and mouth to emit short cries. The father with a deeper, more resonant voice, closely imitates the pitch, duration and intonation of the baby's rudimentary utterances, with matching gentleness of feeling.

They move to one negotiated inter-subjective pulse, playing their parts with flexible synchrony; like two experienced musicians improvising "in the groove" (Gratier and Trevarthen 2008; Kühl 2007; Lee and Schögler 2009; Schögler and Trevarthen 2007; Trevarthen 1986, 1999, 2012a). This inner time sense, and the sensitivity for cooperative contingency of response "in time," make possible the invention of shared meaning from birth. They are abilities of infants that will reach out to understand the world and motivate the development of acquired skills, arts and shared tasks (Dissanayake 2000), including, after many months, those of language to talk with interested others about life and the meaning of its actions and objects (Halliday 1978; Bruner 1990).

The tempo of human moving (Fraisse 1982) that are revealed in this primitive example are also found to give measure to cultivated performances of musical sound, to words and sentences in speech, to intelligence visible in the turns of head and impetuous eyes, in gestures and steps of dance or theatrical performance, in strokes and shapes of drawing, and in workmanlike manipulations of tools to build things. The intended messages of talk are sensed proprioceptively in the bodies of the one who is moving to vocalize, and may be received by another with senses of sight and touch, as well as hearing. They are elements of what Susanne Langer (1942, 1953) has called "forms of feeling," of the musical semantics of Ole Kühl (2007), and of the "vitality dynamics" of Daniel Stern (2010). They characterize the

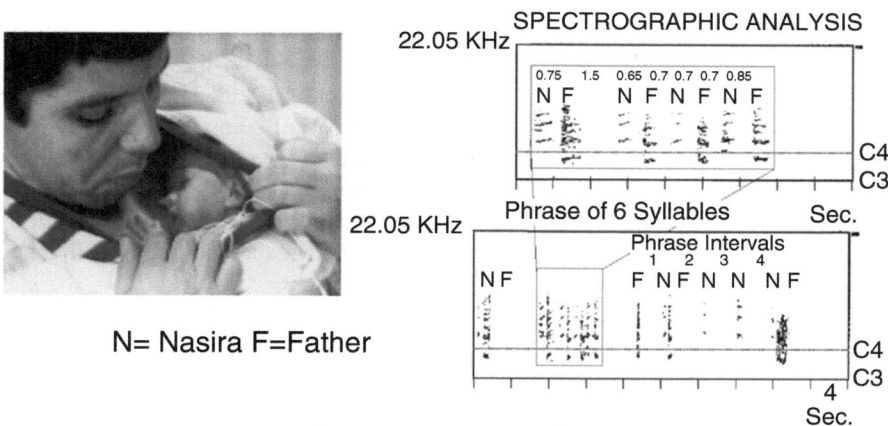

Fig. 12.1 Naseera (N), born 3 months premature, now 2 months premature, exchanges short "coo" sounds with her father (F) who is holding her close to his body, "kangarooing". From a film by Saskia van Rees (van Rees and de Leeuw 1993). Spectrographs produced by Stephen Malloch (1999). They share, with matching precision, the tempo and rhythm of syllables (0.3 s in duration, and separated by 0.7 s) grouped in a phrase (of 4 s). Then they make a sequence of single sounds separated by phrase-length intervals (From Trevarthen and Delafield-Butt 2013a, 172, figure 8.1. By permission of Guildford Press)

way human beings signal intentions and build imagination for meaning by "mimesis" (Donald 2001).

The brain-guided animation of our body parts acting as sensory-motor, proprioceptive organs, is regulated by coordination of innate rhythms created in muscles and among neurons (Sherrington 1906; Buzsaki 2006). Slow walking, *largo*, is 1 per second; comfortable walking, 1 in 700 ms, is *andante*; hurried walking, 1 in 300 ms, is *presto*. The same values animate and give sense to messages of conversation, music or dance. Those particular measures are recognised by musicians and made into visible "scores," useful as symbols to describe measures of experience-in-action with different intensities of purpose in their experiences of sound. But the music is created in the embodied mind, not in the score.

12.2.2 The Progress of Infant Motor Intelligence, from Birth to Language

Forty years ago, using photography and movie film, and with Sony's new video technology, discoveries were published that disproved the theory that infants had no imaginative minds, no sense of self, and therefore no sense of others. Most astonishing, and dismissed with derision by convinced rational constructivists, was the finding that infants habitually move with an exquisite sense of time, and that they use the rhythms of their expressions skilfully, as in the above example, to engage in inter-synchronous imitation with attentive responses from an adult, blending affections (Ammaniti and Gallese 2014; Trevarthen 1974, 1979, 1984, 1986, 2001, 2009a, 2012a, c). Indeed, the paediatrician T. Berry Brazelton, found that the indulgent mother or father could be taught by the infant how to communicate feelings and interests by engaging sympathetically and in time with the gestures, looks, smiles and cooing vocalizations of a newborn (Brazelton 1979). The same principle of sensitivity for the child's intuitive mastery of the art of being "in sympathy" can be used to chart steps in normal development, and to give help when the child is confused and anxious (Brazelton and Sparrow 2006).

In the 1970s the following books with telling titles, mostly edited collections with contributions from many authors, gave detailed evidence of infant intelligence from the new research:

– 1974: *The Growth of Competence* (Bruner and Connolly); *The Integration of a Child into a Social World* (Richards); *The Effect of the Infant on Its Caregiver* (Lewis and Rosenblum).
– 1975: *Child Alive: New Insights Into the Development of Young Children* (Lewin); *Parent-Infant Interaction* (Hofer).
– 1976: *Language and Context: The Acquisition of Pragmatics* (Bates).
– 1977: *Studies in Mother-Infant Interaction* (Schaffer); *Interaction, Conversation and the Development of Language* (Sander); *The First Relationship: Infant and Mother* (Stern).
– 1978: *Action, Gesture and Symbol: The Emergence of Language* (Locke); *The Development of Communication* (Waterson and Snow).

– 1979: *Before Speech: The Beginning of Human Communication* (Bullowa); *The Emergence of Symbols: Cognition and Communication in Infancy* (Bates); *Human Ethology* (von Cranach et al.); *The Origins of Social Responsiveness* (Thomin); *Social Interaction During Infancy: New Directions for Child Development* (Uzgiris); *The Ecology of Human Development: Experiments by Nature and Design* (Bronfenbrenner).

The new infant psychology converged with the insights of the phenomenological philosophy of Europe, which attracted little interest in Anglo-Saxon thinking at that time. Also supported was the idea that the capacity human language has evolutionary origins in the instinctive semiosis, or social signaling, of intelligent social animals (Lenneberg 1967; Maturana et al. 1995; Trevarthen 2012c, 2015). Advance in knowledge of language origins and functions invited a new appreciation of prosody, poetics and metaphor (Bateson 1979; Bullowa 1979a, b; Lenneberg 1967; Stern 1974; Trevarthen 1974, 1979). As Margaret Bullowa explained in the introduction to *Before Speech*, sharing rhythm in subjective time is the key:

> *The Communicative State*: For an infant to enter into the sharing of meaning he has to be in communication, which may be another way of saying sharing rhythm. A great deal of work on interaction with infants during their first half year considers shared attention. This is probably the key to rhythm-sharing underlying also fully elaborated inter-adult communication, even though it is often overlooked in our preoccupation with details of the codes for transmission of messages. (Bullowa 1979a, 15)

Age-related events summarized in Table 12.1, and explored further below, show how the abilities of the newborn become part of an on-going project that educates human collective intelligence by a time-regulated growth process in the young child (Trevarthen 1992; Frank and Trevarthen 2012).

12.2.3 Mimesis in the Neonate

Emese Nagy (2011) has summarized evidence that the first 4 weeks after birth, when medical science recognizes that the human organism must undergo a profound adjustment of physiology and vital functions as both body and brain are become transformed to meet a new environment, should also be recognized as a distinct stage of psychological or mental development. In agreement with Brazelton (1979), she describes "an intentional, intersubjective neonate", citing behavioural and neuroscientific evidence that, "the neonate's early social preferences and responses indicate a unique, sensitive, experience-expectant stage of development."

Considerable mental capacities are now recognized from the first moments, in spite of infant's lack of knowledge of the world. Especially prominent are abilities for sharing intentions, experiences and feelings, abilities that are open to engagement with any other human being, male or female, who is ready to experience the rhythms of intentional time with the infant. They are additional to adaptations of the immature infant for attachment with the mother to care for vital functions with the special resources of her body (Feldman 2012).

Table 12.1 Age-related stages in development of actions-with-awareness, and of communication, in infants and young toddlers

Age	Development
Months 1–2	Primary intersubjectivity: direct sensitivity to the expressions of feeling in intimate contact with an Other. "Dialogic closure" in proto-conversation sustained by two-way transmission of emotions. Identification of familiar affectionate partners (Trevarthen 1979)
Months 3–6	Games I: exploration of surroundings and manipulation of objects. Pleasure in body-action and in object manipulation is shared, and imitated, in play, including musical-poetic play. Laughter, mirror self-awareness and "showing off" as a "social Me" appear (Hubley and Trevarthen 1979; Trevarthen 2005b; Reddy 2008)
Months 6–9	Games II: lively socio-dramatic play and self-confident presentation with family increase, as does fear of strangers. The first ritualized "protosigns" are learned in play. First "emotional referencing" and joint orientation to a locus of interest aided by pointing (Halliday 1978; Hubley and Trevarthen 1979)
Months 9–14	Secondary intersubjectivity: shared interest in tasks and the uses of objects; infant produces "protolanguage." Learning of the conventional meanings of things. Use of objects that others have given value "recreatively," in fantasy play (Halliday 1978; Hubley and Trevarthen 1979; Trevarthen and Hubley 1978)

(From Trevarthen 1992, 125. By permission of Oxford University Press)

Newborn infants are able to exchange displays of interest with felling with adults who use the same times of expression, and they can synchronise gestures of their hands with the syllables and phrases of adult speech (Condon and Sander 1974). Their coherent movements are adjusted with self-perception for comfort and with signals of need for parental care. The organs for this self-awareness and its signalling develop in utero over many months (Trevarthen 1985, 2001; Trevarthen and Delafield-Butt 2013a). They are sensitive to the movements of other persons that are contingent or responsive to what the infant does.

Newborns also imitate forms of expression to animate an exchange, not in an immediate reflex manner and not by chance similarity between their spontaneous actions when "aroused" by the signalling behaviour of the adult. They pay attention and repeat the signal "deliberately," taking time to "reflect" on what is offered as a marked or exaggerated and sustained expression of eyes, face, voice or hands made by the other person (Kugiumutzakis and Trevarthen 2015; Trevarthen 2012b, c) (Fig. 12.2). Importantly, as Nagy discovered, they may repeat the imitated act to "provoke" a responses from the person who had just been imitated (Nagy 2006; Nagy and Molnár 2004). In other words, they engage in imitation reciprocally, to actively "negotiate" the experience of communication, as toddlers do in play (Nadel and Pezé 1993; Nadel 2014).

Right after birth an infant may be ready to respond, not only to the pulse of single movements, as in Fig. 12.1, but they can also respond with imitation and in synchrony to subtle sequential modulations of another person's head movements, eye movements, hand gestures and vocalizations. Figure 12.3, extracted from an engagement when the infant was beginning a "narrative cycle" lasting 30 s (Delafield-Butt and Trevarthen 2015), illustrates matching of the pattern of a

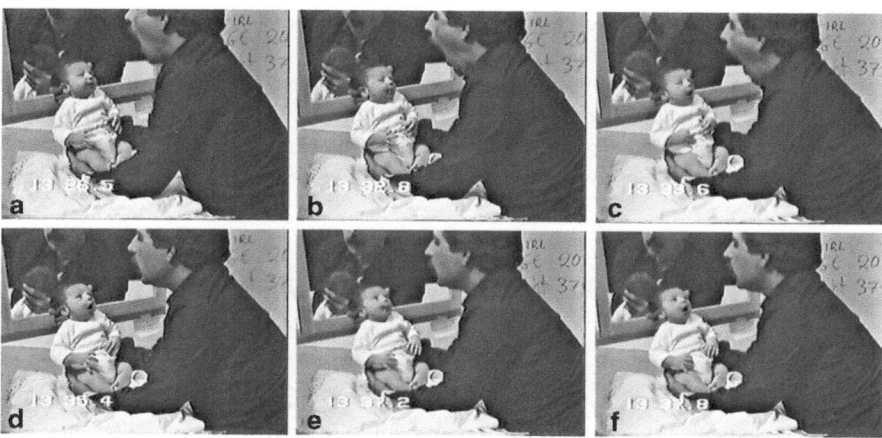

Fig. 12.2 A slow cycle of imitations of Mouth Opening with a female infant 20 min after birth. Recorded a maternity hospital in Herakleion, Crete in 1983 by Giannis Kugiumutzakis for his PhD research at the University of Uppsala. (**a**) (0 s). The researcher presents a wide open mouth for the first time to the attentive infant, focusing on his mouth, and with slightly closed eyes and pursed mouth. (**b**) (6.3 s). The researcher opens his mouth for the fourth time. The neonate continues to observe his mouth with evident interest. The right hand moves up. (**c**) (7.1 s). The researcher opens his mouth for the fifth time. The neonate imitates him once, synchronously while watching his mouth. The right hand closes. (**d**) (8.9 s). The infant imitates a second time, looking up at the researcher's eyes as he waits. (**e**) (10.7 s). Both pause, waiting. The infant is still looking at his eyes. (**f**) (11.3 s). The infant makes a third large imitation while looking at the researcher's mouth (Kugiumutzakis and Trevarthen 2015. By permission of Elsevier)

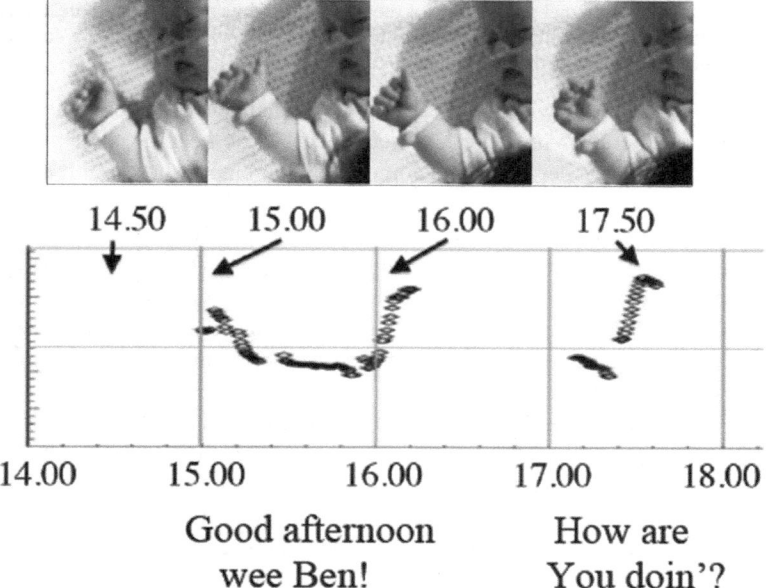

Fig. 12.3 A 1 week premature infant moves his hand in synchrony with intonations of the mother's voice while she encourages him to join in a shared "narrative"

newborn infant's hand openings with the intonations of a mother's voice in a "phrase" enacted over a period of 4 s.

As the mother makes inviting utterances in this "introduction" to the story, the baby's hands open and close in precisely synchronized imitation of her expressions. The infant appears to be "grasping" the mother's greetings with the right hand, with delicate "feeling" for her inviting sounds. Both move with poetic or melodic grace.

12.2.4   "Telling" a "Proto-Conversation" with Gesture and Voice

A few weeks after birth, with maturation of self-regulation and marked improvement in visual awareness, more extensive communicative engagements become common (Trevarthen 2001). These were discovered by Mary Catherine Bateson, a linguist and anthropologist. She described a detailed analysis of filmed and taped interactions between a mother and her 9-week-old infant as follows:

> A study of these sequences established that the mother and infant were collaborating in a pattern of more or less alternating, non-overlapping vocalization, the mother speaking brief sentences and the infant responding with coos and murmurs, together producing a brief joint performance similar to conversation, which I called "proto conversation." The study of timing and sequencing showed that certainly the mother and probably the infant, in addition to conforming in general to a regular pattern, were acting to sustain it or to restore it when it faltered, waiting for the expected vocalization from the other and then after a pause resuming vocalization, as if to elicit a response that had not been forthcoming. These interactions were characterized by a sort of delighted, ritualized courtesy and more or less sustained attention and mutual gaze. Many of the vocalizations were of types not described in the acoustic literature on infancy, since they were very brief and faint, and yet were crucial parts of the jointly sustained performances. (Bateson 1979, 65)

Infants may become very active at this stage of their development, moving their whole body deliberately in expressive sequences while keeping their gaze on the mother's face, seeking her eyes, and sometimes glancing to her mouth (Figs. 12.4 and 12.5).

Figure 12.5 demonstrates the systematic rhythms of a 10-week-old bay girl who is very "earnestly" vocalizing in phrases to her mother who responds with gentle nodding movements and soft sounds of "agreement." The interchange seems like a "lecture" the baby is giving her mother, with intent gaze and a "serious" expression as she utters three phrases made up of five or six short sounds, separated by pauses when the mother responds.

In 1979 I made a recording of a proto-conversation between a 6-week-old girl and her mother. This was subsequently given a comprehensive analysis by Stephen Malloch who reported his findings 20 years later (Malloch 1999). The pitch plot he made of a 27 s narrative sequence, with a transcript of the mothers utterances, is shown in Fig. 12.6. This analysis shows a time structure of proto-conversation that has a clear message. The "poetic form" of this rhythmic and prosodic regulation is confirmed by David Miall and Ellen Dissanayake (2003).

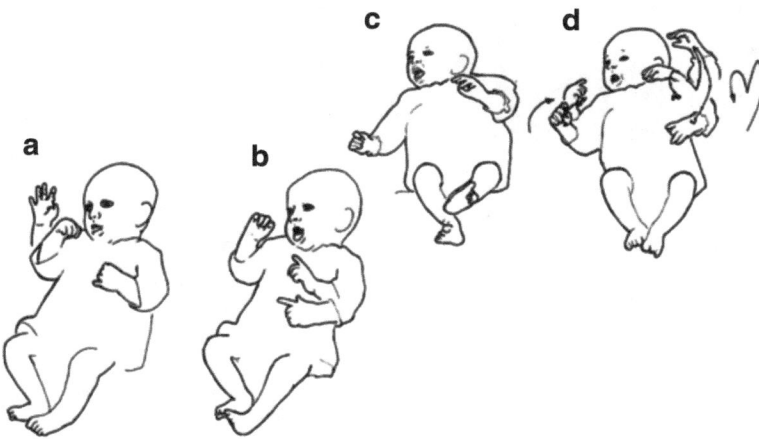

Fig. 12.4 Six-week-olds in face-to-face interaction with their mothers, who are facing them to the left of the pictures. Gestures accompany facial and vocal expression. (**a**) An infant waves the right hand. (**b**) Two seconds later the right hand is closed and the baby extends and raises the left index finger, simultaneously opening the mouth and extending the tongue. In (**c**) an infant raises a left hand with fingers opposed and touches the lower lip with tongue. (**d**) The same infant waves arms rhythmically while vocalizing (From Trevarthen 1984, 251, figure 9)

12.2.5   Animating Play: The Joy of Shared Adventure in Imaginative and Humorous Games

The actions and attentions of infants develop rapidly in the first few months, becoming better controlled by experience, directed selectively toward outside events, and more complex. After 4 months, when the neck and arms are stronger, and looking and reaching out with eyes and hands engage more attentively with objects that may be handled and experimented with, infants tend to pay less attention to proto-conversational interactions. On the other hand, they become more playful; more expressive and more willing to join in action games and songs (Dissanayake 2000; Frank and Trevarthen 2012; Reddy 2008; Trainor 1996). They explore the contingencies of their movements, including reactions of a companion, and can be attracted by activities that invite experimenting with communication for fun, especially if the invitation is rhythmic. Figure 12.7 illustrates typical patterns of baby songs, which mothers everywhere begin to use to entertain their infants after 4 or 5 months (Trevarthen 1986, Trevarthen 1999, Trevarthen 2008).

An infant's intermodal, "proprioceptive" awareness of the body in action, and its sharing of vitality dynamics in "alteroceptive" awareness of the story of a mother's baby-song, are vividly demonstrated in a film by Gunilla Preisler of a 5-month-old baby girl who was born totally blind. Despite this serious sensory loss, she is responding intimately and imaginatively to her mother's singing of a famous Swedish baby song "Mors lille Olle." The infant "conducts" her mother's singing with her left hand, which she has never seen, with graceful movements as of a

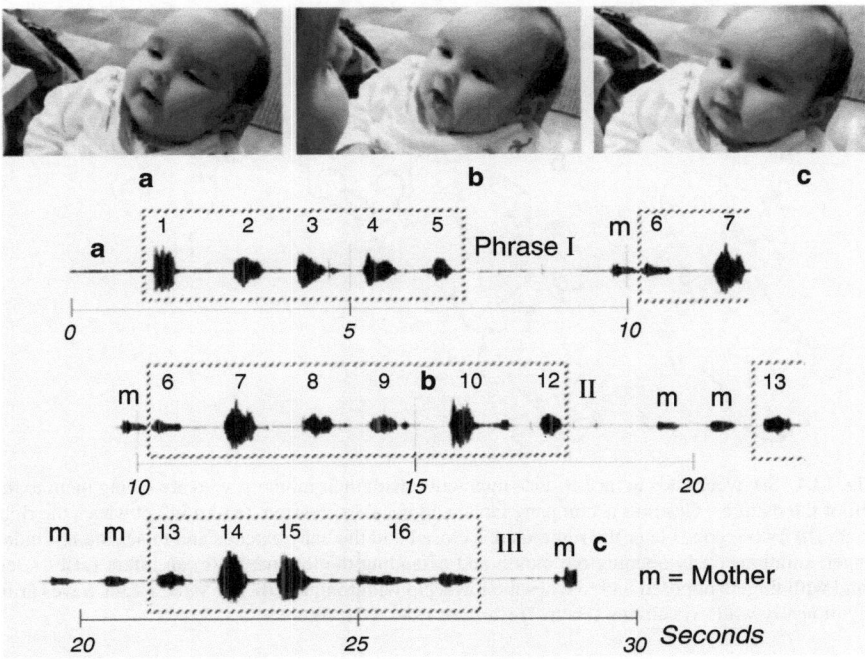

Fig. 12.5 This 10-week-old girl has repeats short energetic vocalisations (1–16), with a pulse of approximately 1 per second, in three groups of five or six beats (I, II and III), to which her mother responds with head nods and gentle sounds of assent (m), as if her daughter was speaking phrases of language. Most of the infant's sounds are open mouthed "calls" or "shouts," but sound number 5 is articulated by closure of lips and tongue to interrupt the vowel with a consonant, making a short two-syllable word, like "d-ba." The mother imitates this sound with her final utterance at 29 s. Clearly the infant is taking the lead in this dialogue, with emphatic timing. The photographs record the infant's expectant look to the mother at the start, watching her mouth (**a**); a major effort at declaration of "pre-speech," (utterance 10 at around 15 s, while her mother leans toward her) (**b**); and watching her mother's final imitative response (**c**)

trained orchestral conductor, marking phrases, shifts of pitch and closure of a verse with delicacy, and at certain significant moment she leads the impulse of the mother's melody by 300 ms (Gratier and Trevarthen 2007, 2008; Schögler and Trevarthen 2007; Trevarthen 1999, 2012c).

The rhythmic stories of baby songs in different cultures have the same times as action games involving shared movements of the hands and body. They can be dance or song, and often are both together (Eckerdal and Merker 2009). In the Kalahari desert of Africa, Akira Takada (2005) has studied how mothers of the !Xun people exercise the playful young bodies of their infants about 6 months when they can just stand, by bouncing them in "baby gymnastics," which prepares them for later dancing with other children.

All these rhythmic and melodic rituals, which give delight to infants, are learned as part of relationships with known people, and the infants anticipate the "proper"

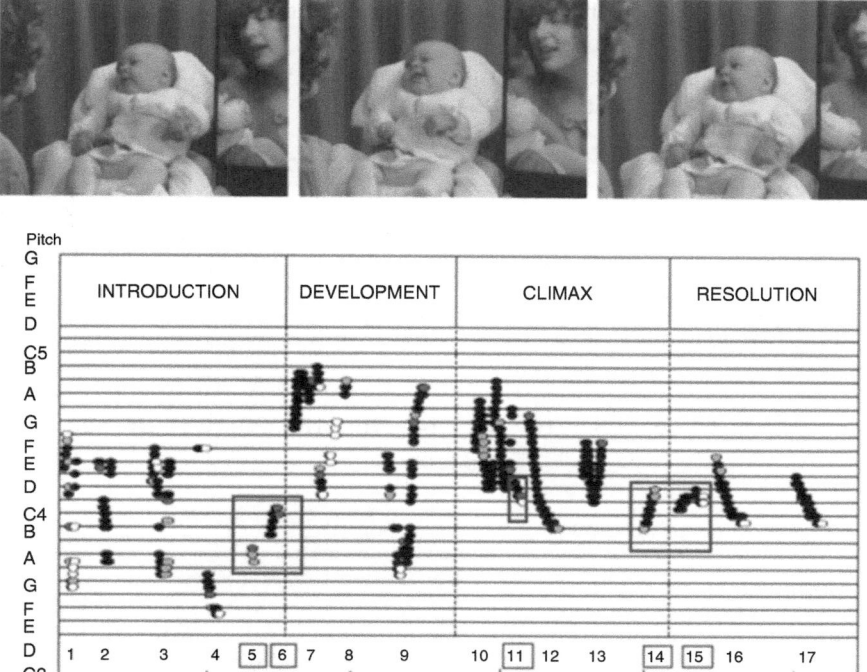

Fig. 12.6 Photos show the expressions of Laura in dialogue; attending to her mother's mouth, smiling, and making a "coo" sound with open mouth and protruding lips. Note the movements of her hands. The pitch plot indicates how the narrative of vocalisations with different values develops in four parts: *Introduction, Development, Climax* and *Resolution*. Utterance numbers appear immediately above the time axis and in the table. The infant's sounds, in boxes, mark the transitions between Introduction and Development, and Climax and Resolution, and utterance 11 animates sharing of the climax with the mother, who follows it with a large glide of nearly one octave. Note that the foundation of the pitch variations for both mother and infant is close to C4, middle C. The infant's final utterance, 15, shows a slight down turn of pitch, anticipating the mothers two closing sounds, which are non-verbal descents (From Malloch and Trevarthen 2009b, 5, figure 1.2. By permission of Oxford University Press)

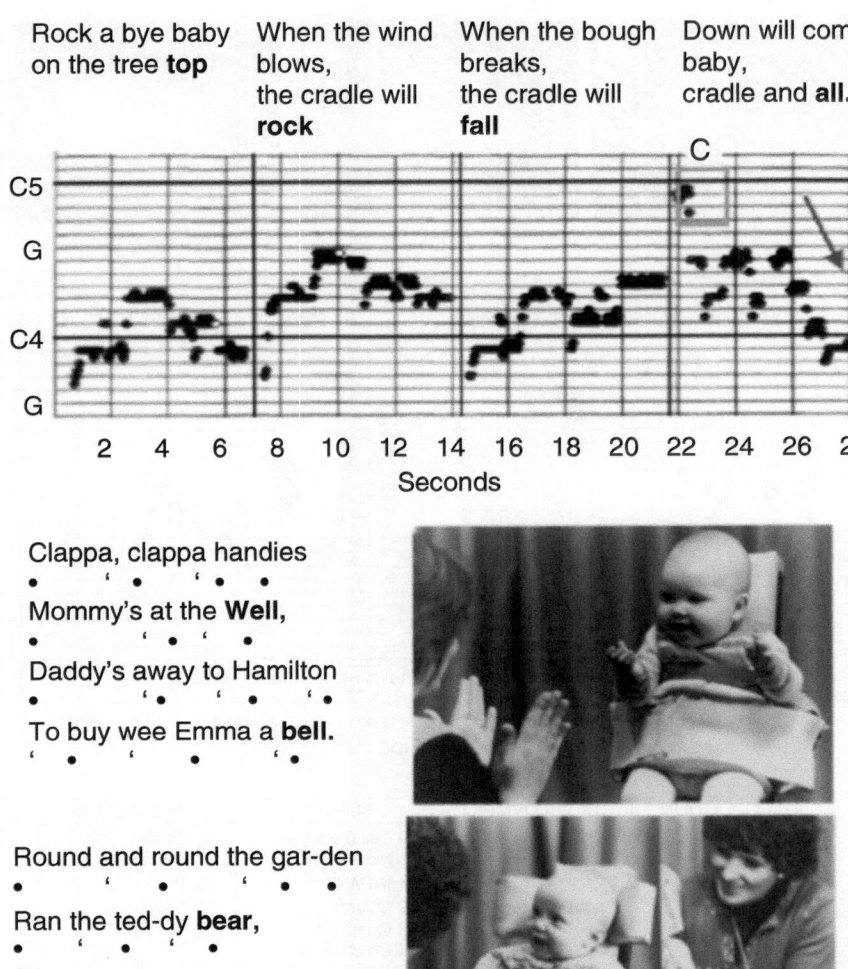

Rock a bye baby   When the wind   When the bough   Down will come
on the tree **top**   blows,   breaks,   baby,
                   the cradle will the cradle will   cradle and **all**.
                   rock   **fall**

Clappa, clappa handies
•   '   •   '   •   •
Mommy's at the **Well**,
•   '   •   '   •
Daddy's away to Hamilton
•   ' •   '   •   ' •
To buy wee Emma a **bell**.
'   •   '   •   ' •

Round and round the gar-den
•   '   •   '   •   •
Ran the ted-dy **bear**,
•   '   •   '   •
One step, two step,
•   •   •   •
And a tickly un-der **there**.
•   ' •   '   •   '   •

Fig. 12.7 *Above*: A pitch plot of a recitation by a female singer of the famous baby song composed in the US as a white person's ironic comment on the Indian custom of hanging a baby in a birch-bark cradle on a tree. The poetry has classical time proportions: each syllable or pause 0.7 s (*andante*); each phrase 3.5 s; each line 7 s, and the verse of 28 s. The pulse marks the steps of the narrative, based on *iambic* poetic "feet," which reproduce the heel to toe rhythm of each comfort-able step as the singer "walks" the story. It is a ritual that tells a drama (Merker 2009a, b; Turner 1982). *Below*: Action songs "Clappa clappa handies" and "Round and round the garden" with 6-month-old infants, showing the four line stanzas and *iambic* feet of the poetic recitation. The first song describes "getting a job done"; the second, approaching slowly on the third line and ending *accelerando*, engages the baby in a teasing game (From Trevarthen and Delafield-Butt 2013a, 177, figure 8.3. By permission of Guildford Press)

way to join in—for example vocalizing with rhyming vowels at the ends of the lines. Merker (2009a, b), making comparison with certain behaviours of social groups of animals, emphasises the special importance of a greatly developed sense of ritual performances for the transmission of human culture. Infants show this sense clearly long before they can share conventions of language.

12.2.6   Cooperative Awareness for Cultural Tasks and Tools: Grasping Meanings

A young psychologist, Penelope Hubley, making a study of the development of a girl Tracy from 3 weeks to 1 year, recorded marked age-related changes in playfulness, and in cooperative use of objects, and discovered a particularly significant transformation about 9 months (Trevarthen and Hubley 1978; Hubley and Trevarthen 1979). Before this she found that Tracy's mother picked up her daughters very strong interest in objects to be grasped and handled, and this turned their play into "person-person-object" games with "toys."

The "age related changes" have been confirmed by subsequent studies with other subjects (Frank and Trevarthen 2012; Trevarthen 2004a, 2012d; Trevarthen and Aitken 2001, 2003) (Fig. 12.8). Boys and girls show some differences, but stages of playfulness and interest in communication and exploration of objects are the same. They are changes not only in movements and alertness but also in "self-consciousness," which means a consciousness of self in relation to others (Trevarthen 2005b). Reddy (2008, 2012) calls this "second person awareness," distinct from "first person awareness," just of the self, or "third person awareness" of impersonal objects or "things." Playfulness experiments with the inventions and challenges of second person awareness, in whatever mode of expression (Trevarthen 1986, 2005b). Developments in appreciation of body action games and songs composed by parents to incite participation with the infant are especially rich in the middle of the first year. They prepare the way to "joint attention" toward objects, or what I prefer to call "shared experience" (Trevarthen 2012d). This identified as a cognitive achievement that begins cultural learning (Tomasello 2008). It is also a development of intersubjectivity (Trevarthen 2004a; Trevarthen and Aitken 2001) that prepares the way for all forms of learning by "intent participation" in purposeful activities (Rogoff 2003; Thelin 2014).

Hubley made a study with five girl infants from 9 to 12 months of age to confirm how they made the transition to Secondary Intersubjectivity or Cooperative Awareness around 40 weeks, and become able and willing to share tasks, and ready to learn the proper routines for use of tools, like books and eating utensils (Hubley and Trevarthen 1979).

These initial steps in the growth of human powers by its own initiative, and with the benefit of intuitive affectionate support, before speech, lay the foundations for a lifetime of conscious movement and talking. They carry evidence of the innate measures of purpose and awareness with feelings of hope in well-being that will

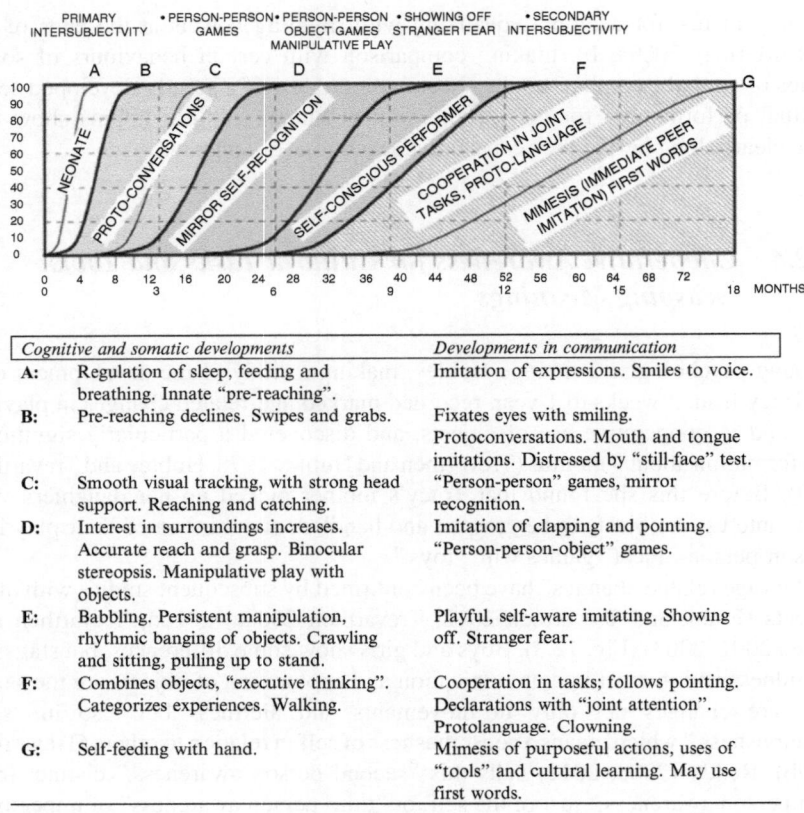

Cognitive and somatic developments	Developments in communication
A: Regulation of sleep, feeding and breathing. Innate "pre-reaching".	Imitation of expressions. Smiles to voice.
B: Pre-reaching declines. Swipes and grabs.	Fixates eyes with smiling. Protoconversations. Mouth and tongue imitations. Distressed by "still-face" test.
C: Smooth visual tracking, with strong head support. Reaching and catching.	"Person-person" games, mirror recognition.
D: Interest in surroundings increases. Accurate reach and grasp. Binocular stereopsis. Manipulative play with objects.	Imitation of clapping and pointing. "Person-person-object" games.
E: Babbling. Persistent manipulation, rhythmic banging of objects. Crawling and sitting, pulling up to stand.	Playful, self-aware imitating. Showing off. Stranger fear.
F: Combines objects, "executive thinking". Categorizes experiences. Walking.	Cooperation in tasks; follows pointing. Declarations with "joint attention". Protolanguage. Clowning.
G: Self-feeding with hand.	Mimesis of purposeful actions, uses of "tools" and cultural learning. May use first words.

Fig. 12.8 Stages in the development of human motives for shared experience and cultural learning before language (From Trevarthen 2001, 849, table I and Trevarthen 2001, 855, part of figure 3. By permission of Springer Verlag)

remain essential in all future attempts to make sense of life with other persons (Trevarthen 2005b). They will, with the benefit of learning, share knowledge of a physical world where experiences of active, "life" time and space can be applied to define abstract measures of "things" and to compare manipulate, and represent them in symbolic ways.

12.3 Brain Science of Time to Move, for Self and in Company

12.3.1 Toward a Natural Science of Active, Innate Self-Awareness

The physiology and psychology of subjective time in movement has had a frustrating history, confused by the "fallacy of misplaced concreteness" (Whitehead 1926); by recourse to explanation in terms of the more "obvious" outside world called "real physical facts", measured by artificial devices (Thelin 2014). But the word "physical" began with an appreciation of the invention of life in movement, of "vitality" and "fertility" as in health, and as also in the growth of plants:

> **physic (n.)** ... from Greek *physike* (episteme) "(knowledge) of nature," from fem. of *physikos* "pertaining to nature," from *physis* "nature," from *phyein* "to bring forth, produce, make to grow" (related to *phyton* "growth, plant," *phyle* "tribe, race," *phyma* "a growth, tumor") from PIE root *bheue-* "to be, exist, grow" (Online Etymology Dictionary, http://www.etymonline.com/)

Animal intelligence, and the intelligence of infants, has been interpreted as made by linking automatic "reactions" of a mindless sensori-motor system in combinations of "reflexes" that are learned by "conditioning" reinforced by associated rewards and punishments that trigger body states of pleasure or pain. In this behaviourist approach, restricted to measurement of responses in artificial experimental situations, the subjects have their freedom to move greatly reduced. Research became more open to acceptance of at least primitive intentions and imaginative emotions by use of "operant" methods, which allowed the subject to make a particular movement "at will," and this led to acceptance that emotions may be causal. For example, operant conditioning of head turns made by infants was applied by Hanuš Papoušek (1967) to study their spontaneous preferences. He found that infants" experience of success or failure in predicting what they would experience was expressed with "human sense" of pleasure or disappointment, by smiles of joy for a "correct" choice, or grimaces and cries of distress and avoidance when the chosen movement was "wrong."

Anticipatory mental functions of action with awareness had been accepted, in the mid nineteenth century, as primary factors in the motivation for learning by looking. Helmholtz, who pioneered sensory psychophysics, proposed that an "unconscious inference" makes it possible for us to perceive a single motionless surround with two moving eyes (Helmholtz 1867). At the beginning of the twentieth century a physiological brain-and-behaviour science of animal intentions, perceptions, self-awareness and feelings of vitality was established. Charles Sherrington, who discovered nerves of "proprioception," self-sensing, that report back to the central nervous system the forces produced inside the body by muscle contractions, founded modern neurophysiology with his *Integrative Action of the Nervous System* (1906). He identified consciousness of objects with knowledge inferred by an imaginative "projicience" that employs the "exteroceptive" distance senses, vision, hearing and touch, to guide movement, seeking confirmation of the anticipatory "proprioception" of the body in action, which he called "the felt Me," with added inferences learned with movements that test the uses of objects. He added that the goal of a movement directed to obtain an object, such as a piece of food or a step forward in walking, included an expectation that a sensory report would be brought back to the body for "affective appraisal" as good or bad, safe or dangerous, and at the right time.

Inspired by Sherrington's studies of the spinal physiology of locomotion, von Holst (1936) confirmed that animal movement with awareness is regulated by a set of "loosely coupled" intrinsic "oscillators" of central nervous activity. He and Mittelstaedt demonstrated the "Reafference Principle" by which an organism distinguishes self-generated sensations from ex-afferent (externally generated) stimuli (von Holst and Mittelstaedt 1950). Sperry (1950) likewise showed that an animal's control of its movements requires a "corollary discharge" of nerve energy that anticipates the "correct" sensory feedback from each movement. In a paper entitled "Neurology and the mind-brain problem," rejecting the behaviourist theory, Sperry (1952) emphasised that perceptions serve movements rather than cause them—that we perceive what we intend. Later, drawing conclusions from his work on the different mental functions of the human cerebral hemispheres dissociated by commissurotomy, for which he received the Nobel Prize in 1981, Sperry (1983) wrote on the creative and unifying power, and values, of a "supervening" consciousness to master the elementary processes of awareness and movement. Like Dewey (1938), he insisted that human Self-awareness, and its development and education, must be imbued with conscious moral purpose.

Prospective motor control by generation of a "motor image" for the movement to achieve a desired effect was rigorously tested in the 1930s by the Russian physiologist Nicolai Bernstein (1967, in English translation) (Fig. 12.9). James Gibson's ecological perception theory (Gibson 1979) explains how practical problems of moving require selective "pick up" of information to guide a person's movements with "prospective control," the mind detecting invariants in sensory information about the world related to how their body may move within it, perceiving "affordances" for use directly, without rational intervention. Bernstein's research and Gibson's theory have been given precise mathematical formulation by David Lee (2009) who identifies a life principle of expectancy in space and time that explains the efficiency of animal movements (Lee et al. 1999), including the movements of infants (Craig and Lee 1999).

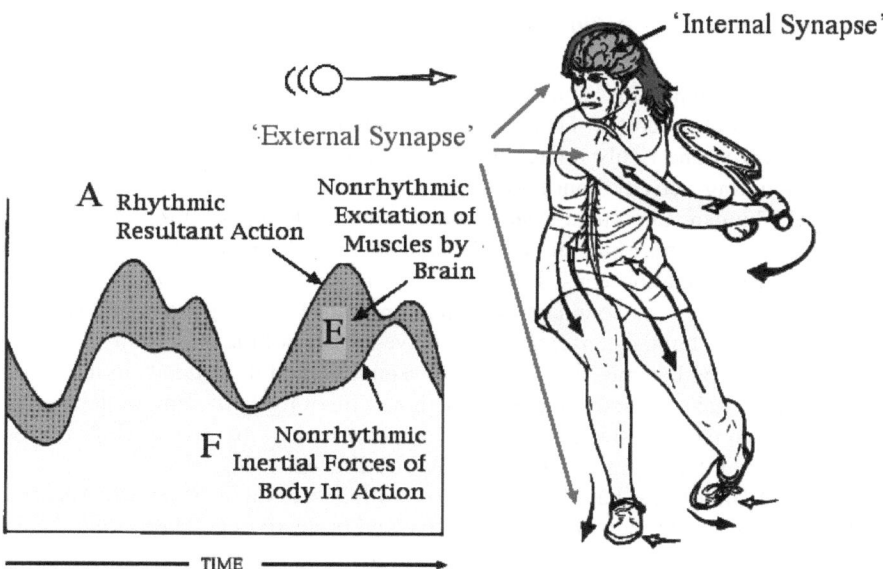

Fig. 12.9 Illustrating Bernstein's theory of prospective motor control for complex movements (Bernstein 1967). *Left*: Bernstein's graph to explain how non-rhythmic forces caused by the masses of moving body parts (*F*), are complemented by continuously adjusted anticipatory central neural processes (*E*) to achieve a smooth rhythmic resultant movement (*A*) with minimum waste of energy. *Right*: A tennis played employs her whole body to bring the racket around to hit the approaching ball with precise control, returning it rapidly to her opponent in the direction she intends. Excitation of muscles (*black arrows*) is monitored by sensory reference (*white arrows*). Sensory-motor control is regulated at two "synapses," at the peripheral engagement of the body with the environment, and centrally in the brain (From Trevarthen 2012c, 16, figure 2. By permission of Peter Lang Publishers)

Clearly the problem of consciousness in action has been, and still is, confused by the theory of the "processing of input," with inadequate attention to the purposes and life-serving functions of creative body activity. The simplest animals must discriminate and use differences between sensations that are caused by their own activity and those that are due to impingement of an external stimulating event. Bjorn Merker (2005) calls this overcoming "the liabilities of mobility", and proposes that, "consciousness arose as a solution to problems in the logistics of decision making in mobile animals with centralized brains, and has correspondingly ancient roots."

The above developments in the psychology of action-with-awareness expose the philosophical divide between "externalist" and "internalist" theories of the causes of consciousness, and a need for an "interactive" approach that accepts that creative Self-related feelings and initiative for adaptive sequences of action are essential properties of organisms that have evolved with powers of movement that have to be regulated "in good time", with anticipation of what may be sensed to be of "interest."

12.3.2   The Brain Stem and the Passionate Self

Modern brain science first sought for centres of human intelligence in a greatly enlarged cerebral neocortex. Then evidence accumulated that areas of the cortex are both specialized and richly interconnected, and, moreover, that lower brain structures, generating activity continuously, are reciprocally linked to it (Trevarthen 2001). The human subcortical systems have comparable organization in other vertebrates. Moruzzi and Magoun (1949) demonstrated that stimulation of the brain stem reticular formation caused "arousal" of the cortical EEG, and behavioural alertness. There is now abundant confirmation that the phylogenetic and ontogenetic impulses of human consciousness involve the core of the brain stem. They are linked to the hypothalamus as gateway between the mind that attends to the world and the emotional system that cares for the inner life of the body. Emotional regulations influence the selective awareness of outside "reality" by giving it life-related values, and by influencing the choice of actions directed to it.

In 1949 Walter Hess shared a Nobel Prize for his work on the subcortical centres of *trophotropic* (parasympathetic, restorative and protective) and *ergotropic* (sympathetic, dynamogenic or action-promoting) regulations mediated by different parts of the hypothalamus and diencephalon in cats. He proved that regulations of inner autonomic functions are coupled with active engagement with the environment, as when a cat made fearful by stimulation of the trophotropic part becomes vigorously active and aggressive toward outside objects by engagement of the ergotropic system (Hess 1954). Co-action of the two systems produces "dynamic equilibrium adapted to the situation at any given moment of the organism as a whole." In other words, a single vital self-consciousness with feelings. Paul MacLean (1990) confirmed the crucial role of limbic system coupling brain stem visceral functions with forebrain intelligence. The importance of emotional processes in generation and regulation of consciousness has been further clarified by the work of Jaak Panksepp (Solms and Panksepp 2012). Commenting on a review of Merker's description of the affectionate and aware behaviours of children lacking cerebral cortex (Merker 2007), Panksepp says the following about "the neurobiology of the soul."

> If we wish to scientifically understand the nature of primary process consciousness, we must study the subcortical terrain where incredibly robust emotional and perceptual homologies exist in all mammalian species. Without work on animal models of consciousness, little progress, aside from the harvesting of correlates, can be made on this topic of ultimate concern. (Panksepp 2007a, 103)

In a recent review of the evidence concerning emotional processes in the creation of conscious states and of experiences in imagination and memory, Vandekerckhove et al. (2014) submit this interpretation of how different levels, and different times, of consciousness come about:

> We argue that this *anoetic* form of evolutionarily refined consciousness constitutes a critical antecedent that is foundational for all forms of knowledge acquisition via learning and memory, giving rise to a knowledge-based, or *noetic*, consciousness as well as higher forms of "awareness" or "knowing consciousness" that permits "time-travel" in the brain-mind.

This leads to

the developmental creation of more subtle higher mental processes such as episodic memory which allows the possibility of *autonoetic* consciousness, namely looking forward and
backward at one's life and its possibilities within the "mind's eye." (Vandekerckhove et al.
2014, 1)

Slower regulatory changes of internal vital state, coupling the neural and the hormonal regulations of well-being or life energy, must have a primary role in the
charting of the plans and memories for meaningful projects and messages over periods of minutes, hours or days, attaching specific events to their "locations" in the
shaped and measured flowing of vitality time, as well as providing the motivation
for common understanding within cultural systems, including language. Both primary episodic memory (Tulving 2002) and elaborated semantic memory depend
upon recruitment of the rhythms of large populations of neurons in the "rational"
neocortex, in collaboration with the allocortex between the oldest and most recent
brain parts (MacLean 1990; Buzsaki 2006). The elements of effort to act and to
know and to categorize the consequences as beneficial or dangerous, and how they
express the fluctuations of vitality of a large and complex body, are the motivation
for development of practical sensory-motor awareness and for cooperative, story-
telling "acts of meaning" (Bruner 1990). They direct the child toward co-construction
of a meaningful spatio-temporal reality, and the functional grammar of talk in a
particular language which has to bring known things to life between people (Halliday
1978; Halliday and Matthiessen 2004).

The total motor brain of a human being, coupling the frontal and parietal cortices
with the intricately structured cerebellum (Glickstein and Doron 2008), a master
time-keeper for intricate and fast motor patterns, has become conspicuously
enlarged in *Homo* with the transformation of clever primate eyes, mouth and hands
used for feeding and grooming into the unique human capacity for speedy communication. We have to exchange subtle intentions, thoughts and feelings by looks
with eyes that have white sclera to heighten perception of the direction of their
regard, by rapid articulations of speech and by subtle gestures of the hands. All
these are active in a newborn and highly effective for soliciting imitative and emo-
tionally enriching engagement with another person.

12.3.3 Affective States and Emotions for Communication: Biochronology of Action and Inter-action

The autonomic nervous system has its own slow rhythms (Delamont et al. 1999),
which are integrated with faster somatic rhythms in the affective nervous system
(Panksepp and Trevarthen 2009) (Table 12.2). Together these regulate the intensity
and goals of action and balance exertions against recuperation. What results is a
hierarchy of times—a biochronological spectrum or chronobiology—by which both
intentional actions and vital functions are kept efficient and well (Trevarthen 2004a,

Table 12.2 Life times, from barely detectable to remembered in life stories

Infants have our inner time, ready to be shared

	A. Somatic, ergo-tropic						B. Visceral, tropo-tropic		
	Preconscious, automatic actions			Conscious sensori-motor control of actions "psychological present"			Imagined and remembered, with emotion		
	Milliseconds						Seconds		
	5–20	30–40	50–100	150–200	300–700	700–1,500	3–6	10–30	30–50
Hz.	200–10			15–0.6			0.3–0.02 (and longer)		
Adult body and brain									
Brain Physiology	Gamma oscillations		Physiological tremor	N200 "Mismatch" wave	Delta waves memory "up-date." Odd-ball effect	Readiness expectancy-wave	Breathing rhythm	Vaso-motor waves	Heart; parasympathetic cycles
Walking		Fast reflex		Step, heal-to-toe	Normal walk	Slow walk			
Manipulating		Twitches	Fiddling	Finger tapping	Fast reach, pickup, grasp. Hammering	Slow reach. Sawing	Manual movement sequence		
Eye and head			Eye-saccade	Inter-saccade interval. Head-turn	Separate head orientations		Eye movement "Scan path"		
Inspecting objects, reading				Short visual fixation	Long visual fixation sight-to-saccade interval				

Adult communications	Utterances, gestures, expressions			Fast gesture or expression. Wink. Laugh, gasp. spasm, Patting fast	Head-shake, nod. Glance eyebrow raise. Hand wave. Burst of laughter Patting gently, stroking. Grimace	Slow smile. Scowl. Stare	
	Conversation timing			Overlaps interruptions	Single utterance	Short turn	Long turn	
	Speech	Voice onset time	Fastest lip and tongue articulation	Articles, prefixes. Counting fast	Syllable, vowel. Slow articulation. Chewing, talking	Word	Phrase. Breath-cycle	
	Music		Trills	Vibrato	Pulse/beat: *presto* to *andante*	*andante* to *largo*	Phrase	
	Poetry			Unstressed syllable	Stressed syllable	Foot	Phrase	Stanza
	Singing			Vibrato	Beat	Bar	Phrase	Verse
	Vitality contour			Fast, bursting	Controlled	Slow, graceful	Slow, sedate	Timeless floating
Infants			Eye saccade	Blink. Quick head turn. Short "coo"	Inter-saccade gap. Slow head turn. Surge in reaching. Gesture longer call, cry	Pre-reaching lift. Sucking. Beat of protoconversation. Breathing	Pause in sucking vocal phrase	

2009b; Osborne 2009a). The intrinsic body sense of time—or "enkinesthesia" (Stuart 2010)—enables one to describe the elaborate prospects of one's purpose in stories, dramas, and strategies that may capture the interest of others and enable them to feel, imagine, and collaborate. Mimetic narrations are informative about how body and mind intend to act and how they sense the pleasures and pains—or benefits and risks—of practical and social enterprises (Donald 2001; Damasio 2010). The poly-rhythmic, poly-sensory brain monitors the internal well-being and safety of the body via a visceral nervous system, components of which have evolved into semiotic organs serving as mediators of one's intimate cooperations, linking the intentional and life-regulating processes of separate selves (Porges 2001, 2003; Porges and Furman 2011). Skilful and sensitive use of the musicality of this psycho-biology can bring benefit to deeply traumatised individuals (Osborne 2009b).

Play is full of energy, fantasy, and self-gratification. With laughter, it brings joy to strengthen relationships (Panksepp and Burgdorf 2003). It can energise the whole body: to run and jump, to shout with excitement, or to share itself with a rising intonation, wink, or shrug. It is a display of self-consciousness, full of invention, that appeals to others, inviting the co-creation of meaning. Communication through play arises by means of extravagant and ritualised motility: using the body, not to perform essential acts within the environment or to use the environment, but to display itself to others, attentive to their response and appreciation in creatively musical ways (Bjørkvold 1992). It is made up of actions—or self-regulations—performed in an exaggerated, non-essential way. It may find new purposes valued for their use in relationship or community. The Self becomes a social Me, sharing symbols of jointly created significance (Mead 1934), playing a game with the "mirrors" of other minds (Trevarthen 2005b; Bråten 2009).

12.3.4   Microkinesics of Self-Synchrony and Inter-synchrony

Daniel Stern's work on the development of the vital powers and awareness of infants in their relations with other persons (Stern 2000) was inspired by a natural science of kinesics developed by Birdwhistell (1970), Schefflen (1973), and Kendon (1980), who measured the way human communicators make expressive movements with self-synchrony between many parts of their own body, and how they easily establish precise inter-synchrony with other persons' movements. Microanalytic description of dynamic features from film (Condon and Ogston 1967) enabled Condon to demonstrate the capacity of a newborn infant to engage in precise inter-synchrony with an adult (Condon and Sander 1974), as we have shown in the vocal exchange of Fig. 12.1.

The subtle dynamics of touch were studied by Clynes a pianist who developed a descriptive system of the emotional delicacy of action that he called "sentics," which applies to control of vocal sounds as well as to hand actions, and all performances of music (Clynes 1978, 1982).

12.3.5   Self-Consciousness and Group Action

Animal gestures transmit purposes and feelings between individuals "musically" (Wallin et al. 2000). They are derivative of self-regulatory movements which have become adapted for acting intersubjectively in social communication (Darwin 1872; MacLean 1990; Porges 2003; Rodriguez and Palacios 2007; Dissanayake 2000, 2009; Panksepp and Trevarthen 2009; Trevarthen 2005b). Neurophysiological studies confirm that intra-synchrony of "proprioception" *within* subjects becomes inter-synchrony of "alteroception" mediated *between* them by "distance senses" of sight and hearing. This is the way habitual "artefacts" of movement in animal groups arise as social habits and rituals, or "fixed action patterns" necessary for efficient cooperation in activities such as mating, feeding of the young, and training for foraging or hunting skills. Joint motor inter-synchrony is seen in the flight of a flock of birds, the swimming of a school of fish, and the movements of swarm of insects, or heard in the singing of birds or the sound chorusing of insects. It is also seen in the behaviour of a team of dancers, or heard in an orchestra of musicians. In human families or working groups, agreed conventions in social time become referred to as semiotic objects or meanings (Trevarthen et al. 2014). They may be learned as conventional units by which we become aware of the rates of physical events that we can employ technically in projects or tools of work, or discuss as elements of abstract theoretical propositions, a function of calculation in language which Halliday calls "mathesis" (Halliday and Matthiessen 2004).

12.3.6   Life times of Sensory-Motor Intelligence before Birth: Movement Comes First

The development of rhythmic limb movements directed away from the body, which have an inbuilt capacity to learn how to guide awareness in intentional and social ways, develop before birth (Trevarthen 2012c; Delafield-Butt and Trevarthen 2013). They are products of autopoiesis of the sensory-motor anatomy, which is mapped for action in space and time in the morphogenesis of an intelligent nervous system (Trevarthen 1985, 2001), not learned.

Human embryos 5 mm. long have beating heart muscles at 1 month after conception, before any motor nerves are formed (Trevarthen 2004b). From 6 weeks— before sensory inputs reach the spinal cord—congregations of motor nerves excite rhythmically paced, whole-body bending.

The first integrative pathways of the brain are in the core of the brain stem and midbrain (Windle 1970), and the earliest complexes of whole body movements, though undifferentiated in their goals, are coherent and rhythmic in time (Lecanuet et al. 1995). When sensory input develops, there is evidence, not of reflex response to stimuli, but of the intrinsic generation of prospective control to complete more individuated actions, before the neocortex is functional. At 2 months of gestation the cortex has no neural cells and thalamo-cortical projections are just starting to grow (Larroche 1981; Hevner 2000), but there is sufficient sensory and motor nervous connectivity for dynamic proprioceptive motor control (Okado 1980).

At 3 months, quantified kinematic analyses indicate that fine rhythmic movements of hands and fingers guided by sensitive touch, show a sequential patterning with modulation of arousal state that may give a grounding for "narrative imagination." Astonishingly, ultrasound recordings of twin fetuses at 4 months show changes of timing that distinguish movements of self-exploration from those directed to a twin, and this is taken to confirm a primary "social awareness" (Castiello et al. 2010). Certainly, by 5 months the kinematic form of the arm movements of single fetuses confirms that "imaginative" and "self-aware" motor planning is operative (Zoia et al. 2007). This natural history of human movement appears to confirm the suggestion by Lashley (1951, 122) that genesis of propositional thought to give order to experiences may grow from the spontaneous syntactic sequencing of movements.

Later in fetal development, more delicate explorations of self and environment can be observed as the hands touch the eyes, the mouth, the uterine wall, and so on. And individual "habits" appear, such as a propensity to fondle the umbilical cord, scratch at the placenta, or to make twin-directed movements (Piontelli 1992, 2002; Jakobovits 2009). Self-touching actions continue throughout life as restless gestural "self adaptors" (Ekman and Friesen 1969), also very evident in animated face-to-face conversation (Kendon 1980). They express a dynamic sense of self that communicates changing states of mind.

Fetal movements are not only exploratory and directed to engage with the external inanimate world or the body of the Self. Affective regulations are also developing. Heart-rate responses to sound appear around 20 weeks (Lecanuet et al. 1995; Parncutt 2006). From this point until term, fetal heart rate slows appreciably and becomes more adaptable, due to increasing parasympathetic maturation. Heart-rate changes are coordinated with phases of motor-activity from 24 weeks (James et al. 1995). This is indicative of the formation of a prospective control of autonomic state, coupled with readiness for episodes of muscular activity of engaging with the environment: a feature of brain function that Jeannerod (1994) cited as evidence of cerebral motor images underlying conscious awareness and purposeful movement.

Facial expressions in fetuses and movements of distress and curious exploration give evidence of emotions of discomfort or pleasure adapted for communicating feelings (Reissland et al. 2011). Maternal hunger with depletion of energy supply to the fetus drives "anxious" patterns of fetal movement. There is consensus in modern paediatrics that by 24 weeks the fetus should be considered a conscious agent deserving the same standard of medical care as adults (Royal College of Obstetricians and Gynaecologists 2010). The mid-term human fetus has the foundations for the space-time defining functions of intention in action, and for the emotional regulation of aesthetic relations with the objective world and moral relations with other persons (Trevarthen 2001, 2005b).

12.3.7 Getting Ready for a Richer Environment

In the last trimester, when the cerebral neocortex is just beginning formation of functional networks, functions are established that anticipate an active post-natal life: especially for collaborating with maternal care (Lecanuet et al. 1995; Feldman 2012; Schore 2012a). Motor coordinations exist for visual exploration, reaching, and grasping; walking; and expressive communication by facial expression and gesture (de Vries et al. 1982; Prechtl 1984). Movements show guidance by touch, by taste and by responses to the sounds of the mother's voice.

The cortex develops its characteristic folds in the final 10 weeks before birth and the patterning of gyri shows differences between the hemispheres characteristic of humans, which reflect asymmetries in sub-cortical self-regulating systems, the right side of the brain being more self-related or proprioceptive and the left being more discriminatory of environmental affordances and eventually directed to learn adaptive articulations of the hands and of vocal activity (Trevarthen 1996, 2001). Areas essential for perceiving and producing sounds of words are identifiable in the left hemisphere at 30 weeks. Complementary enlargements in the right hemisphere are adapted for seeing and hearing other persons' expressions and identity, and for affectionate maternal care (Schore 2012a). The late fetus is in a quiescent state, but can be awakened and can learn. At 7 months it shows cardiac accelerations and startles to sounds. While general body movements decrease, respiratory movements increase, as do face, tongue movements, smiling, eye movements and hand gestures.

Fetuses interact with the mother's movements and uterine contractures and, after 25 weeks, can learn her voice, a process that engages the right cerebral hemisphere (DeCasper and Prescott 2009). There are also motor reactions to rhythmic sounds, such as the bass pulse of dance music, and melodies that the mother attends to frequently or performs on a musical instrument can be learned.

The last trimester is critical for elaboration of asymmetries of cerebral function adapted for cultural learning. First it will be necessary to form an intimate attachment with a caregiver, normally the mother, whose hormonal changes support special affectionate ways of acting that match the newborn's needs. The right hemisphere orbito-frontal system motivates affective communication with the musical prosody of infant-directed parental speech (Schore 2012a), and the left orbitofrontal cortex has a complementary adaptation for generation of expressive signals by the infant (Trevarthen 1996, 2001).

Beneath the cerebral cortex the brain generates fundamental rhythms for self-synchrony of movements of body parts and inter-synchrony in exchanges of signals of motives and emotions from other humans (Buzsáki 2006), including the faster components that become essential for the learning of manipulative skills and the rapidly articulated movements of language or gestural signing (Condon and Sander 1974; Trevarthen et al. 2011). Brain rhythms, while favouring a selective sensitivity to expressive features that identify the mother's voice, also enable the fetus to move in coordination with the sounds of music and to learn certain melodies or musical narrations (Malloch 1999; Gratier and Trevarthen 2008).

12.4 Conclusion: Reconciling the Phenomenal times of Feeling and Fact for Human Agency

12.4.1 The Paradox of Unintended, Dispassionate, Linear Physical Time

Human practical intelligence at its most "cultivated" strives to concentrate on outer factors of experience, to escape from its feelings of vitality in self-related awareness of agency, and from sympathetic emotional communication to share purposes with value. The language of explanation concerning how to master materials and mechanisms becomes a structure of propositions about the necessity of what happens "out there." It becomes timeless, except in the regulation of the assembly of elements of a task, or the meanings of text for communication, the use of conventional forms in media for speaking or writing about knowledge. Yet the discoverer hopes to be believed, to have his or her purposes and experiences valued by others.

Margaret Donaldson, a developmental and educational psychologist, commented on the contradiction between what is believed and motives that wish to be believed, and she noted how our feelings lead us to "pursue our goals with great tenacity." She says,

> This tenacity has a number of sources, but prominent among them is the fact that our purposes are apt to be accompanied by very powerful feelings. Thus they become important to us; and in the extreme case they can become more important than life itself. ... In spite of these facts of experience and observation, a number of serious attempts have been made to account for human behaviour without having recourse to the notion of intention or purpose at all. The notion, however, is one that tends to reappear in some guise or other within psychology, no matter how hard one tries to keep it out. And it is ironic that the attempt to keep it out is generally itself sustained by a passionate aim: the aim of being "scientific" in a manner modeled upon the activity of the physicists. (Donaldson 1992, 8 and 9)

No matter how much we try to put time in events that have no mind and no human purpose, they can only be known by us as metaphors for the time we generate with the aid of memories to give measure to actions and experiences we create and eagerly share—now, in the nostalgic past, and in the hopeful future (James 1890). All our communication requires and uses a shared sense of life in body movement (Trevarthen 2015).

12.4.2 Accepting Consciousness of Purposeful Time, Affective Appraisals and Communication

A philosopher Barbara Goodrich, in a paper entitled "We do, therefore we think: time, motility, and consciousness," presents the problem powerfully, with expert evidence from recent neuroscience of Llinás (2001), who studies how a unified Self is regulated in time and space, and of Buzsaki (2006), who examines the intricate cellular and intercellular rhythms of the brain and how they are organized in maps of the body and its actions to create, by excitation of movements of sensuous agency, action patterns that are both creative and adaptable.

> Most current discussions of consciousness include implicit philosophical presuppositions inherited from the canon of Plato, Aristotle, Descartes, and Kant, e. g. that consciousness is self-reflective, passive, and timeless. Because of this, Llinás's and Buszáki's insights may not be fully appreciated. Western philosophy, however, also includes what might be described as a counter-tradition—and one that is more compatible with empirical biological science than the usual canon. Heraclitus, Spinoza, Schopenhauer, Nietzsche, and especially the 20th century French philosopher and psychologist, Merleau-Ponty, all anticipated aspects of Llinás's and Buszáki's approaches. Their alternative conceptual vocabularies are useful for strengthening Llinás's and Buszáki's approaches, sketching out a notion of consciousness emerging from motility, and generating new hypotheses for neurophysiological research. (Goodrich 2010, 331)

There have been advances in applying such thinking about the mind and the time of movements to explain the inter-subjectivity of communication (Thelin 2014). In *The Laws of Emotion* (2007) Frijda reviews what is known about the importance of the timing of emotional expressions, regulating their intensity and duration in immediate intimate encounters between persons.

The topic of "Vitality Dynamics" is richly explored by Daniel Stern (2010) with scientific clarification of how affective and affecting life time and subtle regulations of "arousal" in action are made, and used to enable communication of experience in "the present moment" (Stern 2004). Like Goodrich, Stern perceives the advantages of a phenomenology that addresses the existence of conscious beings in embodied connection with their environment, and within a sense of rhythm and pace in action and awareness, a philosophy which:

> provides an account of the subjective world experienced *as it is lived*, pre-theoretically, pre-reflectively. The subjective, phenomenal world is as whatever is passing across the "mental stage," right now. It does not concern itself with how the scene got on the mental stage, nor why it got there, nor when, nor whether it is "real" in any objective sense. This current of philosophy ... provides a starting place to look for vitality dynamics or the feel of being alive. William James (1890), Edmund Husserl (1964), and Merleau-Ponty (1962) are the most influential thinkers for this present work. (Stern 2010, 35)

12.4.3 Using the Gifts of Motor Time and Innate Intersubjectivity in Education and Psychotherapy

Donaldson was concerned about the motives for knowledge as a teacher wanting the best for young pupils. The idea that children make and share their experience in playful moving has long been a core principle of a preschool education theory that accepts the natural virtues of early childhood described by revolutionary educators Comenius, Pestalozzi and Froebel (Athey 1990; Bruce 2012). Attention only to how objects may be used or conceived with creative imagination of a rational kind by the individual thinker is insufficient to explain how a young child finds new meanings with the pleasure of human company (Donaldson 1978; Bruner 1996; Rogoff 2003; Erickson 2009; Trevarthen 2011; Trevarthen and Aitken 2001).

Some psychiatrists have express a similar concern for a need to change the practice of their work with distressed or confused patients, finding the concept of emotional defences and the use of verbal interpretation inadequate to account for the shaping of a therapeutic relationship that is mutually rewarding and able to help even the most difficult cases from their confusion and isolation, by engaging intimately with their personal sense of time and purpose, and their emotional apprehensions (Stern et al. 1998; Meares 2004; Meares et al. 2012; Schore 2012b; see Mölder's Chap. 11 of this volume, and Goodrich 2010, who refers to the observations of psychiatrists Melges 1982, and Mo 1990).

There is increasing interest in the use of well-tined and convivial expressive movements of the arts, especially music (Trevarthen 2005b), to aid recovery of self-confidence in autism, schizophrenia and depression, or ADHD in childhood (Pavlicevic et al. 1994; Pavlicevic 1997; Stige 2004; Panksepp 2007b; Pavlicevic and Ansdell 2009; Robarts 2009; Wigram and Elefant 2009; Trevarthen and Delafield-Butt 2013b).

Daniel Stern recognized this shift in the introduction to the 2000 edition of *The Interpersonal World of the Infant* as follows:

> One consequence of the book's application of a narrative perspective to the non-verbal has been the discovery of a language useful to many psychotherapies that rely on the non verbal. I am thinking particularly of dance, music, body, and movement therapies, as well as existential psychotherapies. This observation came as a pleasant surprise to me since I did not originally have such therapists in mind; my thinking has been enriched by coming to know them better. (Stern 2000, xv)

References

Abram, David. 1996. *The spell of the sensuous: Perception and language in a more-than-human world*. New York: Vintage Books.

Ammaniti, Massimo, and Vittorio Gallese. 2014. *The birth of intersubjectivity: Psychodynamics, neurobiology, and the self*. New York: Norton.

Athey, Chris. 1990. *Extending thought in young children: A parent-teacher partnership*. London: Paul Chapman Publishing.

Bateson, Mary C. 1979. The epigenesis of conversational interaction: A personal account of research development. In *Before speech: The beginning of human communication*, ed. M. Bullowa, 63–77. Cambridge: Cambridge University Press.

Beebe, B., J. Jaffe, S. Feldstein, K. Mays, and D. Alson. 1985. Inter-personal timing: The application of an adult dialogue model to mother-infant vocal and kinesic interactions. In *Social perception in infants*, ed. F.M. Field and N. Fox, 217–248. Norwood: Ablex.

Bernstein, N. 1967. *Coordination and regulation of movements*. New York: Pergamon.

Birdwhistell, R. 1970. *Kinesics and context*. Philadelphia: University of Pennsylvania Press.

Bjørkvold, Jon-Roar. 1992. *The muse within: Creativity and communication, song and play from childhood through maturity*. New York: Harper Collins.

Bråten, Stein. 1988. Dialogic mind: The infant and adult in protoconversation. In *Nature, cognition and system*, ed. M. Cavallo, 187–205. Dordrecht: Kluwer Academic Publications.

Bråten, Stein. 1998. Infant learning by alterocentric participation: The reverse of egocentric observation in autism. In *Intersubjective communication and emotion in early ontogeny*, ed. S. Bråten, 105–124. Cambridge: Cambridge University Press.

Bråten, Stein. 2009. *The intersubjective mirror in infant learning and evolution of speech*. Amsterdam: John Benjamins Publishing.

Brazelton, T. Berry. 1979. Evidence of communication during neonatal behavioural assessment. In *Before speech: The beginning of human communication*, ed. M. Bullowa, 79–88. Cambridge: Cambridge University Press.

Brazelton, T. Berry, and Joshua D. Sparrow. 2006. *Touchpoints 0–3: Your child's emotional and behavioral development*, vol. I. Cambridge, MA: DaCapo Press.

Bruce, Tina (ed.). 2012. *Early childhood practice: Froebel today*. London: Sage.

Bruner, Jerome S. 1990. *Acts of meaning*. Cambridge, MA: Harvard University Press.

Bruner, Jerome S. 1996. *The culture of education*. Cambridge, MA: Harvard University Press.

Bullowa, M. 1979a. Introduction. Prelinguistic communication: A field for scientific research. In *Before speech: The beginning of human communication*, ed. M. Bullowa, 1–62. Cambridge: Cambridge University Press.

Bullowa, Margaret (ed.). 1979b. *Before speech: The beginning of human communication*. Cambridge: Cambridge University Press.

Buzsáki, György. 2006. *Rhythms of the brain*. New York: Oxford University Press.

Castiello, Umberto, Cristina Becchio, Stefania Zoia, Cristian Nelini, Luisa Sartori, Laura Blason, Giuseppina D'Ottavio, Maria Bulgheroni, and Vittorio Gallese. 2010. Wired to be social: The ontogeny of human interaction. *PLoS ONE* 5(10): 1–10.

Clynes, Manfred. 1978. *The touch of the emotions*. New York: Doubleday.

Clynes, Manfred (ed.). 1982. *Music, mind, and brain*. New York: Plenum.

Condon, William S., and William D. Ogston. 1967. A segmentation of behavior. *Journal of Psychiatric Research* 5: 221–235.

Condon, William S., and Louis S. Sander. 1974. Neonate movement is synchronized with adult speech: Interactional participation and language acquisition. *Science* 183: 99–101.

Craig, Cathy M., and David N. Lee. 1999. Neonatal control of nutritive sucking pressure: Evidence for an intrinsic tau-guide. *Experimental Brain Research* 124: 371–382.

Damasio, Antonio R. 2010. *Self comes to mind: Constructing the conscious brain*. New York: Pantheon.

Darwin, Charles. 1859. *On the origin of species by means of natural selection*. London: John Murray.

Darwin, Charles. 1872/1998. *The expression of the emotions in man and animals*, 3rd ed. New York: Oxford University Press.

de Vries, J.I.P., Gerard H.A. Visser, and Heinz F.R. Prechtl. 1982. The emergence of fetal behavior, I: Qualitative aspects. *Early Human Development* 7: 301–322.

DeCasper, Anthony J., and Phyllis Prescott. 2009. Lateralized processes constrain auditory reinforcement in human newborns. *Hearing Research* 255: 135–141.

Delafield-Butt, Jonathan, and Colwyn Trevarthen. 2013. Theories of the development of human communication. In *Theories and models of communication: Handbook of communication science*, vol. 1, ed. P. Cobley and P.J. Schultz, 199–221. Berlin: De Gruyter Mouton.

Delafield-Butt, Jonathan, and Colwyn Trevarthen. 2015. The origins of narrative. *Frontiers in Psychology* 6: 1157. doi:10.3389/psyg.2015.01157.

Delamont, R. Shane, Peter O.O. Julu, and Goran A. Jamal. 1999. Periodicity of a noninvasive measure of cardiac vagal tone during non-rapid eye movement sleep in non-sleep-deprived and sleep-deprived normal subjects. *Journal of Clinical Neurophysiology* 16(2): 146–153.

Dewey, John. 1938. *Experience and education*. New York: Macmillan.

Dissanayake, Ellen. 2000. Antecedents of the temporal arts in early mother-infant interaction. In *The origins of music*, ed. N.L. Wallin, B. Merker, and S. Brown, 389–410. Cambridge, MA: MIT Press.

Dissanayake, E. 2009. Root, leaf, blossom or burl: Concerning the origin and adaptive function of music. In *Communicative musicality: Exploring the basis of human companionship*, ed. S. Malloch and C. Trevarthen, 17–30. Oxford: Oxford University Press.

Donald, Merlin. 2001. *A mind so rare: The evolution of human consciousness*. New York: Norton.

Donaldson, Margaret. 1978. *Children's minds*. Glasgow: Fontana/Collins.

Donaldson, Margaret. 1992. *Human minds: An exploration*. London: Allen Lane/Penguin Books.

Eckerdal, Patricia, and Bjorn Merker. 2009. "Music" and the "action song" in infant development: An interpretation. In *Communicative musicality: Exploring the basis of human companionship*, ed. S. Malloch and C. Trevarthen, 241–262. Oxford: Oxford University Press.

Ekman, Paul, and Wallace V. Friesen. 1969. The repertoire of nonverbal behavior: Categories, origins, usage, and coding. *Semiotica* 22: 353–374.

Erickson, Fred. 2009. Musicality in talk and listening: A key element in classroom discourse as an environment for learning. In *Communicative musicality: Exploring the basis of human companionship*, ed. S. Malloch and C. Trevarthen, 449–464. Oxford: Oxford University Press.

Feldman, Ruth. 2012. Parent-infant synchrony: A bio-behavioral model of mutual influences in the formation of affiliative bonds. *Monographs of the Society for Research in Child Development* 77(2): 42–51.

Fraisse, Paul. 1982. Rhythm and tempo. In *The psychology of music*, ed. D. Deutsch, 149–180. New York: Academic.

Frank, Bodo, and Colwyn Trevarthen. 2012. Intuitive meaning: Supporting impulses for interpersonal life in the sociosphere of human knowledge, practice and language. In *Moving ourselves, moving others: Motion and emotion in intersubjectivity, consciousness and language*, ed. A. Foolen, U.M. Lüdtke, T.P. Racine, and J. Zlatev, 261–303. Amsterdam: John Benjamins.

Frijda, Nico H. 2007. *The laws of emotion*. Mahwah: Erlbaum.

Gibson, James J. 1979. *The ecological approach to visual perception*. Boston: Houghton Mifflin.

Glickstein, Mitchell, and Karl Doron. 2008. Cerebellum: Connections and functions. *Cerebellum* 7: 589–594.

Goodrich, Barbara G. 2010. We do, therefore we think: Time, motility, and consciousness. *Reviews in the Neurosciences* 21(5): 331–361.

Gratier, Maya, and Colwyn Trevarthen. 2007. Voice, vitality and meaning: On the shaping of the infant's utterances in willing engagement with culture. *International Journal for Dialogical Science* 2(1): 169–181.

Gratier, Maya, and Colwyn Trevarthen. 2008. Musical narrative and motives for culture in mother-infant vocal interaction. *Journal of Consciousness Studies* 15(10–11): 122–158.

Hadamard, Jacques. 1945. *The psychology of invention in the mathematical field*. Princeton: Princeton University Press.

Halliday, Michael A.K. 1978. Meaning and the construction of reality in early childhood. In *Modes of perceiving and processing information*, ed. H.L. Pick and F. Saltzman, 67–96. Hillsdale: Lawrence Erlbaum.

Halliday, Michael A.K. 1979. One child's protolanguage. In *Before speech: The beginning of human communication*, ed. M. Bullowa, 171–190. Cambridge: Cambridge University Press.

Halliday, Michael A.K., and Christian M.I.M. Matthiessen. 2004. *An introduction to functional grammar*, 3rd ed. London: Arnold.

Helmholtz, Hermann von. 1867. *Handbuch der Physiologischen Optik*. Leipzig: Leopold Voss.

Hess, Walter Rudolf. 1954. *Diencephalon: Autonomic and extrapyramidal functions*. Orlando: Grune and Stratton.

Hevner, Robert F. 2000. Development of connections in the human visual system during fetal midgestation: A Dii-tracing study. *Journal of Neuropathology and Experimental Neurology* 59: 385–392.

Hubley, Penelope, and Colwyn Trevarthen. 1979. Sharing a task in infancy. In *Social interaction during infancy: New directions for child development*, vol. 4, ed. I. Uzgiris, 57–80. San Francisco: Jossey-Bass.

Husserl, Edmund. 1964. *The phenomenology of internal time consciousness*. Trans. J. Churchill. Bloomingtin: Indiana University Press.

Jakobovits, Akos A. 2009. Grasping activity in utero: A significant indicator of fetal behavior. *Journal of Perinatal Medicine* 37: 571–572.

James, William. 1890. *The principles of psychology*. Cambridge, MA: Harvard University Press.

James, David, Mary Pillai, and John Smoleniec. 1995. Neurobehavioral development of the human fetus. In *Fetal development: A psychobiological perspective*, ed. J.-P. Lecanuet, W.P. Fifer, N.A. Krasnegor, and W.P. Smotherman, 101–128. Hillsdale: Erlbaum.

Jeannerod, Marc. 1994. The representing brain: Neural correlates of motor intention and imagery. *Behavioral and Brain Sciences* 17: 187–245.

Kendon, Adam. 1980. Gesticulation and speech: Two aspects of the process of utterance. In *The relationship of verbal and nonverbal communication*, ed. M.R. Key, 207–227. The Hague: Mouton and Co.

Kugiumutzakis, Giannis, and Colwyn Trevarthen. 2015. Neonatal imitation. *Elsevier international encyclopedia of social and behavioural sciences*, Area 2, Entry 23160, In press.

Kühl, Ole. 2007. *Musical semantics*. Bern: Peter Lang.

Langer, Susanne K. 1942. *Philosophy in a new key: A study in the symbolism of reason, rite, and art*. Cambridge MA: Harvard University Press.

Langer, Susanne K. 1953. *Feeling and form*. London: Routledge.

Larroche, Jeanne-Claudie. 1981. The marginal layer in the neocortex of a 7 week-old human embryo: A light and electron microscopy study. *Anatomy and Embryology* 162: 301–312.

Lashley, Karl S. 1951. The problems of serial order in behavior. In *Cerebral mechanisms in behavior*, ed. L.A. Jeffress, 112–136. New York: Wiley.

Lecanuet, Jean-Pierre, William P. Fifer, Norman A. Krasnegor, and William P. Smotherman (eds.). 1995. *Fetal development: A psychobiological perspective*. Hillsdale: Erlbaum.

Lee, David N. 2009. General tau theory: Evolution to date. *Perception* 38: 837–858.

Lee, David N., and Benjamin Schögler. 2009. Tau in musical expression. In *Communicative musicality: Exploring the basis of human companionship*, ed. S. Malloch and C. Trevarthen, 83–104. Oxford: Oxford University Press.

Lee, David N., C.M. Craig, and M.A. Grealy. 1999. Sensory and intrinsic coordination of movement. *Proceedings of the Royal Society of London Series B* 266: 2029–2035.

Lenneberg, Eric. 1967. *Biological foundations of language*. New York: Wiley.

Llinás, Rodolfo. 2001. *I of the vortex: From neurons to self*. Cambridge, MA: MIT Press.

MacLean, Paul D. 1990. *The triune brain in evolution, role in paleocerebral functions*. New York: Plenum Press.

Malloch, Stephen. 1999. Mother and infants and communicative musicality. In *Rhythms, musical narrative, and the origins of human communication. Musicae Scientiae*, Special Issue, 1999–2000, ed. I. Deliège, 29–57. Liège: European Society for the Cognitive Sciences of Music.

Malloch, Stephen, and Colwyn Trevarthen (eds.). 2009a. *Communicative musicality: Exploring the basis of human companionship*. Oxford: Oxford University Press.

Malloch, Stephen, and Colwyn Trevarthen. 2009b. Musicality: Communicating the vitality and interests of life. In *Communicative musicality: Exploring the basis of human companionship*, ed. S. Malloch and C. Trevarthen, 1–11. Oxford: Oxford University Press.

Maturana, Humberto R., and Francisco J. Varela. 1980. *Autopoiesis and cognition*. Dordrecht: Reidel.

Maturana, Humberto, J. Mpodozis, and J. Carlos Letelier. 1995. Brain, language and the origin of human mental functions. *Biological Research* 28(1): 15–26.

Mead, George Herbert. 1934. *Mind, self, and society from the standpoint of a social behaviorist*. Chicago: University of Chicago Press.

Meares, Russell. 2004. The conversational model: An outline. *American Journal of Psychotherapy* 58: 51–66.

Meares, Russell, Nick Bendit, Joan Haliburn, Anthony Korner, Dawn Mears, and David Butt. 2012. *Borderline personality disorder and the conversational model: A clinician's manual*. New York: W.W. Norton and Company.

Melges, Frederick T. 1982. *Time and the inner future: A temporal approach to psychiatric disorders*. New York: Wiley.

Merker, Bjorn. 2005. The liabilities of mobility: A selection pressure for the transition to consciousness in animal evolution. *Consciousness and Cognition* 14(1): 89–114.

Merker, Bjorn. 2007. Consciousness without a cerebral cortex: A challenge for neuroscience and medicine. *Behavioral and Brain Sciences* 30: 63–134.

Merker, Bjorn. 2009a. Ritual foundations of human uniqueness. In *Communicative musicality: Exploring the basis of human companionship*, ed. S. Malloch and C. Trevarthen, 45–60. Oxford: Oxford University Press.

Merker, Bjorn. 2009b. Returning language to culture by way of biology. *Behavioral and Brain Sciences* 32(5): 460.

Merleau-Ponty, Maurice. 1962. *Phenomenology of perception*. London: Routledge and Kegan Paul.

Miall, David S., and Ellen Dissanayake. 2003. The poetics of babytalk. *Human Nature* 14(4): 337–364.

Mo, Suchoon S. 1990. Time reversal in human cognition: Search for a temporal theory of insanity. In *Cognitive models of psychological time*, ed. R.A. Block, 241–254. Hilldale: Lawrence Erlbaum.

Moruzzi, Giuseppe, and Horace W. Magoun. 1949. Brain stem reticular formation and activation of the EEG. *Electroencephalography and Clinical Neurophysiology* 1: 455–473.

Nadel, Jacqueline. 2014. *How imitation boosts development: In typical infants and children with autism*. Oxford: Oxford University Press.

Nadel, Jacqueline, and Ann Pezé. 1993. Immediate imitation as a basis for primary communication in toddlers and autistic children. In *New perspectives in early communicative development*, ed. J. Nadel and L. Camioni, 139–156. London: Routledge.

Nagy, Emese. 2006. From imitation to conversation: The first dialogues with human neonates. *Infant and Child Development* 15: 223–232.

Nagy, Emese. 2011. The newborn infant: A missing stage in developmental psychology. *Infant and Child Development* 20(1): 3–19.

Nagy, Emese, and Peter Molnár. 2004. Homo imitans or homo provocans? The phenomenon of neonatal imitiation. *Infant Behavior and Development* 27: 57–63.

Okado, Nobuo. 1980. Development of the human cervical spinal cord with reference to synapse formation in the motor nucleus. *The Journal of Comparative Neurology* 191: 495–513.

Osborne, Nigel. 2009a. Towards a chronobiology of musical rhythm. In *Communicative musicality: Exploring the basis of human companionship*, ed. S. Malloch and C. Trevarthen, 545–564. Oxford: Oxford University Press.

Osborne, Nigel. 2009b. Music for children in zones of conflict and post-conflict: A psychobiological approach. In *Communicative musicality: Exploring the basis of human companionship*, ed. S. Malloch and C. Trevarthen, 331–356. Oxford: Oxford University Press.

Panksepp, Jaak. 2007a. Emotional feelings originate below the neocortex: Toward a neurobiology of the soul. *Behavioral and Brain Sciences* 30: 101–103.

Panksepp, Jaak. 2007b. Can PLAY diminish ADHD and facilitate the construction of the social brain? *Journal of the Canadian Academy of Child and Adolescent Psychiatry* 16(2): 5–14.

Panksepp, Jaak, and Jeffrey Burgdorf. 2003. "Laughing" rats and the evolutionary antecedents of human joy? *Physiology and Behavior* 79: 533–547.

Panksepp, Jaak, and Colwyn Trevarthen. 2009. The neuroscience of emotion in music. In *Communicative musicality: Exploring the basis of human companionship*, ed. S. Malloch and C. Trevarthen, 105–146. Oxford: Oxford University Press.

Papoušek, Hanuš. 1967. Experimental studies of appetitional behaviour in human newborns and infants. In *Early behaviour: Comparative and developmental approaches*, ed. H.W. Stevenson, E.H. Hess, and H.L. Rheingold, 249–277. New York: Wiley.

Papoušek, Hanuš. 1996a. Musicality in infancy research: Biological and cultural origins of early musicality. In *Musical beginnings: Origins and development of musical competence*, ed. I. Deliège and J. Sloboda, 37–55. Oxford: Oxford University Press.

Papoušek, Mechthild. 1996b. Intuitive parenting: A hidden source of musical stimulation in infancy. In *Musical beginnings: Origins and development of musical competence*, ed. I. Deliège and J. Sloboda, 88–112. Oxford: Oxford University Press.

Parncutt, Richard. 2006. Prenatal development. In *The child as musician*, ed. G.E. McPherson, 1–31. Oxford: Oxford University Press.

Pavlicevic, Mercedes. 1997. *Music therapy in context: Music meaning and relationship*. London: Jessica Kingsley.

Pavlicevic, Mercedes, and Gary Ansdell. 2009. Between communicative musicality and collaborative musicing: A perspective from community music therapy. In *Communicative musicality: Exploring the basis of human companionship*, ed. S. Malloch and C. Trevarthen, 357–376. Oxford: Oxford University Press.

Pavlicevic, Mercedes, Colwyn Trevarthen, and Janice Duncan. 1994. Improvisational music therapy and the rehabilitation of persons suffering from chronic schizophrenia. *Journal of Music Therapy* 31(2): 86–104.

Piontelli, Alessandra. 1992. *From fetus to child*. London: Routledge.

Piontelli, Alessandra. 2002. *Twins: From fetus to child*. London: Routledge.

Porges, Stephen W. 2001. The polyvagal theory: Phylogenetic substrates of a social nervous system. *International Journal of Psychophysiology* 42: 123–146.

Porges, Stephen W. 2003. Social engagement and attachment: A phylogenetic perspective. *Annals of the New York Academy of Sciences* 1008: 31–47.

Porges, Stephen W., and Senta A. Furman. 2011. The early development of the autonomic nervous system provides a neural platform for social behaviour: A polyvagal perspective. *Infant and Child Development* 20: 106–118.

Prechtl, Heinz F.R. 1984. Continuity and change in early human development. In *Continuity of neural functions from prenatal to postnatal life*, ed. H.F.R. Prechtl, 1–13. Oxford: Blackwell.

Reddy, Vasudevi. 2008. *How infants know minds*. Cambridge, MA: Harvard University Press.

Reddy, Vasudevi. 2012. A gaze at grips with me. In *Joint attention: New developments in psychology, philosophy of mind and social neuroscience*, ed. A. Seemans, 137–157. Cambridge, MA: MIT Press.

Reid, Thomas. 1764/1997. An inquiry into the human mind on the principles of common sense, ed. Derek R. Brookes. Edinburgh: Edinburgh University Press.

Reissland, Nadia, Brian Francis, James Mason, and Karen Lincoln. 2011. Do facial expressions develop before birth? *PLoS ONE* 6(8): 1–7.

Robarts, Jacqueline Z. 2009. Supporting the development of mindfulness and meaning: Clinical pathways in music therapy with a sexually abused child. In *Communicative musicality: Exploring the basis of human companionship*, ed. S. Malloch and C. Trevarthen, 377–400. Oxford: Oxford University Press.

Rodrıguez, Cintia, and Pedro Palacios. 2007. Do private gestures have a self-regulatory function? A case study. *Infant Behavior and Development* 30: 180–194.

Rogoff, Barbara. 2003. *The cultural nature of human development*. Oxford: Oxford University Press.

Royal College of Obstetricians and Gynaecologists. 2010. *Fetal awareness: Review of research and recommendations for practice*. London: Royal College of Obstetricians and Gynaecologists.

Schefflen, Albert E. 1973. *Communicational structure: Analysis of a psychotherapy transaction*. Bloomington: Indiana University Press.

Schögler, Benjaman, and Colwyn Trevarthen. 2007. To sing and dance together. In *On being moved: From mirror neurons to empathy*, ed. S. Bråten, 281–302. Amsterdam: John Benjamins.

Schore, Allan. 2012a. Bowlby's "environment of evolutionary adaptiveness": Recent studies on the interpersonal neurobiology of attachment and emotional development. In *Evolution, early experience and human development: From research to practice and policy*, ed. D. Narvaez, J. Panksepp, A. Schore, and T. Gleason, 31–67. New York: Oxford University Press.

Schore, Allan. 2012b. *The science of the art of psychotherapy*. New York: W.W. Norton.

Sebeok, Thomas A. 1994. *Signs: An introduction to semiotics*. Toronto: University of Toronto Press.

Sherrington, Charles S. 1906. *The integrative action of the nervous system*. New Haven: Yale University Press.

Smith, Adam. 1777/1982. Of the nature of that imitation which takes place in what are called the imitative arts. In *Essays on philosophical subjects*, ed. W.P.D. Wightman, and J.C. Bryce. Indianapolis: Liberty Fund.

Solms, Marc, and Jaak Panksepp. 2012. The "Id" knows more than the "Ego" admits: Neuropsychoanalytic and primal consciousness perspectives on the interface between affective and cognitive neuroscience. *Brain Sciences* 2: 147–175.

Sperry, Roger Wolcott. 1950. Neural basis of the spontaneous optokinetic response produced by visual inversion. *Journal of Comparative and Physiological Psychology* 43: 482–489.

Sperry, Roger Wolcott. 1952. Neurology and the mind-brain problem. *American Scientist* 40: 291–312.

Sperry, Roger Wolcott. 1983. *Science and moral priority*. New York: Columbia University Press.

Stern, Daniel N. 1974. Mother and infant at play: The dyadic interaction involving facial, vocal and gaze behaviours. In *The effect of the infant on its caregiver*, ed. M. Lewis and L.A. Rosenblum, 187–213. New York: Wiley.

Stern, Daniel N. 2000. *The interpersonal world of the infant: A view from psychoanalysis and development psychology*, 2nd ed. New York: Basic Books.

Stern, Daniel N. 2004. *The present moment in psychotherapy and everyday life*. New York: W.W. Norton.

Stern, Daniel N. 2010. *Forms of vitality: Exploring dynamic experience in psychology, the arts, psychotherapy and development*. Oxford: Oxford University Press.

Stern, Daniel N., Kristine MacKain, and Susan Spieker. 1982. Intonation contours as signals in maternal speech to prelinguistic infants. *Developmental Psychology* 18: 727–735.

Stern, Daniel N., Lynne Hofer, Wendy Haft, and John Dore. 1985. Affect attunement: The sharing of feeling states between mother and infant by means of inter-modal fluency. In *Social perception in infants*, ed. T.M. Field and N.A. Fox, 249–268. Norwood: Ablex.

Stern, Daniel N., Louis W. Sander, Jeremy P. Nahum, Alexandra M. Harrison, Karlen Lyons-Ruth, Alexander C. Morgan, Nadia Bruschweiler-Stern, and Edward Z. Tronick. 1998. Non-interpretive mechanisms in psychoanalytic therapy: The "something more" than interpretation. *International Journal of Psychoanalysis* 79: 903–921.

Stige, Brunholf. 2004. Community music therapy: Culture, care and welfare. In *Community music therapy*, ed. M. Pavlicevic and G. Ansdell, 91–113. London: Jessica Kingsley.

Stuart, Susan A.J. 2010. Enkinaesthesia, biosemiotics and the ethiosphere. In *Signifying bodies: Biosemiosis, interaction and health*, ed. S. Cowley, J.C. Major, S. Steffensen, and A. Dini, 305–330. Braga: Portuguese Catholic University.

Takada, Akira. 2005. Mother-infant interactions among the !Xun: Analysis of gymnastic and breastfeeding behaviors. In *Hunter-gatherer childhoods: Evolutionary, developmental, and cultural perspectives*, ed. B.S. Hewlett and M.E. Lamb, 289–308. New Brunswick: Aldine Transaction Publishers.

Thelin, Nils. 2014. *On the nature of time: A biopragmatic perspective on language, thought and reality*. Uppsala: Uppsala University.

Tomasello, Michael. 2008. *Origins of human communication*. Cambridge, MA: MIT Press.

Trainor, Laurel J. 1996. Infant preferences for infant-directed versus non-infant-directed play songs and lullabies. *Infant Behavior and Development* 19: 83–92.

Trehub, Sandra E. 1990. The perception of musical patterns by human infants: The provision of similar patterns by their parents. In *Comparative perception, Vol. 1: Mechanisms*, ed. M.A. Berkley and W.C. Stebbins, 429–459. New York: Wiley.

Trevarthen, Colwyn. 1974. The psychobiology of speech development. In *Language and brain: Developmental aspects*, ed. E.H. Lenneberg, 570–585. Boston: Neurosciences Research Program.

Trevarthen, Colwyn. 1979. Communication and cooperation in early infancy: A description of primary intersubjectivity. In *Before speech: The beginning of human communication*, ed. M. Bullowa, 321–347. Cambridge: Cambridge University Press.

Trevarthen, Colwyn. 1984. How control of movements develops. In *Human motor actions: Bernstein reassessed*, ed. H.T.A. Whiting, 223–261. Amsterdam: Elsevier (North Holland).

Trevarthen, Colwyn. 1985. Neuroembryology and the development of perceptual mechanisms. In *Human growth*, 2nd ed, ed. F. Falkner and J.M. Tanner, 301–383. New York: Plenum.

Trevarthen, Colwyn. 1986. Development of intersubjective motor control in infants. In *Motor development in children: Aspects of coordination and control*, ed. M.G. Wade and H.T.A. Whiting, 209–261. Dordrecht: Martinus Nijhof.

Trevarthen, Colwyn. 1990. Signs before speech. In *The semiotic web*, ed. T.A. Sebeok and J. Umiker-Sebeok, 689–755. Amsterdam: De Gruyter.

Trevarthen, Colwyn. 1992. An infant's motives for speaking and thinking in the culture. In *The dialogical alternative: Towards a theory of language and mind* (Festschrift for Ragnar Rommetveit), ed. A.H. Wold, 99–137. Oslo/Oxford: Scandinavian University Press/Oxford University Press.

Trevarthen, Colwyn. 1996. Lateral asymmetries in infancy: Implications for the development of the hemispheres. *Neuroscience and Biobehavioral Reviews* 20(4): 571–586.

Trevarthen, Colwyn. 1999. Musicality and the intrinsic motive pulse: Evidence from human psychobiology and infant communication. In *Rhythms, musical narrative, and the origins of human communication. Musicae Scientiae*, Special Issue, 1999–2000, ed. I. Deliège, 157–213. Liège: European Society for the Cognitive Sciences of Music.

Trevarthen, Colwyn. 2001. The neurobiology of early communication: Intersubjective regulations in human brain development. In *Handbook on brain and behavior in human development*, ed. A.F. Kalverboer and A. Gramsbergen, 841–882. Dordrecht: Kluwer.

Trevarthen, Colwyn. 2004a. Language development: Mechanisms in the brain. In *Encyclopedia of neuroscience*, 3rd ed, ed. G. Adelman and B.H. Smith. Amsterdam: Elsevier Science.

Trevarthen, Colwyn. 2004b. Brain development. In *Oxford companion to the mind*, 2nd ed, ed. R.L. Gregory, 116–127. Oxford: Oxford University Press.

Trevarthen, Colwyn. 2005a. First things first: Infants make good use of the sympathetic rhythm of imitation, without reason or language. *Journal of Child Psychotherapy* 31(1): 91–113.

Trevarthen, Colwyn. 2005b. Stepping away from the mirror: Pride and shame in adventures of companionship. Reflections on the nature and emotional needs of infant intersubjectivity. In *Attachment and bonding: A new synthesis*, ed. C.S. Carter, L. Ahnert, K.E. Grossmann, S.B. Hrdy, M.E. Lamb, S.W. Porges, and N. Sachser, 55–84. Cambridge, MA: MIT Press.

Trevarthen, Colwyn. 2008. *The musical art of infant conversation: Narrating in the time of sympathetic experience, without rational interpretation, before words*, 11–37. Special Issue: Musicae Scientiae.

Trevarthen, Colwyn. 2009a. The functions of emotion in infancy: The regulation and communication of rhythm, sympathy, and meaning in human development. In *The healing power of emotion: Affective neuroscience, development, and clinical practice*, ed. D. Fosha, D.J. Siegel, and M.F. Solomon, 55–85. New York: W.W. Norton.

Trevarthen, Colwyn. 2009b. Human biochronology: On the source and functions of "musicality". In *Music that works: Contributions of biology, neurophysiology, psychology, sociology, medicine and musicology*, ed. R. Haas and V. Brandes, 221–265. Berlin: Springer.

Trevarthen, Colwyn. 2011. What young children give to their learning, making education work to sustain a community and its culture. *European Early Childhood Education Research Journal* 19(2): 173–193.

Trevarthen, Colwyn. 2012a. Communicative musicality: The human impulse to create and share music. In *Musical imaginations: Multidisciplinary perspectives on creativity, performance, and perception*, ed. D.J. Hargreaves, D.E. Miell, and R.A.R. MacDonald, 259–284. Oxford: Oxford University Press.

Trevarthen, Colwyn. 2012b. Born for art, and the joyful companionship of fiction. In *Evolution, early experience and human development: From research to practice and policy*, ed. D. Narvaez, J. Panksepp, A. Schore, and T. Gleason, 202–218. New York: Oxford University Press.

Trevarthen, Colwyn. 2012c. Embodied human intersubjectivity: Imaginative agency, to share meaning. *Cognitive Semiotics* 4(1): 6–56.

Trevarthen, Colwyn. 2012d. The generation of human meaning: How shared experience grows in infancy. In *Joint attention: New developments in philosophy, psychology, and neuroscience*, ed. A. Seemann, 73–113. Cambridge, MA: MIT Press.

Trevarthen, Colwyn, and K.J. Aitken. 2001. Infant intersubjectivity: Research, theory, and clinical applications. *The Journal of Child Psychology and Psychiatry and Allied Disciplines, Annual Research Review* 42(1): 3–48.

Trevarthen, Colwyn, and K.J. Aitken. 2003. Regulation of brain development and age-related changes in infants' motives: The developmental function of "regressive" periods. In *Regression periods in human infancy*, ed. M. Heimann, 107–184. Mahwah: Erlbaum.

Trevarthen, Colwyn. 2015. Infant Semiosis: The psycho-biology of action and shared experience from birth. In *Semiotics and cognition in human development*, eds. Chris Sinha, Göran Sonesson and Sara Lenninger, Special Issue in Cognitive Development (in press).

Trevarthen, Colwyn, and J. Delafield-Butt. 2013a. Biology of shared experience and language development: Regulations for the inter-subjective life of narratives. In *The infant mind: Origins of the social brain*, ed. M. Legerstee, D. Haley, and M. Bornstein, 167–199. New York: Guildford Press.

Trevarthen, Colwyn, and J. Delafield-Butt. 2013b. Autism as a developmental disorder in intentional movement and affective engagement. *Frontiers in Integrative Neuroscience* 7(49): 1–16.

Trevarthen, Colwyn, and Penelope Hubley. 1978. Secondary intersubjectivity: Confidence, confiding and acts of meaning in the first year. In *Action, gesture and symbol*, ed. A. Lock, 183–229. London: Academic.

Trevarthen, Colwyn, J. Delafield-Butt, and B. Schögler. 2011. Psychobiology of musical gesture: Innate rhythm, harmony and melody in movements of narration. In *New perspectives on music and gesture*, ed. A. Gritten and E. King, 11–43. Burlington: Ashgate.

Trevarthen, Colwyn, M. Gratier, and N. Osborne. 2014. The human nature of culture and education. *Wiley Interdisciplinary Reviews: Cognitive Science* 5: 173–192.

Tulving, Endel. 2002. Episodic memory: From mind to brain. *Annual Review of Psychology* 253: 1–25.

Turner, Victor. 1982. *From ritual to theatre: The human seriousness of play*. New York: Performing Arts Journal Publications.

Turner, Robert, and Andreas A. Ioannides. 2009. Brain, music and musicality: Inferences from neuroimaging. In *Communicative musicality: Exploring the basis of human companionship*, ed. S. Malloch and C. Trevarthen, 147–181. Oxford: Oxford University Press.

van Rees, Saskia, and Richard de Leeuw. 1993. Born too early: The kangaroo method with premature babies. Video by Stichting Lichaamstaal, Scheyvenhofweg 12, 6093 PR. Heythuysen.

Vandekerckhove, Marie, Luis C. Bulnes, and Jaak Panksepp. 2014. The emergence of primary anoetic consciousness in episodic memory. *Frontiers in Behavioral Neuroscience* 7(210): 1–8.

von Holst, Erich. 1936. Versuche zur Theorie der relativen Koordination. *Archives für Gesamte Physiologie* 236: 93–121.

von Holst, Erich, and Horst Mittelstaedt. 1950. Das Reafferenzprinzip. *Naturwissenschaften* 37: 256–272.

von Uexküll, Jakob. 1957. A stroll through the worlds of animals and men: A picture book of invisible worlds. In *Instinctive behavior. The development of a modern concept*, ed. Claire H. Schiller, 5–80. New York: International Universities Press.

Wallin, Nils L., Bjorn Merker, and Steven Brown (eds.). 2000. *The origins of music*. Cambridge, MA: MIT Press.

Whitehead, Alfred North. 1926. *Science and the modern world. Lowell Lectures, 1925*. Cambridge: Cambridge University Press.

Wigram, Tony, and Cochavit Elefant. 2009. Therapeutic dialogues in music: Nurturing musicality of communication in children with autistic spectrum disorder and Rett syndrome. In *Communicative musicality: Exploring the basis of human companionship*, ed. S. Malloch and C. Trevarthen, 423–445. Oxford: Oxford University Press.

Windle, William F. 1970. Development of neural elements in human embryos of four to seven weeks gestation. *Experimental neurology* (Suppl.) 5: 44–83.

Zoia, Stefania, Laura Blason, Giuseppina D'Ottavio, Maria Bulgheroni, Eva Pezzetta, Aldo Scabar, and Umberto Castiello. 2007. Evidence of early development of action planning in the human foetus: A kinematic study. *Experimental Brain Research* 176: 217–226.

Index

© Springer International Publishing Switzerland 2016
B. Mölder et al. (eds.), *Philosophy and Psychology of Time*, Studies
in Brain and Mind 9, DOI 10.1007/978-3-319-22195-3